D1762117

Springer
Proceedings in Physics 33

Springer Proceedings in Physics

Managing Editor: H. K. V. Lotsch

Volume 30 *Short-Wavelength Lasers and Their Applications*
Editor: C. Yamanaka

Volume 31 *Quantum String Theory*
Editors: N. Kawamoto and T. Kugo

Volume 32 *Universalities in Condensed Matter*
Editors: R. Jullien, L. Peliti, R. Rammal, and N. Boccara

Volume 33 *Computer Simulation Studies in Condensed Matter Physics: Recent Developments*
Editors: D. P. Landau, K. K. Mon, and H.-B. Schüttler

Volumes 1 – 29 are listed on the back inside cover

Computer Simulation Studies in Condensed Matter Physics

Recent Developments

Proceedings of the Workshop,
Athens, GA, USA, February 15–26, 1988

Editors: D. P. Landau, K. K. Mon,
and H.-B. Schüttler

With 101 Figures

Springer-Verlag Berlin Heidelberg New York
London Paris Tokyo

Professor David P. Landau, Ph. D.
Professor K. K. Mon, Ph. D.
Professor Heinz-Bernd Schüttler, Ph. D.

Center for Simulational Physics, The University of Georgia, Athens, GA 30602, USA

ISBN 3-540-50449-4 Springer-Verlag Berlin Heidelberg New York
ISBN 0-387-50449-4 Springer-Verlag New York Berlin Heidelberg

Library of Congress Cataloging-in-Publication Data. Computer simulation studies in condensed matter physics: recent developments: proceedings of the workshop. Athens, GA, USA, February 15–26, 1988 / editors, D. P. Landau, K. K. Mon, and H.–B. Schüttler. (Springer proceedings in physics; v. 33) Includes index.
(U.S.: alk. paper)
1. Condensed matter – Computer simulation – Congresses. I. Landau, David P. II. Mon, K. K. (Kin K.), 1950–. III. Schüttler, Heinz-Bernd, 1956–. IV. Series. QC173.4.C65C65 1988 530.4'1–dc 19 88-30707

Printing: Weihert-Druck GmbH, D-6100 Darmstadt
Binding: J. Schäffer GmbH & Co. KG., D-6718 Grünstadt
2154/3150-543210 – Printed on acid-free paper

Preface

Computer simulation studies in condensed matter physics represent a rapidly changing field, making significant contributions in techniques and in new results to important physical problems. The workshop on "Recent Developments in Computer Simulation Studies in Condensed Matter Physics", held at the Center for Simulational Physics at the University of Georgia, February 14–26, 1988, was an attempt to bring together some of the practitioners in this field and to provide a forum for the presentation and exchange of new ideas and recent developments. These proceedings are a record of the workshop and are published with the goal of timely dissemination of the papers to a wider audience.

Although by their very nature workshops are limited in their scope of coverage, a broad range of current topics is discussed. The papers present new techniques and results on both static and dynamical phenomena of classical and quantum mechanical systems. The reader will find descriptions of studies of static properties as well as time-dependent ones where time may be real time, stochastic time or imaginary (path integral) time.

A substantial portion of the material, the first part of the proceedings, deals with simulations of classical systems. Monte Carlo simulations are presented for simple Ising models. On the one hand, the reader will find here new approaches suggested for avoiding critical slowing down, and on the other, there are descriptions of the study of critical slowing down using standard techniques. Both standard and novel Monte Carlo methods have been used to study static properties at phase transitions. New results for classical spin dynamics are discussed. New material is presented for the formation of aggregates. Molecular dynamics and/or Monte Carlo approaches have been developed for studying fluid flow as well as for polymer systems, adsorbed monolayers and crystal growth.

The second part of the proceedings is devoted to the simulation of quantum systems. New results for interacting boson and fermion systems, adsorbed ^{4}He monolayers and the electronic structure of clusters have been obtained by numerical functional and path integral techniques. Novel variational wave functions and minimization techniques are discussed in the context of solid ^{4}He, and the electronic structure of real atoms and solids. Promising new hybrid methods have been developed by combining more conventional quantum methods, such as quantum Monte Carlo or density functional theory, with well-established classical techniques, such as molecular dynamics. The workshop also included several presentations about emerging computer technologies and programming methods. but for the most part no written record is available.

We hope that the readers will benefit from papers in their own special field as well as gain new ideas from other related specialties.

This workshop was made possible through the generous support of the Advanced Computational Methods Center, the Department of Physics and Astronomy, and the Vice President for Research at the University of Georgia.

Athens, GA, USA
April 1988

D.P. Landau
K.K. Mon
H.-B. Schüttler

Contents

Part II Quantum Systems

Part III Computer Graphics

Part IV Contributed Papers

Introduction

D.P. Landau, K.K. Mon, and H.-B. Schüttler

Center for Simulational Physics, University of Georgia, Athens, GA 30602, USA

Computer simulations have become recognized as a new approach to research in physics; they provide great flexibility in the choice of parameters to be included, the ensemble to be used, and even the questions to be answered. Simulations provide not only numerical data but also allow the visualization of microscopic behavior (through graphics output) which helps flesh out our intuitive understanding of the processes at work. Early work was done using computers which were considerably slower than today's personal computers, and it was only in the late 1960's that computer speed had increased to the point that detailed studies of physical problems became possible. A broad range of problems which have already been attacked using Monte Carlo methods has been summarized in two earlier volumes in the Springer-Verlag series on Topics in Current Physics and we refer the newcomer to the field to these books [1, 2]. The proceedings of this workshop concentrate instead on the most recent developments in the field of computer simulations in condensed matter physics.

Monte Carlo methods determine the properties of interacting many-particle systems by generating (carefully chosen) sample configurations of the system and using these states to estimate the actual distribution of the total number of states. New configurations are generated from preceding ones using "single site" changes, but "multi–spin" methods may speed up the computation substantially either through the implementation of vector operations or through modification of the fundamental dynamic time scale. Increasing accuracy is of course obtainable by simply using larger systems and generating more states. A "standard" unit for measuring the lengths of Monte Carlo runs is the MCS/site or Monte Carlo Step per site, which represents one examination of every site in the system. Whereas two decades ago only a few hundred MCS/site were generally possible for classical systems, today run lengths for simple models may be measured in millions of MCS/site. The speed of modern computers has also made it possible to examine more realistic models, which of course consume far more computer time than do simple theoretical systems. In these proceedings you will find data obtained using both Cyber 205 and CRAY/XMP supercomputers; on the other hand there are also very high resolution data for simple, small systems which have been produced by using a totally dedicated micro–Vax.

Whereas Monte Carlo methods may be applied directly to collections of classical particles, they may be used only indirectly for the exploration of quantum systems. Quantum behavior can be produced only with the inclusion of an extra degree of freedom (or added dimension) before simulations are introduced and the second part of these proceedings presents some novel ways in which this may be done.

A second simulation approach, Molecular Dynamics, involves the integration of coupled equations of motion which govern the time development of classical systems. The resultant (true) dynamics is different than the (stochastic) kinetic behavior resulting from Monte Carlo simulations although it is presently possible to extend such studies to only very short periods of physical time. In these proceedings the reader will find descriptions of the application of molecular dynamics techniques to larger and to more complex collections of interacting particles than is usually found in the literature.

The representaion of simulations data through computer graphics is gaining in importance and the state of the art in computer graphics is discussed in these proceedings by Follin.

Springer Proceedings in Physics Vol. 33: **Computer Simulation Studies in Condensed Matter Physics**
Editors: D. P. Landau • K. K. Mon • H-B. Schüttler

The first part of the proceedings is devoted to the study of classical systems. The first two papers provide a readable review and instructive summary of recent advances in a number of methods designed to overcome the long–standing problem of critical slowing down in Monte Carlo simulations. Sokal presents an overview of these new numerical algorithms, which involve collective–mode updates, and Loh discusses a method which may eliminate critical slowing down completely.

The use of a Gaussian ensemble in Monte Carlo simulations is discussed in the paper by Challa and Hetherington. They show examples where this method is particularly useful at first–order phase transitions.

Wysin, Gouvea, Bishop and Mertens consider the classical spin dynamics in the two–dimensional anisotropic Heisenberg model with numerical solutions of the equations of motion and approximate ideal gas phenomenology. These results yield detailed information on the dynamic structure factors which are then compared to Monte Carlo–Molecular Dynamics simulations.

Mountain considers different applications of computer simulation, i.e. to help clarify the amount of information available from a particular measurement technique. He considers light scattering experiments on the agglomeration process and models it by Langevindynamics. A detailed picture of diffusion and growth with clusters characterized by fractal dimension is presented. In a paper on a related topic, Meakin discusses some recent directions in the computer simulation of non–equilibrium growth and aggregation models. In particular, he shows how the simple DLA (diffusion limited aggreation) model has influenced the development of realistic models for mechanical failure in thin film deposits. Family considers in his paper the fractal growth and pattern formation in a Laplacian field. This has important applications to a wide range of physical processes of fundamental and applied interest, including crystal shapes and snowflake formation.

There are two papers on the important applications of simulations to the study of polymers. Grest and Kremer consider a molecular dynamics approach to dense polymer melts and cover the entire regime from Rouse to reptation dynamics. Binder describes the impact of Monte Carlo simulations on polymer physics with emphasis on three examples; a test of the classical Flory–Huggins theory for polymer mixtures, configurational statistics and dynamics of chains confined to cylindrical tubes, and the adsorption of polymers at walls.

Rapaport considers the application of Molecular Dynamics to the study of hydrodynamics and describes detailed simulations involving hundreds of thousands of particles producing realistic pictures of eddy formation and wake oscillations. The development of Rayleigh–Benard phenomena is presented and illustrates the capabilities of molecular dynamics techniques in dealing with complex hydrodynamics phenomena.

Landman reviews applications of molecular dynamics simulations in material science and condensed matter physics. Important examples are taken from studies of solid–melt interfaces of silicon and crystal growth, amorphous film growth, cluster aggregates , tip–substrate interaction in atomic force microscopy and metalic systems.

The last paper of the first part of the proceedings is by Flurchick and Etters on simulations of oxygen monolayer and bilayer systems. A detailed picture of various possible phases is presented for realistic models of oxygen–substrate interactions.

The second part of the proceedings is devoted to the simulation of quantum systems. New results for interacting boson and fermion systems, adsorbed monolayers and the electronic structure of clusters have been obtained by numerical functional and path integral techniques. Novel variational wavefunctions and minimization techniques are discussed in the context of solid ^{4}He, and the electronic structure of real atoms and solids. Promising new hybrid methods have been developed by combining more conventional quantum methods, such as quantum Monte Carlo or density functional theory with well–established classical techniques, such as molecular dynamics.

The first paper, by Abraham, presents a study of adsorbed monolayers of krypton and helium on graphite. Employing, in the case of krypton, classical molecular dynamics and, in the case of helium, path integral quantum Monte Carlo simulations, he has obtained unique physical insights into the complicated phase diagram of such systems. In particular, the nature of commensurate and incommensurate phases and transitions between them has been explored in detail.

The paper by Landman discusses the quantum path integral molecular dynamics and time–dependent self–consistent field method. The objective here is to simulate not only the static properties of quantum (in particular, electronic) systems, but also, to obtain information about dynamical (time–dependent) correlations and, hence, excited state properties. A variety of applications, involving electron localization in small ionic and molecular clusters, is presented.

Using path integral quantum Monte Carlo techniques, Peters and Alder have studied the two–dimensional Bose gas with Coulomb interaction, which they propose as a possible model for high–T_c superconductors. They suggest that, due to the long–range nature of the interaction, Bose condensation occurs at finite temperatures, similar to the three–dimensional Bose systems. Transition temperatures in the range of 100 to 200 K are estimated. The behavior of this system near the transition is contrasted with the Kosterlitz–Thouless transition predicted for a two–dimensional Bose system with short–range interactions.

The paper by Scaletter, Bickers and Scalapino discusses recent progress in the simulation of lattice fermion systems, using a novel hybrid method that combines quantum Monte Carlo with molecular dynamics simulation. They present results for the superconducting and charge density correlations in the two–dimensional version of the Holstein electron–phonon model. The simulational data are compared with approximate analytical results from Eliashberg theory. The competition between charge density wave and superconducting order and the breakdown of Eliashberg theory in the strong–coupling regime are explored.

A new type of wavefunctions for solid ^{4}He, referred to as "shadow wavefunctions", has been studied by Vitiello, Runge and Kalos. Such a wavefunction has the advantage of describing the crystalline correlations of the quantum solid without violating the requirements of translational symmetry of free space or Bose symmetry under particle exchange. They demonstrate that the shadow wavefunction gives a lower variational energy than previously used trial functions in which translational and exchange symmetry are explicitly broken, whereas other physical properties are comparable to those obtained from non–symmetric trial states. The authors propose to study the interesting question of Bose condensation in the solid phase by means of shadow wavefunctions.

Benedek, Min, Woodward and Garner discuss recent developments in the application of Car–Parrinello–type dynamical non–linear optimization methods in electronic structure calculations. Such methods are based on density functional or related variational approaches to electronic structure, but avoid explicit diagonalization of large matrices. Rather, one attempts to minimize the total energy of the electronic (and sometimes, simultaneously, ionic) degrees of freedom by use of molecular dynamics techniques. The present paper discussed possibilities of implementing such an energy minimization by use of first order (steepest descent) equations of motion instead of the original (second order) simulated annealing technique.

Umrigar, Wilson and Wilkins explore an efficient method for optimizing trial wavefunctions in variational electronic structure calculations. The method is based on the observation that, for a true eigenstate Ψ of a (many–body) Hamiltonian H, the so–called local energy, $H\Psi/\Psi$, is a constant in configuration space. Hence, instead of minimizing, for a given trial state, the energy expectation value, one may rather try to minimize the variance of this local energy of the trial state, sampled at a certain number of fixed points in configuration space. The paper discusses the advantages of such a method. It presents results obtained with this method for small atomic systems and gives a comparison with conventional variational techniques.

REFERENCES

1. Monte Carlo Methods in Statistical Physics, 2nd edition, ed. K. Binder (Springer-Verlag, Heidelberg, Berlin, 1986).
2. Applications of the Monte Carlo Method in Statistical Physics, 2nd edition, ed. K. Binder (Springer-Verlag, Heidelberg, Berlin, 1987).

Part I

Classical Systems

New Numerical Algorithms for Critical Phenomena (Multi-Grid Methods and All That)

A.D. Sokal

Department of Physics, New York University, 4 Washington Place, New York, NY 10003, USA

ABSTRACT

Monte Carlo computations in statistical mechanics and quantum field theory have been greatly hampered by critical slowing-down. We review recent progress in devising new Monte Carlo algorithms having radically reduced critical slowing-down. We discuss "collective-mode" algorithms (multi-grid and Swendsen-Wang) for spin systems and lattice field theories, and algorithms for the self-avoiding walk. We also discuss an algebraic multi-grid algorithm for solving Kirchhoff's equations in the random-resistor problem.

1. Introduction

This talk will be an overview of recent progress in devising new and more efficient numerical algorithms for problems in quantum field theory and the statistical mechanics of critical phenomena. The first question is, therefore: why should anyone care? Here are two reasons:

- *Quantum chromodynamics*. We think that QCD is the correct theory of the strong interactions (albeit not the Theory of Everything). But we are unable to compare QCD's predictions with experiment because we are unable to compute what its predictions are — its strongly nonlinear equations defy analytic solution. We can study QCD numerically by Monte Carlo, but current calculations are very crude: a really adequate computation of hadron masses in QCD with dynamical fermions, using conventional algorithms, would take around 10^{14} seconds of CPU time on a Cray X-MP supercomputer — that is, a few million years.* On the other hand, if the exciting new algorithms that I'll describe in Sections 3 and 5 turn out to be applicable to QCD — and it is far from clear that they will be — then this estimate could be reduced by a factor of 10 or even 100.

- *Hyperscaling in the $d = 3$ Ising model*. This has been a key controversy in the theory of critical phenomena for over 20 years [1], and a convincing resolution (either analytical or numerical) is still lacking. A good Monte Carlo computation, using conventional algorithms, might be performed in as little as 10^9 seconds of Cray time (roughly 30 years). But that's still a bit slow for graduate students trying to complete their theses, or junior faculty facing tenure decisions.

Clearly, improvements in computer technology will be essential to future progress in these fields, but brute force will not gain us a factor of 10^6: there is an urgent need for new and more efficient Monte Carlo algorithms. Now this is not primarily a question of clever programming, but of fundamental physics: for as I shall argue, the efficiency of

*Be warned that this estimate is highly controversial. Other physicists' estimates may be a factor of 10^6 higher or lower.

Springer Proceedings in Physics Vol. 33: **Computer Simulation Studies in Condensed Matter Physics**
Editors: D. P. Landau • K. K. Mon • H-B. Schüttler

an algorithm depends primarily on the *insight into the physical behavior of the model* which is implicit in the algorithm.

My goal in this talk, therefore, is to give an overview of some of the key difficulties arising in numerical studies of quantum field theory and critical phenomena — notably the problem of *critical slowing-down* — and of some recent progress in devising new algorithms which (partly) alleviate these difficulties. Section 2 contains a brief introduction to dynamic Monte Carlo methods. Section 3 describes Monte Carlo algorithms, old and new, for spin models and lattice field theories (*e.g.* Ising, ϕ^4, lattice gauge theories, etc.). Section 4 describes Monte Carlo algorithms for the self-avoiding walk. Section 5 describes some very recent work on the random resistor problem.

Some important topics will *not* be discussed here: purely computer improvements (*e.g.* vectorization, multi-spin coding); fermions (whether quenched or dynamic); and the choice of observables to be measured (*e.g.* Monte Carlo renormalization group).

2. Dynamic Monte Carlo Methods

All Monte Carlo work has the same general structure: given some probability measure μ (on some configuration space S), we wish to generate many random samples from μ. How is this to be done?

Static Monte Carlo methods are those which generate a sequence of *independent* samples from the distribution μ. These techniques are widely used in Monte Carlo numerical integration [2]. But they are unfeasible for most applications in statistical physics and quantum field theory, where μ is the Gibbs measure of some rather complicated system (many coupled degrees of freedom).

The idea of *dynamic* Monte Carlo methods is to invent a *Markov chain* with state space S having μ as its unique equilibrium distribution. In other words, we invent (somehow or other) a transition probability $P(\phi \to \phi')$ that leaves μ invariant, *i.e.*

$$\int d\mu(\phi)\, P(\phi \to \phi') \quad = \quad d\mu(\phi') \tag{2.1}$$

and also satisfies an ergodicity condition ("it is possible to get from every state to every other state"). In physical terms, we are inventing a *stochastic time evolution* for the given system. Note, however, that this time evolution *need not correspond to any real "physical" dynamics*: rather, the dynamics is simply a numerical algorithm, and it is to be chosen, like all numerical algorithms, on the basis of its computational efficiency.

So we simulate this Markov chain on the computer, starting from an arbitrary initial configuration $\phi^{(0)}$. The general principles of Markov-chain theory [3-6] ensure that time averages converge (with probability 1) to μ-averages. Thus, with sufficient computer time, one can in principle measure expectation values $\langle A \rangle_\mu$ to arbitrary precision.

The trouble in practice, of course, is that the samples $\phi^{(1)}, \phi^{(2)}, \ldots$ from the Markov chain are not statistically independent, but are in general strongly correlated. Crudely speaking, the statistical error in Monte Carlo work behaves as $\sim N^{-1/2}$, where N is the number of "effectively independent" samples. Thus, the statistical error in dynamic Monte Carlo is of order $\sim (\tau/T)^{1/2}$, where T is the total number of samples (run length) and τ is an "autocorrelation time" of the Markov chain. As we shall see, τ can get very large!

More precisely, let A be an observable, and let

$$\rho_{AA}(t) \quad = \quad \frac{\langle A(s)\, A(s+t)\rangle \; - \; \langle A(s)\rangle \langle A(s+t)\rangle}{\langle A(s)^2\rangle \; - \; \langle A(s)\rangle^2} \tag{2.2}$$

be its normalized time-autocorrelation function in the *stationary* Markov process (*i.e.* "in equilibrium"). Typically $\rho_{AA}(t)$ decays exponentially ($\sim e^{-|t|/\tau}$) for large t; we define the *exponential autocorrelation time*

$$\tau_{exp,A} = \limsup_{t\to\infty} \frac{t}{-\log|\rho_{AA}(t)|} \tag{2.3}$$

and

$$\tau_{exp} = \sup_A \tau_{exp,A} . \tag{2.4}$$

Thus, τ_{exp} is the relaxation time of the slowest mode in the system. (It might be $+\infty$!)

On the other hand, the statistical errors in Monte Carlo estimates of $\langle A\rangle_\mu$ are controlled by the *integrated autocorrelation time*

$$\tau_{int,A} = \frac{1}{2}\sum_{t=-\infty}^{\infty} \rho_{AA}(t) . \tag{2.5}$$

If $\rho_{AA}(t)$ decays approximately as a pure exponential, then $\tau_{exp,A} \approx \tau_{int,A}$; but in general $\tau_{exp,A}$ may be much larger than $\tau_{int,A}$.

Now let μ be the Gibbs measure of some statistical-mechanical system at inverse temperature β. What typically occurs is that as β approaches the critical temperature β_c, the autocorrelation time τ diverges to *infinity* — so that the computational efficiency tends to *zero*! This behavior is called *critical slowing-down*; it plagues Monte Carlo studies of critical phenomena in statistical mechanics and of the continuum limit in quantum field theory. My goal in the remainder of this lecture is to explain the physical causes of critical slowing-down in conventional Monte Carlo algorithms, and some recent progress in devising new Monte Carlo algorithms with radically reduced critical slowing-down.

3. Monte Carlo Methods for Spin Models and Lattice Field Theories

Consider, for purposes of illustration, a ϕ^4 model on a finite lattice $\Lambda \subset \mathbf{Z}^d$ of linear size L. The Hamiltonian is

$$H = -\frac{J}{2}\sum_{|i-j|=1} \phi_i\phi_j + \sum_i (A\phi_i^2 + \lambda\phi_i^4) \tag{3.1}$$

and the Gibbs measure is

$$d\mu(\phi) = Z^{-1} e^{-H(\phi)}\, d\phi . \tag{3.2}$$

A typical Monte Carlo method for this problem is the *single-site heat-bath algorithm*, defined as follows: We visit the sites $i \in \Lambda$ in some order (systematic or random); when working on site i we erase the old spin value ϕ_i and replace it by a new value ϕ_i' taken from the Gibbs conditional distribution of spin i given its neighbors,

$$P(\phi_i' \mid \{\phi_j\}_{j\neq i}) = \text{const}(\phi) \times \exp\left[J\phi_i' \sum_{j:\,|j-i|=1} \phi_j - A\phi_i'^{\,2} - \lambda\phi_i'^{\,4}\right] d\phi_i' \;; \tag{3.3}$$

all other spins are left unchanged.

Why does the single-site heat-bath algorithm have critical slowing-down? The key fact is that the updates are *local*: in a single step of the algorithm, "information" is transmitted only to nearest neighbors. One might guess that this "information" executes a random walk around the lattice. In order for the system to reach an "effectively new" configuration, the "information" has to travel a distance of order ξ, the (static) correlation length. One would guess, therefore, that $\tau \sim \xi^2$ near criticality. This guess is correct for the Gaussian model (ϕ^4 model with $\lambda = 0$). It is only roughly correct for other models: in general, $\tau \sim \xi^z$ where z is a dynamic critical exponent. For ϕ^4 and Ising models, we have $z \approx 2$ in all dimensions [7]. Note that the same heuristic argument applies to *all* Monte Carlo algorithms (*e.g.* Metropolis, Langevin) with purely *local* updates.

It follows that the computational work needed to get *one* "independent" sample behaves as $\sim L^d \xi^z \gtrsim \xi^{d+z}$. For accurate statistics one might want 10^6 "independent" samples. The reader is invited to plug in $L = 100$, $d = 3$ (or $d = 4$ if you're an elementary-particle physicist) and get depressed. Note that the factor ξ^d is inherent in *all* Monte Carlo algorithms for spin models (but not for self-avoiding walks, see Section 4). The factor ξ^z could, however, conceivably be reduced or eliminated by a more clever algorithm.

What is to be done? Our knowledge of the *physics* of critical slowing-down tells us that the slow modes are the long-wavelength modes, if the updating is purely local. The natural solution is therefore to speed up those modes by *collective-mode* (nonlocal) updating [8, 9]. Of course, this is easier said than done: one must decide exactly *which* collective modes to speed up and *how*; and the algorithm must not be so nonlocal that its computational cost outweighs the reduction in critical slowing-down. Specific implementations of the collective-mode idea are thus highly model-dependent. At least three such algorithms have been invented so far:

- Fourier acceleration [9-11]
- Multi-grid Monte Carlo (MGMC) [12-17]
- Swendsen-Wang algorithms [18,19]

Fourier acceleration and MGMC are very similar in spirit (though quite different technically); their performance is probably roughly similar. I'll describe MGMC very briefly.

One way of looking at the single-site heat-bath algorithm is that it proposes to add an unspecified number t to the spin ϕ_i, and then chooses t according to the appropriate conditional probability distribution. MGMC generalizes this by proposing updates in which a single unspecified number t is to be added simultaneously to *all* the spins in a subset $B \subset \Lambda$; again, t is chosen according to the appropriate conditional distribution. In MGMC, the subset B is taken successively to run through single-element subsets, cubes of side 2, cubes of side 4, etc. in a suitable order. Thus, MGMC updates *collective modes* (indicator functions of the blocks B) on *all length scales*. (In truth, this is not how the computation is performed, but it is mathematically equivalent and probably easier to understand physically.) I realize that this description of MGMC is so sketchy as to be probably incomprehensible. A somewhat different incomprehensible description can be found in [13]; a more detailed and hopefully comprehensible treatment will soon be available [14].

Variants on this algorithm can be devised for nonlinear σ-models and for $U(1)$ lattice gauge theories with or without *bosonic* matter fields [13-16]. We don't yet know how to do non-Abelian gauge theories, but we're working on it!

Note that MGMC is philosophically very similar to the block-spin renormalization group (RG), but technically very different: RG *integrates out* the short-wavelength degrees of freedom, while MGMC *conditions* on them.

How well does MGMC perform? The answer is highly model-dependent:

- For the *Gaussian model*, it can be proven rigorously [13, 17] that τ is *bounded* as criticality is approached (empirically $\tau \approx 1-2$); the gain in efficiency over traditional algorithms thus grows to *infinity* near the critical point. This behavior is to be expected: the correct collective modes in the Gaussian model are sine waves, and the upshot of the MGMC convergence proof is that piecewise-constant waves are nearly as good.

• For the ϕ^4 *model*, numerical experiments [13, 14] show that τ diverges with the *same* dynamic critical exponent as in the heat-bath algorithm; the gain in efficiency thus approaches a *constant* factor, empirically ≈ 10. This behavior can be understood [14] as due to the double-well nature of the ϕ^4 potential, which makes MGMC ineffective on large blocks. Thus, the correct collective modes at long length scales are nonlinear excitations *not* well modelled by $\phi \rightarrow \phi + t\chi_B$; understanding what these excitations *are* would be the key to inventing a better collective-mode algorithm for the ϕ^4 model.

• For the $d = 2$ *XY model*, preliminary data [15] show a more complicated behavior: As the critical temperature is approached from above, τ seems to diverge with approximately the same dynamic critical exponent as in the heat-bath algorithm; thus, the gain in efficiency approaches a constant factor, of order ≈ 10. On the other hand, below the critical temperature, τ seems to be very small ($\approx 2-3$); since for the heat-bath algorithm τ is infinite in this regime, the gain in efficiency is infinite as well. This behavior can also be understood physically: in the low-temperature phase the main excitations are spin waves, which are well handled by MGMC (as in the Gaussian model); but near the critical temperature the important excitations are widely separated vortex-antivortex pairs, which are apparently not easily created by the MGMC updates.

A very different type of collective-mode Monte Carlo algorithm was proposed by Swendsen and Wang [18] for Potts models, and later generalized by Edwards and myself to arbitrary models [19]. Basically, the idea is to augment the given model by means of *auxiliary variables*, and then to simulate this augmented model. It goes as follows:

Consider an arbitrary statistical-mechanical model with dynamical variables $\{\phi\}$ and probability distribution

$$d\mu(\phi) = Z^{-1} \left(\prod_b W_b(\phi) \right) d\mu_0(\phi) , \tag{3.4}$$

where the Boltzmann weight is decomposed as a product of terms W_b which we assume normalized to $0 \leq W_b \leq 1$, and $d\mu_0$ is some *a priori* measure (which is usually a product measure, though it need not be). In most applications the terms W_b in (3.4) would be associated with sites, links or plaquettes of a regular lattice, but they need not have any geometric significance at all. We now introduce, for each b, a variable κ_b taking values in the interval $[0,1]$, and use the trivial identity $W_b = \int_0^1 \theta(W_b - \kappa_b)\, d\kappa_b$. This leads us to define a joint (interacting) model of the fields $\{\phi\}$ and $\{\kappa\}$, with probability distribution

$$d\mu_{joint}(\phi, \kappa) = Z_{joint}^{-1} \left(\prod_b \theta(W_b(\phi) - \kappa_b) \right) d\mu_0(\phi) \prod_b d\kappa_b , \tag{3.5}$$

where $d\kappa_b$ is Lebesgue measure on $[0,1]$. It is immediately obvious that the joint model (3.5) has the following properties:

a) The marginal distribution on the $\{\phi\}$ variables (*i.e.* integrating out the $\{\kappa\}$) is precisely the original model $d\mu$.

b) The conditional distribution of the $\{\kappa\}$ given the $\{\phi\}$ is as follows: independently for each "bond" b, κ_b is a real number uniformly distributed in the interval $[0, W_b(\phi)]$.

c) The conditional distribution of the $\{\phi\}$ given the $\{\kappa\}$ is the restriction of the *a priori* measure $d\mu_0$ to the set of $\{\phi\}$ satisfying the constraints $W_b(\phi) \geq \kappa_b$ for all b.

The Swendsen-Wang (SW) algorithm then simulates the joint model (3.5) by alternately applying the conditional distributions (b) and (c) — that is, by alternately generating new $\{\kappa\}$ values (independent of the old ones) given the "spins" $\{\phi\}$, and new spin variables (independent of the old ones, if this is feasible) given the $\{\kappa\}$.

The SW algorithm becomes particularly simple for the ferromagnetic Potts model, in which the partial Boltzmann factors W_b take only the two values 1 and $w_{b,min}$: then all relevant information in the real-valued variables κ_b is contained in the dichotomous variables $n_b \equiv \theta(\kappa_b - w_{b,min})$. The conditional distribution (c) can then be described elegantly as follows: find the maximal clusters of sites which are connected by "occupied" bonds ($n_b = 1$), and assign a new spin value randomly to each cluster.

It is certainly plausible that the SW algorithm might have less critical slowing-down than the conventional algorithms: the reason is that a local move in one set of variables can have highly nonlocal effects in the other — for example, setting $n_b = 0$ on a single bond may disconnect a cluster, causing a big subset of the spins in that cluster to be flipped simultaneously. In some sense, therefore, the SW algorithm is a collective-mode algorithm in which the collective modes are *chosen by the system* rather than imposed from the outside as in multi-grid.

How well does the SW algorithm perform? For ferromagnetic Potts models, the performance is nothing short of extraordinary. Table 1 shows some preliminary data [20] on a two-dimensional Ising model at the bulk critical temperature. These data are consistent with $\tau_{int,\mathcal{E}} \sim L^{\approx 0.3}$ [18], and it is even conceivable that $\tau_{int,\mathcal{E}}$ is *bounded* as $L \to \infty$, *i.e.* no critical slowing-down at all! On the 512×512 lattice, the SW algorithm is over 1000 times more efficient than a single-site heat-bath (or Metropolis) algorithm. For the 3- and 4-state Potts models in two dimensions and the Ising model in three dimensions, the performance of the SW algorithm is only slightly less spectacular. On the other hand, for the two-dimensional *XY* model the SW algorithm performs [19] qualitatively like MGMC: there is significant critical slowing-down on the high-temperature side of criticality, but none on the low-temperature side of criticality. Unfortunately, I have at present no good physical understanding of this behavior.

Table 1. Susceptibility χ and autocorrelation time $\tau_{int,\mathcal{E}}$ ($\mathcal{E}$ = energy $\approx$ slowest mode) for two-dimensional Ising model at criticality, using Swendsen-Wang algorithm. Error bars are $\pm$ one standard error.

L	χ	$\tau_{int,\mathcal{E}}$
64	1575 ± 10	5.25 ± 0.30
128	5352 ± 53	7.05 ± 0.67
256	17921 ± 109	6.83 ± 0.40
512	59504 ± 632	7.99 ± 0.81

Collective-mode Monte Carlo is thus a general philosophy, not a cut-and-dried recipe. For each model, the challenge is to find the correct collective modes, and then to use this physical knowledge (and some ingenuity) to invent an efficient Monte Carlo algorithm. The field is wide open!

4. Monte Carlo Methods for the Self-Avoiding Walk

An *N-step self-avoiding walk* ω on a lattice $\mathcal{L}$ is a sequence of *distinct* points $\omega_0, \omega_1, \ldots, \omega_N$ in $\mathcal{L}$ such that each point is a nearest neighbor of its predecessor. Let c_N [resp. $c_N(x)$] be the number of N-step SAWs starting at the origin and ending anywhere [resp. ending at x]. Let $\langle \omega_N^2 \rangle$ be the mean-square end-to-end distance of an N-step SAW. These quantities are believed to have the asymptotic behavior

$$c_N \sim \mu^N N^{\gamma - 1} \tag{4.1}$$

$$c_N(x) \sim \mu^N N^{\alpha_{\text{sing}} - 2} \qquad (x \text{ fixed} \neq 0) \tag{4.2}$$

$$\langle \omega_N^2 \rangle \sim N^{2\nu} \tag{4.3}$$

as $N \to \infty$; here γ, α_{sing} and ν are critical exponents, while μ (the connective constant of the lattice) is the analogue of a critical temperature. The SAW has direct application in polymer physics [21], and is indirectly relevant to ferromagnetism and quantum field theory by virtue of its equivalence with the $n \to 0$ limit of the n-vector model [22].

The SAW has some advantages over spin systems for Monte Carlo work: Firstly, one can work directly with SAWs on an infinite lattice; there are no systematic errors due to finite-volume corrections. Secondly, there is no L^d (or ξ^d) factor in the computational work, so one can go closer to criticality. Thus, the SAW is an exceptionally advantageous "laboratory" for the numerical study of critical phenomena.

Different aspects of the SAW can be probed in three different ensembles:

- Free-endpoint grand canonical ensemble (variable N, variable x)
- Fixed-endpoint grand canonical ensemble (variable N, fixed x)
- Canonical ensemble (fixed N, variable x)

In the remainder of this section we survey some typical Monte Carlo algorithms for these ensembles.

- *Free-endpoint grand canonical ensemble.* Here the configuration space S is the set of all SAWs, of arbitrary length, starting at the origin and ending anywhere. The grand partition function is

$$\Xi(\beta) = \sum_{N=0}^{\infty} \beta^N c_N \tag{4.4}$$

and the Gibbs measure is

$$\mu(\omega) = \Xi(\beta)^{-1} \times \beta^{|\omega|}. \tag{4.5}$$

The "monomer activity" β is a user-chosen parameter satisfying $0 \leq \beta < \beta_c \equiv \mu^{-1}$. As β approaches the critical activity β_c, the average walk length $\langle N \rangle$ tends to infinity.

A dynamic Monte Carlo algorithm for this ensemble was proposed by Berretti and Sokal [23]. Its elementary moves are to delete the last link of the chain ($\Delta N = -1$) or to append one link to the chain ($\Delta N = +1$); the relative probabilities are chosen so as to leave the Gibbs measure μ invariant. This algorithm does have critical slowing-down: heuristic arguments and numerical evidence suggest that $\tau \sim \langle N \rangle^{\approx 2}$, and slightly weaker bounds have recently been proven [24,25]. It is worth comparing the computational work required for SAW versus Ising simulations: $\langle N \rangle^2 \sim \xi^{2/\nu} = \xi^{\approx 3.4}$ for the $d=3$ SAW, versus $\xi^{d+z} = \xi^{\approx 5.0}$ (resp. $\xi^{\approx 3.8}$) for the $d=3$ Ising model using the Metropolis (resp. Swendsen-Wang) algorithm. This vindicates our assertion that the SAW is an advantageous model for Monte Carlo studies of critical phenomena.

• *Fixed-endpoint grand canonical ensemble*. The configuration space S is the set of all SAWs, of arbitrary length, starting at the origin and ending at the fixed site x ($\neq 0$). The ensemble is as in the free-endpoint case, with c_N replaced by $Nc_N(x)$.

A dynamic Monte Carlo algorithm for this ensemble was proposed by Berg and Foerster [26] and Aragão de Carvalho, Caracciolo and Fröhlich [22, 27] (BFACF). The elementary moves are local deformations of the chain, with $\Delta N = 0, \pm 2$. The critical slowing-down in the BFACF algorithm is quite subtle. A heuristic argument [28, 29] suggests that $\tau \sim \langle N \rangle^{2+2\nu}$, but this is very likely wrong! Indeed, Sokal and Thomas [30] have recently proven the surprising result that $\tau_{exp} = +\infty$ for *all* $\beta \neq 0$. On the other hand, numerical experiments [29] show that $\tau_{int,N} \sim \langle N \rangle^{3.0 \pm 0.4}$ (in $d=2$). Clearly, the BFACF dynamics is not well understood at present: further work, both theoretical and numerical, is needed.

• *Canonical ensemble*. Algorithms for this ensemble, based on local deformations of the chain, have been used by polymer physicists for more than 25 years [31, 32]. So the recent proof [33] that all such algorithms are nonergodic comes as a slight embarrassment. Fortunately, there does exist a *non-local* algorithm which is ergodic: the "pivot" algorithm, invented by Lal [34] and independently reinvented by MacDonald *et al.* [35] and by Madras [36]. The elementary move is as follows: choose at random a pivot point k along the walk ($1 \leq k \leq N-1$); choose at random a non-identity element of the symmetry group of the lattice (rotation or reflection); then apply the symmetry-group element to $\omega_{k+1}, \ldots, \omega_N$ using ω_k as a pivot. The resulting walk is accepted if it is self-avoiding; otherwise it is rejected and the walk ω is counted once more in the sample. It can be proven [36] that this algorithm is ergodic and preserves the equal-weight probability distribution.

At first thought the pivot algorithm sounds terrible (at least it did to me): for N large, nearly all the proposed moves will get rejected. This is in fact true: the acceptance fraction behaves $\sim N^{-p}$ as $N \to \infty$, where $p \approx 0.19$ in $d=2$ [36]. On the other hand, the pivot moves are very radical: after very few (5 or 10) accepted moves the SAW will have reached an "essentially new" conformation. One conjectures, therefore, that $\tau \sim N^p$. Actually it is necessary to be a bit more careful: for *global* observables A (such as the end-to-end distance ω_N^2) one expects $\tau_{int,A} \sim N^p$; but *local* observables (such as the angle between the 17th and 18th bonds of the walk) are expected to evolve a factor of N more slowly: $\tau_{int,A} \sim N^{1+p}$. Thus, the *slowest* mode is expected to behave as $\tau_{exp} \sim N^{1+p}$. For the pivot algorithm applied to *ordinary* random walk one can calculate the dynamical behavior exactly [36]: for *global* observables A the autocorrelation function behaves roughly like

$$\rho_{AA}(t) \sim \sum_{i=1}^{n} (1 - \frac{i}{n})^t , \tag{4.6}$$

from which it follows that

$$\tau_{int,A} \sim \log N \tag{4.7}$$

$$\tau_{exp,A} \sim N \tag{4.8}$$

— in agreement with our heuristic argument modulo logarithms. For the SAW, it is found numerically [36] that $\tau_{int,A} \sim N^{\approx 0.20}$ in $d = 2$, also in close agreement with the heuristic argument.

A careful analysis of the computational complexity of the pivot algorithm [36] shows that one "effectively independent" sample (at least as regards *global* observables) can be produced in a computer time of order N. This is a factor $\sim N$ more efficient than the Berretti-Sokal algorithm, a fact which opens up exciting prospects for high-precision Monte Carlo studies of critical phenomena in the SAW. Thus, with a modest computational effort (300 hours on a Cyber 170-730), Madras and I found $\nu = 0.7496 \pm 0.0007$ (95% confidence limits) for 2-dimensional SAWs of lengths $200 \leq N \leq 10000$ [36]. We hope to carry out soon a convincing numerical test of hyperscaling in the three-dimensional SAW.

5. Multi-Grid Algorithm for the Random Resistor Problem

The random resistor problem [37, 38] is a simplified model of electrical conduction in composite materials. Consider a lattice of $L + 1$ sites in the vertical direction (free boundaries) and L sites in each of the $d - 1$ horizontal directions (periodic boundary conditions). Let each nearest-neighbor bond be a 1-ohm resistor with probability p, or an infinite resistor with probability $1 - p$ (independently on each bond). Now attach conducting plates to the top and bottom faces, connect a 1-volt battery, and measure the current flow (*i.e.* the conductance G_L). Then

$$\Sigma = \lim_{L \to \infty} \langle G_L \rangle / L^{d-2} \tag{5.1}$$

defines an "effective conductivity" of the random medium; we would like to know how Σ depends on p. Clearly Σ vanishes whenever p is less than the percolation threshold p_c; and it can be proven that Σ is nonvanishing for $p > p_c$ [39]. It is thus reasonable to expect that

$$\Sigma \sim (p - p_c)^t \quad \text{as } p \downarrow p_c \tag{5.2}$$

for some critical exponent t which we would like to compute.

One obvious approach to this problem is direct Monte Carlo simulation combined with finite-size scaling. Indeed, the Monte Carlo aspect of this problem is trivial, since it is simply independent bond percolation (no critical slowing-down here!). The hard part is the *deterministic* problem of computing the conductance for a given resistor configuration. This boils down to solving Kirchhoff's equations for the current flow, or equivalently a Laplace-like equation for the node voltages — a large, sparse, and highly ill-conditioned system of linear equations.

Direct methods of solution, such as sophisticated variants of Gauss elimination, have been tried with modest success [38, 40], but the computational labor grows as L^{3d-3} when $p = 1$ (albeit more slowly when $p \approx p_c$) [41]. Iterative methods of solution require a labor of only order L^d per iteration (when $p = 1$), but typically suffer from critical slowing-down, so that the number of iterations required grows as a power of L. In fact, the critical slowing-down is much more severe in disordered problems ($p \approx p_c$) than in regular problems ($p = 1$).

In recent years much effort has been devoted to devising improved iterative algorithms with reduced critical slowing-down. In particular, Fourier preconditioning [42, 10] has been shown to reduce appreciably the critical slowing-down in the random-resistor problem [43], but severe critical slowing-down still remains. This is because the preconditioning operator mimics the *ensemble-averaged* current flow in the lattice, but the current flow in any *particular* realization of the random resistor network is affected strongly by the local and global topology of interconnections. This reasoning suggests that an improved strategy must take account of the topology of the particular resistor network at hand.

The multi-grid method [44] is known to be an extraordinarily effective approach for solving wide classes of large linear systems arising from the discretization of elliptic partial differential equations (PDEs). Usually the coarse grids and interpolation operators are defined geometrically, *e.g.* cubical (2×2) blocks with piecewise-constant or piecewise-linear interpolation. This approach, which is suitable for PDEs with smooth coefficients, will clearly *not* be appropriate in disordered systems such as the random-resistor network: just because two sites are close geometrically does not mean that they are close in the topology of the resistor network and hence in voltage. Rather, a successful multi-grid algorithm will have to define its coarse grids and interpolation operators in accordance with the connection structure of the particular resistor network.

For the past year I have been trying to come up with an algorithm for doing this, without much success. Very recently, however, I realized that an identical philosophy has been pursued for a number of years under the name of *algebraic multi-grid* (AMG) [45]: these are "black-box" solvers which, given an arbitrary matrix (of a certain type), attempt to choose appropriate coarse grids and interpolation operators. Of course, there is no guarantee as to how well the algorithm will perform on any given problem. Heretofore the AMG codes have been applied to partial differential equations with strongly discontinuous coefficients [45, 46], but not to problems as singular as the random-resistor network.

So we decided to try out a "canned" AMG code (AMG1R4 [47]) on the random resistor problem. We expected that it would reduce, but not eliminate, the critical slowing-down, and we hoped to gain some insight into how to exploit the specific features of the problem to obtain further improvements. To our surprise, we have found that AMG1R4 succeeds in eliminating entirely the critical slowing-down in the two-dimensional random-resistor problem. Some preliminary results [48] are shown in Table 2.

Here the "convergence factor" is the (worst-case) factor by which the norm of the error is reduced at each iteration; critical slowing-down would correspond to this factor becom-

Table 2. Mean convergence factors for algebraic multigrid (AMG) algorithm on two-dimensional random resistor problem at percolation threshold, computed on either connected cluster or current-carrying backbone. Error bars are ± one standard error.

L	Convergence Factor	
	Cluster	Backbone
100	0.348 ± 0.002	0.264 ± 0.002
200	0.403 ± 0.001	0.319 ± 0.001
400		0.362 ± 0.001

ing very close to 1. Thus, for the 400×400 lattice we can obtain the solution to six-decimal-place accuracy in about 10 iterations of the AMG method, while it would take roughly 10 *million* iterations using the Gauss-Seidel method!

We are now trying to devise AMG algorithms to deal with the fermion problems which arise in elementary-particle physics (Dirac propagator in a random gauge field) and in condensed-matter physics (Hubbard model).

This research was supported in part by NSF grants PHY-8413569 and DMS-8705599 and John Von Neumann Supercomputer Center grant NAC-705.

REFERENCES

1. G.A. Baker Jr., Phys. Rev. **B15**, 1552 (1977); M.E. Fisher and J.-H. Chen, J. Physique **46**, 1645 (1985); A.J. Guttmann, Phys. Rev. **B33**, 5089 (1986).
2. J.M. Hammersley and D.C. Handscomb, *Monte Carlo Methods* (Methuen, London, 1964), chap. 5.
3. J.G. Kemeny and J.L. Snell, *Finite Markov Chains* (Springer, New York, 1976).
4. M. Iosifescu, *Finite Markov Processes and Their Applications* (Wiley, Chichester, 1980).
5. K.L. Chung, *Markov Chains with Stationary Transition Probabilities*, 2nd ed. (Springer, New York, 1967).
6. E. Nummelin, *General Irreducible Markov Chains and Non-Negative Operators* (Cambridge Univ. Press, Cambridge, 1984).
7. G.F. Mazenko and O.T. Valls, Phys. Rev. **B24**, 1419 (1981); J.K. Williams, J. Phys. **A18**, 49 (1985); R.B. Pearson, J.L. Richardson and D. Toussaint, Phys. Rev. **B31**, 4472 (1985); S. Wansleben and D.P. Landau, J. Appl. Phys. **61**, 3968 (1987).
8. M. Kalos, in Proceedings of the Brookhaven Conference on Monte Carlo Methods and Future Computer Architectures, May 1983 (unpublished).
9. G. Parisi, in *Progress in Gauge Field Theory* (1983 Cargèse lectures), ed. G. 't Hooft *et al.* (Plenum, New York, 1984).
10. G.G. Batrouni *et al.*, Phys. Rev. **D32**, 2736 (1985).
11. J.B. Kogut, Nucl. Phys. **B275 [FS17]**, 1 (1986).
12. A. Brandt, D. Ron and D.J. Amit, in *Multigrid Methods II* (Lecture Notes in Mathematics #1228), ed. W. Hackbusch and U. Trottenberg (Springer, Berlin, 1986).
13. J. Goodman and A.D. Sokal, Phys. Rev. Lett. **56**, 1015 (1986).
14. J. Goodman, A.D. Sokal and D. Zwanziger, in preparation.
15. R.G. Edwards, J. Goodman, A.D. Sokal and D. Zwanziger, in preparation.
16. R.G. Edwards, J. Goodman, D. Ritchie, A.D. Sokal and D. Zwanziger, in preparation.
17. J. Goodman and A.D. Sokal, in preparation.
18. R.H. Swendsen and J.-S. Wang, Phys. Rev. Lett. **58**, 86 (1987).

19. R.G. Edwards and A.D. Sokal, NYU preprint (1988).

20. R.G. Edwards and A.D. Sokal, in preparation.

21. P.G. DeGennes, *Scaling Concepts in Polymer Physics* (Cornell Univ. Press, Ithaca NY, 1979).

22. C. Aragão de Carvalho, S. Caracciolo and J. Fröhlich, Nucl. Phys. **B215 [FS7]**, 209 (1983).

23. A. Berretti and A.D. Sokal, J. Stat. Phys. **40**, 483 (1985).

24. G.F. Lawler and A.D. Sokal, Trans. Amer. Math. Soc. (to appear).

25. A.D. Sokal and L.E. Thomas, to be submitted to J. Stat. Phys.

26. B. Berg and D. Foerster, Phys. Lett. **106B**, 323 (1981).

27. C. Aragão de Carvalho and S. Caracciolo, J. Physique **44**, 323 (1983).

28. A.D. Sokal, Comparative analysis of Monte Carlo methods for the self-avoiding walk, preliminary draft (1984).

29. S. Caracciolo and A.D. Sokal, J. Phys. **A19**, L797 (1986).

30. A.D. Sokal and L.E. Thomas, J. Stat. Phys. (to appear).

31. M. Delbrück, in *Mathematical Problems in the Biological Sciences* (Proc. Symp. Appl. Math., vol. 14), ed. R.E. Bellman (American Math. Soc., Providence RI, 1962).

32. P.H. Verdier and W.H. Stockmayer, J. Chem. Phys. **36**, 227 (1962).

33. N. Madras and A.D. Sokal, J. Stat. Phys. **47**, 573 (1987).

34. M. Lal, Molec. Phys. **17**, 57 (1969).

35. B. MacDonald *et al.*, J. Phys. **A18**, 2627 (1985).

36. N. Madras and A.D. Sokal, J. Stat. Phys. **50**, ___ (1988).

37. S. Kirkpatrick, Rev. Mod. Phys. **45**, 574 (1973).

38. S. Kirkpatrick, in *Ill-Condensed Matter* (Les Houches 1978), ed. R. Balian *et al.* (North-Holland, Amsterdam, 1979).

39. J.T. Chayes and L. Chayes, Commun. Math. Phys. **105**, 133 (1986).

40. D.J. Frank and C.J. Lobb, Phys. Rev. **B37**, 302 (1988).

41. R.J. Lipton, D.J. Rose and R.E. Tarjan, SIAM J. Numer. Anal. **16**, 346 (1979).

42. P. Concus, G.H. Golub and D.P. O'Leary, in *Sparse Matrix Computations*, ed. J.R. Bunch and D.J. Rose (Academic Press, New York, 1976).

43. G.G. Batrouni, A. Hansen and M. Nelkin, Phys. Rev. Lett. **57**, 1336 (1986).

44. A. Brandt, Math. Comp. **31**, 333 (1977); W. Hackbusch and U. Trottenberg, eds., *Multigrid Methods* (Lecture Notes in Mathematics #960) (Springer, Berlin, 1982); W. Hackbusch, *Multi-Grid Methods and Applications* (Springer, Berlin, 1985); W.L. Briggs, *A Multigrid Tutorial* (SIAM, Philadelphia, 1987).

45. J. Ruge and K. Stüben, in *Multigrid Methods for Integral and Differential Equations*, ed. D.J. Paddon and H. Holstein (Clarendon Press, Oxford, 1985); A. Brandt, Appl. Math. Comput. **19**, 23 (1986); J. Ruge and K. Stüben, in *Multigrid Methods*, ed. S.F. McCormick (SIAM, Philadelphia, 1987).

46. R.E. Alcouffe *et al.*, SIAM J. Sci. Statist. Comput. **2**, 430 (1981).

47. R. Hempel, private communication via J. Ruge.

48. R.G. Edwards, J. Goodman and A.D. Sokal, in preparation.

Multigrid Monte Carlo Methods

E. Loh, Jr.

Theoretical Division and Center for Nonlinear Studies,
Los Alamos National Laboratory, Los Alamos, NM87545, USA

This paper is intended to be a tutorial on multigrid Monte Carlo techniques, illustrated with two examples. Path-integral quantum Monte Carlo is seen to take only a finite amount of computer time even as the paths are discretized on infinitesimally small scales. A method for eliminating critical slowing down completely — even for models with discrete degrees of freedom, as in Potts models, or discrete excitations, such as isolated vortices in the XY model — is presented.

1. Introduction

Monte Carlo methods have been used quite successfully to investigate many-body systems in condensed-matter physics. Stochastic simulations offer the theorist a tool to examine models in very-high-dimensional spaces in parameter regimes which are inaccessible to analytical methods. Yet many of the interesting phenomena in many-body problems occur at or near critical points in parameter space where the critically slow physical dynamics, which make these points interesting, affect the simulational dynamics as well. Hence, critical phenomena have been a particularly challenging source of difficulties for traditional Monte Carlo techniques.

A variety of attempts have been made to beat such critical slowing down. I would like to mention two other speakers at this workshop in particular. The very first talk was given by Bob Swendsen, who spoke about non-universal dynamics in Monte Carlo simulations. Swendsen and coworkers utilized a mapping between Potts models, such as the Ising model, and percolation models, which do not suffer from critical slowing down, to change the dynamics of Potts simulations. This mapping was first described by FORTUIN and KASTELEYN [1], used for Monte Carlo simulations by SWEENY [2], and adapted differently by SWENDSEN and WANG [3]. The resultant simulation flips disconnected, stochastically defined, clusters at random. Since, at criticality, these clusters are present at all length scales, the dynamics of the simulation slow down much less dramatically with system size than for local-update algorithms.

Also, Alan Sokal described multigrid methods which would block all variables in a local domain together into a single coarse-grid variable. Coarsening deterministically in this way, these simulations would perform lattice updates on every length scale and so, again, change the dynamics of the simulation. Such an approach proves to eliminate critical slowing down completely for "trivial" Hamiltonians such as the gaussian model. For systems with discrete excitations, on the other hand, the acceptance ratio goes down exponentially with the length scale of the Monte Carlo move. Thus, these multigrid methods can only offer savings in computer time by collecting degrees of freedom and processing them together and by thermalizing continuous excitations. The discrete excitations, in contrast,

Springer Proceedings in Physics Vol. 33: **Computer Simulation Studies in Condensed Matter Physics**
Editors: D. P. Landau • K. K. Mon • H-B. Schüttler

still suffer from critical slowing down. This has been observed in simple-blocking simulations of the XY model above the Kosterlitz-Thouless temperature, where the discrete excitations are the isolated vortices, and in a ϕ^4 scalar-field model in a double-well potential [4]. Other, self-admittedly disappointing, attempts [5] have been made to generalize the Fortuin-Kasteleyn mapping to arbitrary models. And, of course, there have been many other efforts to address critical slowing down that have not been represented at this workshop.

In this talk, I will try to give a tutorial on multigrid methods by discussing two very simple models: a single quantum-mechanical particle in a harmonic well and the Potts model. From the harmonic oscillator, we will see how high-order interpolation schemes can be used to accelerate simulations of systems with continuous degrees of freedom. The thermodynamics of a quantum particle may be studied by summing over paths of the particle in imaginary time, discretized in nonzero time steps $\Delta\tau$. We will see that the study can achieve the accuracy of an arbitrarily small discretization step even in a finite amount of computer time through the use of multigrid methods. I will also discuss a stochastic coarsening procedure, which was proposed by BRANDT [6] and studied by KANDEL et al.[7]. For the Potts model, the procedure reduces to SWENDSEN and WANG's use of FORTUIN and KASTELEYN's mapping to percolation models. Using the coarsening procedure to employ multigrid ideas provides a means for eliminating critical slowing down completely even in the presence of discrete excitations. This is seen in multigrid simulations on the two-dimensional Ising model.

Most of all, multigrid ideas are only starting to be used in Monte Carlo simulations. What I hope to do, then, is simply to invite others to think about multigridding, using natural geometrical considerations in improving simulational methods, especially in the study of critical phenomena.

2. Quantum mechanical harmonic oscillator

We first consider a single quantum-mechanical particle in a harmonic well. The Hamiltonian describing the motion of the particle is $H = T + V$ with

$$T = \frac{p^2}{2m} \qquad\qquad V = \frac{m\omega^2 x^2}{2} \quad , \tag{2.1}$$

where the quantum-mechanical nature of the particle arises from the fact that $[p, x] = -i\hbar \neq 0$. Of course, the natural solution to this model comes from rewriting the position and momentum coordinates in terms of the creation and annihilation operators $b^\dagger$ and b, resulting in the diagonal, second-quantized Hamiltonian $H = \hbar\omega(b^\dagger b + 1/2)$.

Instead, anticipating numerical solutions of less tractable quantum Hamiltonians, we adopt Feynmann's path-integral formalism. While other speakers have already developed this formalism, I present the essential features here once more. Expanding the partition function

$$Z = \mathrm{Tr}\ e^{-\beta H} = \mathrm{Tr}\ e^{-\Delta\tau\ H}\ e^{-\Delta\tau\ H}\ \cdots\ e^{-\Delta\tau\ H} \tag{2.2}$$

as a product of L identical factors, we may now approximate $\exp(-\Delta\tau\ H) \approx \exp(-\Delta\tau\ T)\ \exp(-\Delta\tau\ V)$ if $\Delta\tau = \beta/L$ is made suitably small. Working in a position-coordinate basis, the factor $\exp(-\Delta\tau\ V)$ is diagonal and easy to handle. The kinetic-energy factor, $\exp(-\Delta\tau\ T)$, is simply the free-particle propagator in imaginary time; it is the gaussian

$$< x'|e^{-\Delta\tau\ T}|x >= \sqrt{\frac{m}{2\pi\Delta\tau\hbar^2}}\ \exp(-\Delta\tau\frac{m(x'-x)^2}{2\ (\Delta\tau\hbar)^2}) \quad , \tag{2.3}$$

which, of course, becomes a delta function in the limit $\Delta\tau \rightarrow 0$. In this limit, the resultant expression for the partition function becomes a path integral of the exponential $\exp(-S)$ of the action over all paths in imaginary time. For numerical simulations, we will use the approximate expression for Z arising from nonzero $\Delta\tau$. This so-called Trotter approximation has been well studied [8] and is the foundation of most quantum Monte Carlo work.

People speak of the quantum particle as being represented by a polymer or by a world line in imaginary time. In the latter picture, the world line is described only at discretized times $\Delta\tau$, $2\Delta\tau$, ... β. At each of the discretized times, the world-line coordinate feels an external force from the harmonic potential as well as spring forces from its neighbors in imaginary time. In the limit $\Delta\tau \rightarrow 0$, these spring forces become infinitely strong. Hence, the world line is continuous in imaginary time. Quantum mechanically, however, while such lines are continuous, they are not differentiable. This physical property manifests itself in simulations as statistical noise in measurements of observables which depend on short imaginary-time scales. As one goes to the zero-temperature limit, $\beta \rightarrow \infty$, the "polymer" becomes infinitely long with correlations on the imaginary-time scale $\tau \sim 1/\hbar\omega$. In the classical limit $\hbar \rightarrow 0$, the correlation time diverges and the world lines become completely straight.

Using the Trotter breakup, we have reduced the quantum-mechanical partition function to a high-dimensional sum amenable to classical Monte Carlo techniques. At this point, one may treat the sum using a local-update algorithm [9], accepting or rejecting proposed moves $x(\tau) \rightarrow x(\tau)+\delta$ according to the Metropolis algorithm. Unfortunately, such "wiggles" of the world lines require $\delta \sim \sqrt{\Delta\tau}$ for reasonable acceptance ratios, meaning that Monte Carlo moves must become very small. Furthermore, the Trotter formula assumes $\Delta\tau\ \hbar\omega \ll 1$, meaning that each local move can only affect a segment of the world line which is very short compared to the correlation time. Of course, this situation is exacerbated in the treatment of many-particle systems, for which the correlation times can grow much larger, especially if the system goes critical.

One solution to this difficulty was alluded to by Farid Abraham in his discussion of quantum He on graphite. One may "Fourier accelerate" the simulation by considering nonlocal moves of the form $x(\tau) \rightarrow x(\tau) + \delta_\Omega \cdot \cos(\Omega \cdot (\tau - \tau_0))$ for all points on the world line at once. Not only may one use much larger step sizes δ_Ω for small Ω, accelerating movement through phase space for the long time-scale modes, but one may also sample the various modes with different sampling frequencies. Fourier-accelerated Langevin simulations take advantage of this flexibility by assigning different "masses" to the various fourier modes of the system. For example, if one is interested in measuring only the particle's mean-square displacement, which depends only on $\Omega = 0$ characteristics, then it is straightforward to show that the high-Ω modes should be sampled with frequency $\sim\ \Omega^{-4}$ for optimal statistics. In contrast to most quantum Monte Carlo algorithms, for which more lattice sweeps are needed as $\Delta\tau \rightarrow 0$ to achieve the same quality statistics, this procedure does not slow down since only a finite number of lattice sweeps will be required for high-Ω modes, even as the number of these modes diverges. Put another way, the mean-square displacement depends somehow only on the very longest time scale. The Ω^{-4} rule only gives one a sense of how often the other modes must be sampled to ensure ergodicity — that is, to ensure that the system can sample all of phase space. Of course, the simulation does slow down in the sense that the computer time required for a single sweep increases as $\Delta\tau$ vanishes.

(In contrast to long-time-scale observables, consider measurements of the particle's kinetic energy. Naively, one would measure the expectation value of

$$< x(\tau+\Delta\tau)|T\exp(-\Delta\tau\ T)|x(\tau) > / < x(\tau+\Delta\tau)|\exp(-\Delta\tau\ T)|x(\tau) > \quad , \tag{2.4}$$

giving one the kinetic-energy estimator

$$<T> = \frac{1}{2\,\Delta\tau} - <\frac{m\,(x(\tau+\Delta\tau)-x(\tau))^2}{2\,(\Delta\tau\hbar)^2}> \quad . \tag{2.5}$$

The nondifferentiability of the world line requires the $1/2\Delta\tau$ term for convergence. The estimator (2.5) depends on all modes of the world line and would require one to sample one value of Ω as often as the next. As the imaginary time variable is discretized on a finer and finer scale, more and more lattice sweeps would be required and yet the measurements would still be more noisy.)

In most Monte Carlo simulations, one updates only one degree of freedom at a time, holding all others fixed for that update. Ergodicity is achieved by subsequently updating other degrees of freedom as well. In our multigrid approach, let us hold fixed interpolations of the world line while updating a position coordinate $x(\tau)$. Consider a short time segment $\tau_0-\Delta\tau<\tau<\tau_0+\Delta\tau$ of the world line, which is defined by its coordinates at the discretized times, $x(\tau_0-\Delta\tau)$, $x(\tau_0)$, and $x(\tau_0+\Delta\tau)$. We fix a particular interpolation of this segment by fixing the displacement $x(\tau_0)-(x(\tau_0-\Delta\tau)+x(\tau_0+\Delta\tau))/2$ of $x(\tau_0)$ from the linearly interpolated position $(x(\tau_0-\Delta\tau)+x(\tau_0+\Delta\tau))/2$. Higher- and lower-order interpolations are possible.

To multigrid the simulation, then, we first define the world line of the particle on a very fine grid — that is, the imaginary time is discretized on a very fine time scale. We then "decimate" the position coordinate at every other time step by memorizing its displacement from an interpolated value which is defined in terms of coordinates that are not decimated. Since degrees of freedom have been eliminated from the description of the world line, the cost of updates on the coarser levels decreases inversely with the time scale.

In order to perform updates on the coarser time scale, we must write expressions for the renormalized action of the path. There are two contributions to the action over the segment $\tau_0-\Delta\tau$, τ_0, $\tau_0+\Delta\tau$ — one due to the kinetic energy and the other due to the external potential. Writing $x_\pm = x(\tau_0\pm\Delta\tau)$ and $x_0=x(\tau_0)$ and fixing $c=x_0-(x_++x_-)/2$, get

$$\begin{aligned} S_{\mathrm{T}} &= \Delta\tau\frac{m((x_+-x_0)^2+(x_0-x_-)^2)}{2\,(\Delta\tau\,\hbar)^2} \\ &= \Delta\tau\frac{m((x_+-(\frac{x_++x_-}{2}+c))^2+((\frac{x_++x_-}{2}+c)-x_-)^2)}{2\,(\Delta\tau\,\hbar)^2} \\ &= (2\,\Delta\tau)\frac{m(x_+-x_-)^2}{2\,(2\,\Delta\tau\hbar)^2}+\Delta\tau\frac{mc^2}{(\Delta\tau\,\hbar)^2}. \end{aligned} \tag{2.6}$$

Hence the kinetic-energy contribution to the action on the coarser time scale has the same form as that on the finer scale with $\Delta\tau \to 2\Delta\tau$. The second term is simply a contribution to the action which depends only on the particular interpolation, c, and not on the coarse degrees of freedom.

The potential energy contribution to the action is $S_V = \Delta\tau \sum_\tau V(x(\tau))$. The contribution from the segment $\tau_0 - \Delta\tau,\ \tau_0,\ \tau_0 + \Delta\tau$ is therefore

$$(2\Delta\tau)\ \tilde{V}(x_+, x_-) = \Delta\tau(V(x_+, x_0) + V(x_0, x_-)) \quad , \tag{2.7}$$

where $\tilde{V}$ is the renormalized potential which depends on the particular interpolation and x_0, of course, is fixed with relation to the interpolated value $(x_+ + x_-)/2$. While $\tilde{V}$ grows in complexity as it is coarsened repeatedly, this is not true in several interesting cases. In particular, for the harmonic potential, V remains quadratic no matter how many times it is coarsened.

The multigrid algorithm, then, is composed of coarsenings, local Monte Carlo updates on the various time scales, and uncoarsenings. To go back to a finer time scale, one simply interpolates between the discretized times and adds back in the fixed displacements. Again, the number of times a fine time scale must be updated decreases rapidly with the time scale — in a quartic fashion $\sim\ \Omega^{-4}$ for the harmonic oscillator, for example — while the processing cost scales only with the inverse of the time scale. The total cost of processing the very fine time scales, then, is finite even as the finest level is discretized for a smaller and smaller $\Delta\tau$! This is in contrast to fourier-accelerated algorithms, whose processing time grows linearly with $1/\Delta\tau$, and to local-update Monte Carlo, for which processing grows algebraically with $1/\Delta\tau$ even faster than for the fourier moves.

Multigridding is expected to be essential up to the correlation time of the world line — in our harmonic oscillator, this time is $\sim\ 1/\hbar\omega$. For a system of many quantum particles, this correlation time could be longer and could even diverge critically. Again, the short time scales (Ω large) require only negligibly frequent sampling.

To make the connection to other collective moves, it should be noted that fourier moves correspond to adding cosine waves to the world line. In contrast, GOODMAN and SOKAL's [4] approach may be thought of as adding square pulses. The linear-interpolation scheme mentioned here corresponds to adding triangular pulses whose widths are the time scales at which the pulses are added.

In Fig. 1, the potential energy is plotted as a function of the logarithm G of the number of levels for a single quantum particle of unit mass at inverse temperature $\beta = 5$ in a harmonic well of level spacing $\hbar\omega = 1$. In the limit of few levels, G small, the error due to using a nonzero Trotter parameter $\Delta\tau = \beta \cdot 2^{-G}$ results in large deviations from the $\Delta\tau \to 0$ limit (the dotted line). Three curves are plotted, each representing the same number of processings at the very coarsest time scale, at which measurements of the potential energy were made. While the coarsest level was always sampled the same number of times, finer levels were only visited with relative sampling frequency $\sim\ \Omega^{-p}$, where, drawing from the fourier picture, Ω is the imaginary-time frequency corresponding to the time scale $\sim\ 1/\Omega$. The three curves are for $p = 1, 2, 3$. Notice that the three data sets are essentially indistinguishable. Both the data points and the error bars are independent of p; hence, it is not important to visit the fine time scales often. In

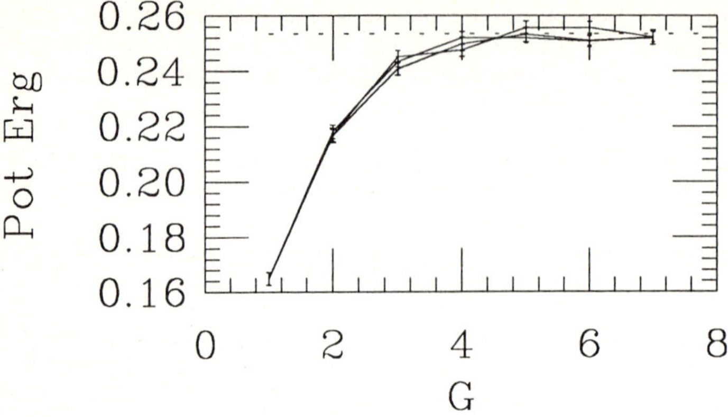

Fig. 1. Potential energy as a function of logarithm of the number of levels for three different sampling frequencies of the short time scales

particular, for $p = 3$, only a finite amount of computer time would be spent on the finest scales even while the number of such levels goes to infinity. Meanwhile, it is also clear from the systematic errors at small G that it is crucial to include these fine time scales to reduce the error due to the Trotter breakup. Notice that this statement is equivalent to the need for ergodicity — including fine time scales is of no consequence if these levels are never sampled. Our rule of thumb from the harmonic oscillator, again, is that $p \approx 4$ offers the best statistics. In our illustration, processing time grows exponentially, 2^G, for $p = 1$, linearly, G, for $p = 2$, and not at all for $p > 2$ as a function of the number, G, of levels.

To close out this section on the quantum particle, we may summarize by saying that multigrid methods allow one to study the limit $\Delta\tau \rightarrow 0$ in a finite amount of computer time. (If one chose to store all the path interpolations, however, the memory requirements would grow.) These methods offer efficient numerical solutions for quantum problems — indeed, this single-particle illustration was motivated by work in progress on anharmonic lattice dynamics, whose treatment by these methods is quite straightforward. Multigridding allows one to make large-scale moves through configuration space by stochastically eliminating degrees of freedom on smaller scales.

Nevertheless, one shortcoming, reflective more of path-integral formulations rather than multigrid Monte Carlo itself, must be mentioned. We have discussed measurements of long-time-scale observables, such as the potential energy of the system. Unfortunately, many interesting quantities — the specific heat and kinetic energy, as examples — depend on short-time behavior. As discussed above, however, the nondifferentiability of quantum world lines makes measurement of such quantities difficult. Inclusion of short-time contributions is required to remove systematic errors in the measured averages. Short time scales contribute little to the averages but substantially to the noise. Sampling these fluctuations more often is an unattractive tact as this would be quite time consuming, even in a multigrid approach. Several short-time-dependent observables, fortunately, can be expressed in terms of long-time averages. The specific heat, as many Monte Carlo workers well know, is often best measured by differentiating the energy, which is less dependent on short times than the specific heat itself, than by measuring fluctuations in the energy, which is what a specific-heat operator would do. The kinetic energy, in turn, can be measured using the virial theorem. As a generalization, we note that the expectation value of the commutator of the Hamiltonian $H = T + V$ with any operator A is zero. Thus, $< [T, A] >= - < [V, A] >$. Since $T \sim p^2$, $[T, A]$ has one fewer factor of x and one more factor of p than does A. On the other hand, V is composed only of position variables and so $[V, A]$ has one more factor of x and one fewer factor of p than does A. This identity, therefore, relates the averages of quantities with different numbers of p's, which have great short-time dependences, and x's, which have only long-time dependence. For example, $A = xp$ gives the virial theorem $< p^2/2m >=< xV'/2 >$. In practice, long-time estimators do not exist for all observables and when they do they are not always well behaved. Alternatively, one can express measurements as sums of contributions from each of the separate time scales as we saw in (2.6), for example, for the kinetic-energy contribution S_T to the action. The contributions, of course, become progressively more noisy as one goes to shorter time scales. On the other hand, in the limit of short time scales, one may construct estimates of the contributions that depend only on long-time averages. For example, for S_T, the contribution from the shortest time scale is $< m(x_0 - (x_+ + x_-)/2)^2/(\Delta\tau\ \hbar)^2 >$. Since correlations such as $< x(\tau + \Delta\tau)x(\tau) >$ and $< x(\tau + \Delta\tau)x(\tau - \Delta\tau) >$ can be written for small $\Delta\tau$ solely in terms of functions of position coordinates (functions of x, V, and its derivatives), long-time estimates of short-time contributions to averages can be constructed. It is not clear how useful this approach is in practice.

Of course, one could simply live with small $\Delta\tau$ errors and settle for using multigridding only to beat critical slowing down, which may arise in many-body problems.

3. Potts models

Now let us turn to the question of problems with discrete excitations. In the case of Potts models, for example, changes in the energy clearly must come in finite quanta which are not small on an energy scale set by the critical temperature. I include not only systems described by discrete degrees of freedom, but also models with continuous variables which have discrete excitations. In the classical XY model, for instance, the spin variables are continuous and give rise to continuous excitations such as spin waves. They also allow discrete excitations, however, such as individual vortices, which characterize the phase transition at the Kosterlitz-Thouless temperature. Our strategy is to design Monte Carlo moves on all length scales with energy changes always of the same, hopefully small, scale. Up to now we have discussed only moves which are everywhere gradual. Unfortunately, for discrete-variable models, such moves are no longer possible. For continuous-variable models, such Monte Carlo moves are possible, but they are ineffectual in thermalizing the discrete excitations. Alternatively, one could consider making non-gradual moves over domains of variables — flipping a prescribed domain of Ising spins, for example, or rotating a selected domain of XY spins. These moves suffer from exponentially decreasing acceptance ratios as the domains grow, meaning that in practice the simulation still does not incorporate large-scale moves.

Here I will describe a stochastic coarsening procedure proposed by BRANDT [6] which allows performing local updates on a coarser length scale even without increasing the scale of energy changes (which would decrease the acceptance ratio). Used in conjunction with multigridding techniques, this procedure allows the simulation of many-body problems without any critical slowing down.

We eliminate the finest length scales and so reduce the number of degrees of freedom with a stochastic coarsening procedure. Disregarding geometrical considerations for the moment, consider a model whose thermodynamics are governed now by a classical Hamiltonian $H = H_0 + V$, where factors of $-\beta = -1/k_B T$ have been absorbed into H. Here, k_B and T are Boltzmann's constant and the temperature, respectively, and H_0 is somehow easier to simulate than the original Hamiltonian. The probability of finding the system in some state Q is proportional to the Boltzmann weight $\exp(H(Q))$, where $H(Q)$ is the energy of the system in state Q. We may "kill" the contribution V to the Hamiltonian stochastically by either "deleting" it with probability $p_d = c_V \exp(-V(Q))$ or by "freezing" it with probability $p_f = 1 - p_d$. If the interaction is frozen, only states Q' with $V(Q') = V(Q)$ are considered in the ensuing simulation. If the interaction is deleted, no such restriction is placed on the states. In either case, the thermodynamics are subsequently governed only by the simplified Hamiltonian H_0. The coefficient c_V must be chosen so that $p_d, p_f \in [0,1]$ — p_d and p_f must be probabilities. The largest choice of c_V produces the best statistics. By assumption, H_0 is easier to study than H and so the simulation will proceed more efficiently than before.

Clearly, this procedure is strongly ergodic since there is always a nonzero probability that no restriction will be placed on the simulation, allowing nonzero transition probabilities between all states. It also satisfies detailed balance. To see this, first consider two states Q and Q' with $V(Q) \neq V(Q')$. Then, a transition from one state to the other can take place only if V has been deleted:

$$T(Q \longrightarrow Q') = c_V \, e^{-V(Q)} \cdot \frac{e^{H_0(Q')}}{Z_0} \quad , \tag{3.1}$$

where Z_0 is the partition function for the reduced Hamiltonian. Now,

$$\frac{T(Q \longrightarrow Q')}{T(Q' \longrightarrow Q)} = \frac{e^{-V(Q)} \, e^{H_0(Q')}}{e^{-V(Q')} \, e^{H_0(Q)}} = \frac{e^{H_0(Q')+V(Q')}}{e^{H_0(Q)+V(Q)}} = \frac{e^{H(Q')}}{e^{H(Q)}} \quad . \tag{3.2}$$

Alternatively, if $V(Q) = V(Q')$, the interaction V may either be deleted or frozen:

$$\begin{aligned} T(Q \longrightarrow Q') &= c_V \, e^{-V(Q)} \cdot \frac{e^{H_0(Q')}}{Z_0} + (1 - c_V \, e^{-V(Q)}) \cdot \frac{e^{H_0(Q')}}{Z_0^f} \\ &= \left(\frac{c_V \, e^{-V(Q)}}{Z_0} + \frac{1 - c_V \, e^{-V(Q)}}{Z_0^f}\right) e^{H_0(Q')} \quad , \end{aligned} \tag{3.3}$$

where Z_0^f is the partition function for the reduced Hamiltonian over the restricted space. Then, using $V(Q') = V(Q)$, we find

$$\frac{T(Q \longrightarrow Q')}{T(Q' \longrightarrow Q)} = \frac{e^{H_0(Q')}}{e^{H_0(Q)}} = \frac{e^{H_0(Q')+V(Q)}}{e^{H_0(Q)+V(Q)}} = \frac{e^{H_0(Q')+V(Q')}}{e^{H_0(Q)+V(Q)}} = \frac{e^{H(Q')}}{e^{H(Q)}} \quad , \tag{3.4}$$

completing the proof of detailed balance.

In practice, H_0 is still nontrivial to simulate efficiently and so additional terms of the Hamiltonian must be killed. After killing all the interactions in H, one arrives at a system which is completely decoupled — and so is trivial to simulate — but it is subject to an arbitrary set of restrictions on its states.

As an example, consider the Ising model $H = \sum_{<ij>} K_{ij} s_i s_j$. The optimal probability for deletion is $p_d = \exp(-K_{ij}(1 + s_i s_j))$. Interactions between antiparallel spins will always be deleted; only parallel spins can be frozen together in ferromagnetic models. We may kill the interactions $K_{ij} s_i s_j$ one at a time until we are left with irregular, fractal blocks of spins which are completely decoupled from one another. This coarsened system is trivial to study and so statistics over many such coarsenings are easily gathered. This procedure differs from standard block-spin projection methods in that here blocks are generated in a stochastic manner and are generally of irregular shape. As one can see, for Potts models the coarsening procedure is identical to that used by SWENDSEN and WANG [3]. Due to the restrictions that are introduced, however, Swendsen and Wang still observe critical slowing down, albeit with a considerably reduced dynamical exponent.

To eliminate critical slowing down completely, we incorporate the stochastic blocking procedure as part of a multigrid scheme. Instead of coarsening the system until all that remains are large, decoupled blocks of spins, we coarsen out only the very finest length scales at each level of the multigrid algorithm. The coarsened system is composed of small blocks, typically all of some small length scale b, which interact according to a reduced Hamiltonian. We have explored several ways of killing only fine length scales; perhaps the most transparent, though perhaps also not the best, is one in which some fraction of the bonds selected at random are killed. The coupling between two coarse blocks is the sum of the living couplings that connect fine-lattice spins frozen to those blocks. Notice that long-range interactions may be generated, but they are improbable and their presence does not influence the convergence of the algorithm. The resultant "simplified" system is stored. It is studied by further coarsening.

While Swendsen and Wang coarsened their Potts lattices completely for each iteration of their simulations, we return to intermediate length scales. In particular, we coarsen the lattice γ times at each level of processing before returning to the next finer level. In this language, Swendsen and Wang, in effect, use $\gamma = 1$. To "uncoarsen" the system, decoupled blocks should be set to some arbitrary value of spin, all fine-lattice spins should be set to the block spin to which they were frozen, and the fine-lattice couplings should be restored. Metropolis updates may be performed at any length scale by using a standard Metropolis algorithm on the block spins at that length scale using the Hamiltonian appropriate to those blocks.

Our Monte Carlo method "cycles" through all the various length scales.[4] At each intermediate length scale, the system is coarsened γ times before it is uncoarsened. Each time the system reaches the coarsest level, at which all the blocks are decoupled, measurements may be made and the system is immediately uncoarsened. The cycle ends each time the finest level is reached. A few Metropolis sweeps are performed between coarsenings and uncoarsenings. These Metropolis sweeps are not essential to defeating critical slowing down and, in practice, performing more than one such sweep at a time is ineffective in accelerating the procedure.

We carried out simulations of the $d = 2$ Ising model on square lattices from 4^2 to 128^2 sites with periodic boundary conditions, using a cycle of $\gamma = 2$ with rescaling factor $b = 2$. Starting from fully magnetized states, we measured the decay of the energy to its equilibrium value for an ensemble of configurations. The energy relaxation was described by an exponential decay with, surprisingly, no discernable short-time transients. The relaxation times were $\tau(L) = 3.6 \pm 0.5$, independent of linear size L for $L \geq 16$. In contrast, other Monte Carlo algorithms show critical slowing down — $\tau(L) \sim L^z$ — with dynamical exponents $z \approx 2.1$ for standard single-spin-flip Monte Carlo and $z \approx 0.35$ for Swendsen and Wang's method. Similar measurements for the magnetic susceptibility at criticality showed the relaxation time saturating at $\tau = 7 \pm 2$, again at $L = 16$, for our algorithm.

But how does our approach eliminate critical slowing down? The first answer is simply that it does and not every algorithm that allows large-scale Monte Carlo moves achieves this. If the coarsening procedure tended to create coarse lattices with higher connectivities or stronger bonds than those of the fine lattices, the acceptance ratios for the large-scale moves would become prohibitively small. On the other hand, if the connectivities tended to be lower or the couplings weaker, blocks would become decoupled at short length scales and large-scale flips would not be possible. Thus, in addition to the empirical evidence that our multigrid Monte Carlo algorithm produces a dynamical exponent $z = 0$, it is important to note that we find that our coarsening procedure yields similar distributions of bonds and connectivities in lattices at different length scales.

Swendsen and Wang also produce Monte Carlo moves that flip blocks at all length scales. Why then do they not achieve $z = 0$? Consider, again, the multigrid coarsening procedure, which essentially treats interactions first at the finest length scales and then at increasingly longer length scales. The reason this is only "essentially" true is that the various length scales cannot be treated completely independently — to some extent, whenever an interaction is "frozen" at a fine length scale, restrictions on the allowed states are introduced at all coarser length scales as well. For example, if interactions are killed (more importantly, frozen) up to some length scale, then the coarsened Hamiltonian were simulated infinitely fast, and finally the fine-lattice spins restored and the procedure iterated, there would still be a correlation time. That time would grow with the number of levels that had been frozen and from scaling arguments we would conclude that the rate of growth would be independent of level number. Physically, it would be nice to decouple all the various length scales. In practice, we do not know how to design stochastic coarsening procedures that will do this for nontrivial models. Results on $\gamma = 1$ cycles tell us to what extent a particular coarsening procedure interlocks the length scales, measuring the amount of correlations that have built into the simulation by killing interactions over some number of length scales. Thus, each level must be visited $\gamma_{\min}$ times more often than the next finer level, where $\gamma_{\min} = b^{z_1}$, b is the length rescaling factor which describes the degree of coarsening that takes place between consecutive levels, and z_1 is the dynamical exponent for $\gamma = 1$ cycles. We find that z_1 is in fact scale invariant and, for the coarsening procedure that we and Swendsen and Wang use, is $z_1 = 0.35$ for the $d = 2$ Ising model.

Exponentially many coarsenings must take place at the coarser levels. This increased processing is acceptable since processing at coarser levels is cheaper due to the smaller number of degrees of freedom. Indeed, the work function is still proportional only to lattice size so long as the increase in the amount of processing is smaller than the amount by which the number of degrees of freedom is reduced: $\gamma < b^d$, where d is the dimensionality of the system. This is not a practical difficulty since the dynamical exponent for the $\gamma = 1$ cycle is typically much smaller than the dimensionality of the system: for the Ising ($q = 2$ Potts) model, complete elimination of critical slowing down for a finite amount of processing time per site requires $b^{0.35} < \gamma < b^2$ in two dimensions and $b^{0.75} < \gamma < b^3$ in three dimensions. For the 3-state Potts model in two dimensions, $b^{0.6} < \gamma < b^2$. In our simulations for the two-dimensional Ising model, we chose $\gamma = 2$ and $b = 2$, which is clearly within the regime of no critical slowing down.

Aside from geometrical considerations, it is clear that it is most efficient to perform most of the processing at the coarsest levels, for which there are the fewest degrees of freedom. The role of multigrid ideas and the above scaling arguments is to suggest coarsening schemes and parameter regimes for which the processing time is not only lowered but, in fact, critical slowing down is eliminated completely.

4. Conclusions

In multigrid methods, one coarsens the degrees of freedom to reduce the computational complexity of the problem and to enable local processing at coarse levels to effect large-scale changes on the fine lattice. In multigrid Monte Carlo, random processes are used both to represent coarsened interactions stochastically, but exactly, and also to update the coarsened degrees of freedom with local Monte Carlo moves.

In the longer range, the application of multigrid Monte Carlo to frustrated systems would be interesting. Uniform frustration presents no special difficulties. In contrast, it is not clear how best to model spin glasses, which have random frustration.

Some of this work was motivated by studies of dynamics. While the advantage of techniques that beat critical slowing down is that they change the dynamics of the system, relating the dynamics of a d-dimensional model to the equilibrium statistical mechanics of a $d+1$-dimensional problem [10] allows one to employ multigridding for nonequilibrium studies.[11]

Finally, a number of interesting technical questions remain. How, for instance, does one optimize the coarsening procedures to reduce statistical noise and to minimize the number of visits to the finest levels? What ways are there of constructing estimators which have minimum noise and depend only on very large scales? How can one further simplify complicated effective potentials on coarse length scales stochastically? And what detailed tests can be performed of the scaling conjectures presented above?

But most of all, multigrid Monte Carlo is still a very young class of techniques. Clearly, a great deal of work in the near future will consist simply of knocking down straw men — multigridding straight-forward Monte Carlo simulations in order to draw up a more lengthy list of credentials. Among such systems are models with both continuous and discrete excitations, such as the XY model, with its spin waves and independent vortices, and many-boson Hamiltonians, which are described by continuous world lines as well as discrete exchange effects.

1. P.W. Kasteleyn and C.M. Fortuin, J. Phys. Soc. Jpn. Suppl. **26s**, 11 (1969); and C.M. Fortuin and P.W. Kasteleyn, Physica (Utrecht) **57**, 536 (1972).

2. M. Sweeny, Phys. Rev. **B27**, 4445 (1983).

3. R.H. Swendsen and J.S. Wang, Phys. Rev. Lett. **58**, 86 (1987).

4. J. Goodman and A.D. Sokal, Phys. Rev. Lett. **56**, 1015 (1986).

5. R.G. Edwards and A.D. Sokal, unpublished.

6. A. Brandt, Multilevel Computations: Review and Recent Developments, in Preliminary Proceedings of the Third Copper Mountain Conference on Multigrid Methods (April 1987).

7. D. Kandel, E. Domany, D. Ron, A. Brandt, and E. Loh, Jr., submitted to Physical Review Letters.

8. See, for example, M. Suzuki, J. of Math. Phys. **26**, 601 (1985).

9. M. Creutz and B. Freedman, unpublished.

10. See, for example, E. Domany, Phys. Rev. Lett. **52**, 871 (1984).

11. D. Kandel, to be published.

Monte Carlo Simulations Using the Gaussian Ensemble

M.S.S. Challa and J.H. Hetherington*

Department of Physics and Astronomy, Michigan State University, East Lansing, MI48824, USA

The Gaussian ensemble [1-3] consists of a sample of size N (whose properties we are seeking) in contact with a finite bath of size N' and having a specific functional form for its entropy. This ensemble is intermediate to the microcanonical (N'=0) and canonical (N'=∞) ensembles and can be made to attain these limiting cases by continuously varying N'. If N' is small enough, the ensemble allows us to sample the intermediate states (with negative heat capacity/compressibility) at first-order transitions. Such states are then manifest through van der Waals' loops in the temperature-energy or pressure-volume curves of the sample and permit a relatively easy diagnosis of the order of a phase transition. This is an advantage vis-a-vis canonical-ensemble Monte Carlo simulations using the method of Metropolis et al. [4] where the intermediate states have very low probability and make the distinction between first- and second-order transitions difficult [5]. An appealing property of this ensemble is that it can be easily implemented as a Monte Carlo process so that models such as the Ising model, which lack equations of motion, can be readily simulated in the microcanonical ensemble (The molecular dynamics method, which also uses the microcanonical ensemble, is not feasible for such stochastic models). Further, by examining the behaviour of quantities such as the average energy of the sample as a function of N', the method allows us to investigate systematically the effects of the heat bath on static properties of the sample.

In what follows, we first describe the analysis of the method. The validity of the analysis is then demonstrated by simulations of Potts models [6] which are a convenient testing ground for numerical methods since many of their properties are exactly known.

Theory

We shall restrict ourselves here to the case of the sample and the bath being in thermal contact, the combination forming a closed system. The temperature is then a derived quantity, being determined by thermal equilibrium obeying the constraint of constant total energy. (Application of the method to obtain other intensive quantities is straight-forward. Thus, if we consider the temperature fixed and impose the condition that the total magnetization is a constant, then the magnetic field is a derived quantity. Similarly, pressure may be derived using the constraint of constant total volume of the system).

Let S and E be the energy of the sample and let primes denote the corresponding quantities of the bath. All quantities are extensive unless otherwise stated. The (constant) total energy is $E_t = E+E'$ and the total entropy is $S_t = S+S'$. The Gaussian ensemble is defined by taking $S' = -aE'^2 = -a(E_t - E)^2$ with a (>0) being the constant curvature of S'. Using the postulate of equal

* Research supported by the Center for Fundamental Materials Research, Michigan State University.

Springer Proceedings in Physics Vol. 33: **Computer Simulation Studies in Condensed Matter Physics**
Editors: D. P. Landau · K. K. Mon · H-B. Schüttler

apriori probabilites, the energy distribution is $P(E) \propto \rho(E)\rho'(E') \propto \exp(S_t)$ where ρ,ρ' are the densities of states.

The entropy of the bath, $S'=-aE'^2$, may be "justified" as follows. Consider the bath to be a Heisenberg paramagnet in a magnetic field; S' will then have a maximum S'_{max} at some E'_{max}. Shifting the origin to (E'_{max},S'_{max}) and retaining only the 2nd-order term in the expansion of S' about this origin, we obtain the form $S'= -aE'^2$. The classical nature of the spins ensures the continuity of S'. Further, since these are non-interacting spins, both S' and E' are extensive and we can identify $a \propto 1/N'$. We shall sometimes refer to a instead of N' as a parameter.

For a given set of input parameters $\{a,E_t,N\}$ equilibrium occurs at some $E=\tilde{E}$ where $S_t(\tilde{E})$ is a maximum. Expanding $S_t(E)$ about $\tilde{E}$ upto 2nd order in $E-\tilde{E}$ and defining $\beta=\partial S/\partial E$ and $C = -\beta^2/\frac{\partial^2 S}{\partial E^2}$ we obtain [1,2]

$$\langle E\rangle=\tilde{E} \quad , \tag{1a}$$

$$\langle \beta\rangle=\tilde{\beta}= 2a(\langle E\rangle-E_t) \tag{1b}$$

and

$$\langle C\rangle= \tilde{C} = \frac{\tilde{\beta}^2 G_2}{1 - 2a\,G_2} \quad , \tag{2}$$

where we have introduced the notation $G_n=\langle(E-\langle E\rangle)^n\rangle$, $n=2,3,4,\ldots$ The tildes indicate that the quantities are the true (microcanonical) values and are to be evaluated at $\tilde{E}$.

We can easily verify that (2) is equivalent to $\langle C\rangle= -\langle\beta\rangle^2\frac{\partial\langle E\rangle}{\partial\langle\beta\rangle}$ as follows. We define $Q_a(E_t)$, the analog of the partition function, through

$$Q_a= \Sigma_i \exp[\ -a(E_i-E_t)^2] \quad , \tag{3}$$

where the subscript a denotes that we are considering a given bath. All quantities are now functions of E_t so that

$$\frac{\partial\ \ell n Q_a}{\partial E_t} = \langle\beta\rangle = 2a(\langle E\rangle-E_t) \tag{4}$$

$$\frac{\partial\langle E\rangle}{\partial E_t} = 2a\ G_2 \tag{5}$$

and

$$\frac{\partial\langle\beta\rangle}{\partial E_t} = 2a(2aG_2-1) \quad . \tag{6}$$

Using (5) and (6) in $\langle C\rangle = -\langle\beta\rangle^2\{\frac{\partial\langle E\rangle}{\partial E_t} \ / \ \frac{\partial\langle\beta\rangle}{\partial E_t}\}$ recovers (2).

The equilibrium condition also requires that

$$2a + \frac{\tilde{\beta}^2}{\tilde{C}} \geq 0 \quad . \tag{7}$$

(7) shows that states with $\tilde{C}<0$ can be equilibrium states provided that a is large enough; thus the van der Waals' loops are allowed in the microcanonical ensemble. Conversely, such states will have very low probability in the

canonical ensemble (a=0). An alternative explanation for the loops follows if we invert (2). We see that $G_2 \to 0$ as $a \to \infty$ so that the loops are allowed for small fluctuations. This is why the loops appear in mean field theories where quantities like the order parameter have a fixed value [7].

A graphical interpretation of the method is given in Fig. 1. Define $S'' = -S' = a(E-E_t)^2$. For a given E_t, $\tilde{E}$ is then defined by the maximum in $S_t = S - S''$, i.e. , the point where the slopes of S and S" are equal. By a suitable choice of an additive constant in S" we can arrange that the curves then touch at $\tilde{E}$. Since $\partial S''/\partial E$ is a single-valued function of E, we can evaluate β at any point on S(E) by a proper choice of E_t. As shown in Fig. 1, the curvature of S" (or S') is a $\propto 1/N'$. The limit $a \to 0$, $-2aE_t \to \tilde{\beta}$ yields the canonical ensemble while the $a \to \infty$, $E_t \to \tilde{E}$ limit is the microcanonical case; thus the Gaussian ensemble interpolates smoothly between these extremes. In the canonical limit, S" is characterized only by its constant slope $\tilde{\beta}$ and coincides with the tangent as shown. Then the probability of any state is given by $P(E) \propto \exp(S-\tilde{\beta}E) \propto \exp(-\tilde{\beta}F)$ where F is the free energy of the state. As a final example of the interpolating properties of the ensemble, note that Eq. (2) recovers the canonical ensemble definition of $\langle C \rangle$ in the limit $a \to 0$. Fig.2 illustrates the manner in which the method allows us to sample the intermediate states at first-order transitions.

Equations (1), which result from a Gaussian approximation to P(E), are good for $\beta \to 0$ and ∞. Near phase transitions, however, fluctuations are large and the term of order $(E-\tilde{E})^3$ in $S_t(E)$ becomes significant and introduces N'-dependent corrections in $\langle E \rangle$ and $\langle \beta \rangle$. We consider approximately the effect of this cubic term as follows. Define $\tilde{S}_3 = \frac{1}{3!}\left(\frac{\partial^3 S}{\partial E^3}\right)_{\tilde{E}}$, $\sigma^2 = \frac{\tilde{C}}{2a\tilde{C}+\tilde{\beta}^2}$ and $z = 3\tilde{S}_3\sigma^4$. Now approximate $\exp[\tilde{S}_3(E-\tilde{E})^3] \approx 1+\tilde{S}_3(E-\tilde{E})^3$ in P(E). This retains the Gaussian form of P(E) and we obtain readily

$$z^3 + z(G_2/2) - (G_3/4) = 0 , \qquad (8)$$

$$\langle E \rangle = \tilde{E} + z \qquad (9a)$$

and

$$\langle \beta \rangle = \tilde{\beta} + 2az \quad . \qquad (9b)$$

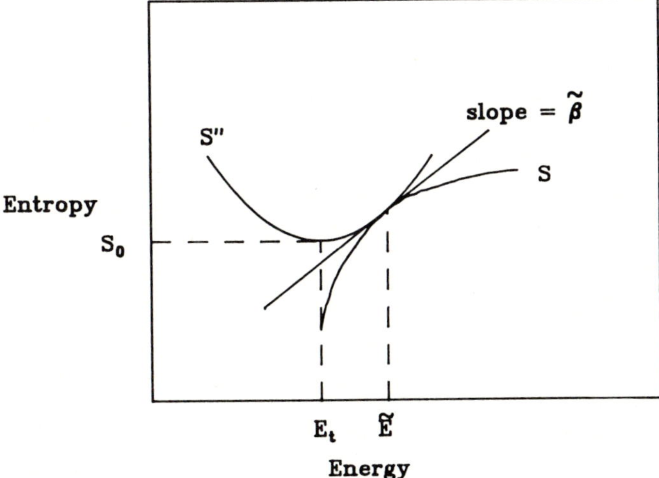

Fig. 1. The interpretation of $\tilde{\beta}$ in the Gaussian ensemble. S"= -S' where S' is the entropy of the bath and $\tilde{E}$ is the location of the maximum in $S_t = S - S''$. $\tilde{\beta}$ is the slope of the straight line which is simultaneously tangent to S and S".

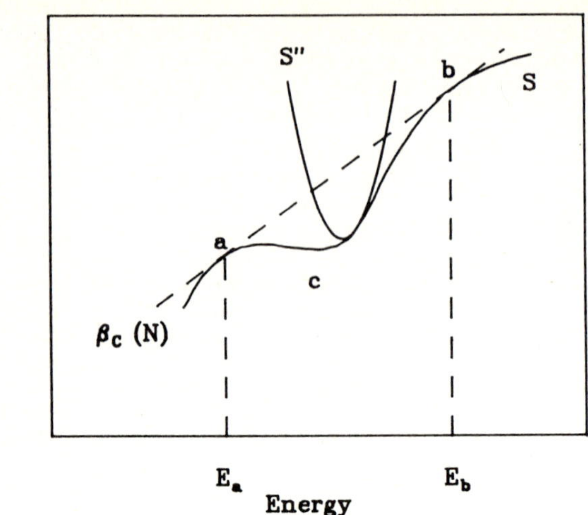

Fig. 2. Schematic diagram explaining the sampling of the intermediate states at first-order transitions using the Gaussian ensemble. The notation is that of Fig.1. The acb portion of S is due to the interface effects in a finite sample. As $N\to\infty$, these effects become negligible and acb approaches the straight dashed line ab. $\beta_c(N)$ is the slope of ab and is defined as the transition temperature of the sample in the canonical ensemble because of the equality of the free energies at a and b. States near c become equilibrium states in the Gaussian ensemble when the curvature of S" is large.

(9a,b) then show that z is a measure of the deviations of $\langle E\rangle$ and $\langle \beta\rangle$ from the true values. These deviations can be estimated from (8) since (8) has only one real root and G_2 and G_3 are "measurable" quantities.

One of the useful features of (8) is that $G_3=0$ implies z=0 and vice-versa. But $z \propto \left(\frac{\partial^2 \beta}{\partial E^2}\right)_{\tilde{E}}$ and hence inflection points in $\tilde{\beta}(\tilde{E})$ are independent of N'. The curves of $\langle\beta\rangle$ vs. $\langle E\rangle$ for various a cross at these points, denoted by (β^*, E^*), which are thus non-trivial fixed points with respect to N'. $\beta=0$ and ∞ then represent trivial fixed points since P(E) approaches a δ-function at these extremes $(G_3\to 0)$. Here we take β^* as alternate choice for the transition temperature even at first-order transitions where it is an alternative to the equal-area construction.

Equations (9) also allow a form of scaling. From $\langle\beta\rangle$ vs. $\langle E\rangle$ for a given a we can estimate $\tilde{E}$ and $\tilde{\beta}$ and thus all the curves for various a (N fixed) should collapse onto a single curve - the microcanonical one. Of course, in view of the approximations made in obtaining these equations, we expect the results to be good only for large values a (small fluctuations). This is confirmed by the simulation results.

If we denote a knowledge of $\tilde{\beta}(\tilde{E})$ as complete information, the preceeding arguments show that, for finite, non-zero N and N', information is inevitably lost during statistical averaging. This fact is also illustrated by the following theorem.

Let $\beta_a(E_t)$ denote the curve $\langle\beta\rangle$ vs. $\langle E\rangle$ for a given a. Then if we know $\beta_a(E_t)$, we can find $\beta_{a'}(E_t)$ for any other a'<a.
Proof.

Rewrite (3) as

$$Q_a(E_t) = \int_{-\infty}^{+\infty} \rho(E)\exp\{-a(E-E_t)^2\}\,dE \quad . \tag{10}$$

Using (4) we can write

$$Q_a(E_t) = \exp\left[\int^{E_t} \beta_a(E_t')\,dE_t'\right] \quad . \tag{11}$$

Consider the integral,

$$I = \int_{-\infty}^{+\infty} \exp\{-b(E_t - E_t')^2\}Q_a(E_t')\,dE_t' \quad , \tag{12}$$

which becomes, by (10),

$$I = \int_{-\infty}^{+\infty} \rho(E)dE \int_{-\infty}^{+\infty} dE_t' \exp[-\{b(E_t-E_t')^2 + a(E-E_t')^2\}]$$

$$= (\pi/a+b)^{1/2} \int_{-\infty}^{+\infty} \rho(E)\exp\left[-\frac{ab}{a+b}(E-E_t)^2\right]dE \quad . \tag{13}$$

Defining $a' = \frac{ab}{a+b}$, the integral is simply $Q_{a'}(E_t)$ so that (12) and (13) may be combined into

$$Q_{a'}(E_t) = (a^2/\pi(a-a'))^{1/2} \int_{-\infty}^{+\infty} \exp\left\{\frac{-aa'}{(a-a')}(E_t-E_t')^2\right\} Q_a(E_t')\,dE_t' \quad . \tag{14}$$

$Q_a(E_t')$ can be numerically evaluated through (11) and thus $\beta_{a'}(E_t)$ can be predicted through (14) and (4). We require, however, that $a>a'$ in order that that integral in (14) be tractable. Therefore it is not possible to construct the density of states of a finite system using canonical ensemble averages.

Simulations of Potts models

We illustrate the validity of the method by simulating q-state Potts models on square and simple-cubic lattices of side L with periodic boundary consitions. ($N=L^d$ where d is the lattice dimensionality). In a q-state Potts model the spin at the i-th site, σ_i, can take one of q different values, say, the numbers 1 to q. The (zero-field) ferromagnetic Hamiltonian used in our simulations is given by

$$H = -J\sum_{(ij)}\delta_{\sigma_i\sigma_j} \quad , \tag{15}$$

where δ is the Kronecker delta, the sum is over all the nearest-neighbour pairs and J (>0) is the interaction strength. We shall use units where energy is in units of J and β is in units of J^{-1}. The following notation is used. $\beta_c(\infty)$ is the inverse of the infinite-lattice transition temperature. Energies per site at the transitions in the infinite lattice are denoted by $E_c(\infty)$ if the transition is second-order and by E_- and E_+ if the transition is first-order. The notation "$\beta^*(L)$" emphasizes the L-dependence of β^*.

The nature of the transition in these models is q-dependent and many of the properties are well-known [6,8,9]. For d=2 Baxter [8] has shown that the model has a second-order transition for q<5 and a first-order transition for q≥5. We can obtain E_+ and E_- (or $E_c(\infty)$ at the second-order transitions) for d=2 by combining Baxter's results with those of Kihara et al [9]. The situation is not exactly known for d=3 although approximate calculations suggest that the transition is first-order for q>2. (Note that q=2 is the Ising model).

Monte Carlo realization of the Gaussian ensemble is straightforward. Phase space is sampled by going to each spin on the lattice in turn and then testing it for "flipping" from a configuration $\{\mu\}$ to a new configuration $\{\nu\}$. If P_μ and P_ν be the probabilities for two configurations of energies E_μ and E_ν, their relative probability is given by

$$\frac{P_\nu}{P_\mu} = \frac{\exp\{-a(E_\nu - E_t)^2\}}{\exp\{-a(E_\mu - E_t)^2\}} \tag{16}$$

and the process is implemented on the computer by comparing the ratio to a random number as usual. One pass over all the spins in the lattice constitutes a Monte Carlo step/site, abbreviated to MCS, and is taken as the unit of computer time. An important difference from canonical-ensmeble Monte Carlo is that E_ν, E_μ are now the total energies of the lattice and not just the energies of the particular spin in question, although this does not pose a problem computationally.

Fig. 3 shows typical $\langle \beta \rangle$ vs. $\langle E \rangle$ plots at the first-order transition in the d=2, q=10 model. Notice that the loops appear only for large a. As shown in [2], P(E) changes continuously from a double-peaked distribution for small a to a single peaked one for large a. Data for G_3 [3] confirm that this quantity vanishes at the fixed point. It can be shown [2,3] that the N'-effects vanish when $N \to \infty$; the loops thus shrink to a horizontal straight line for large lattices. Notice the good agreement between the curves, which are predictions based on the theorem above (10-14), and the data points.

Figure 4 shows the "scaled" curves at the second-order transition. Note the absence of loops. The successful scaling for large a demonstrates the correctness of the theory. (The results are equally good at first-order transitions). There are dramatic differences in the slopes of the curves near β^*. These N'-dependencies in the specific heat are fully evident in Fig.5 where the specific heat data coresponding to Fig.4 are plotted as a function of temperature.

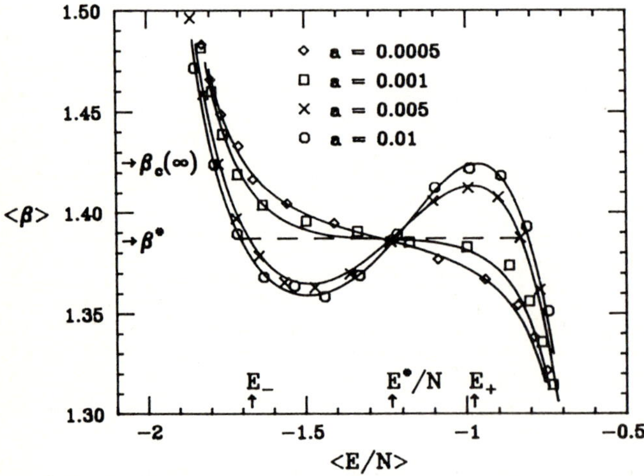

Fig.3. Typical $\langle \beta \rangle$ vs. $\langle E \rangle/N$ curves at a first-order transition. Symbols represent data for the q=10 model on an 8x8 lattice and represent averages over 200,000 MCS each. The curves are predictions based on the data for a=0.05 (not shown) and using (10-14). The horixontal dashed line is the equal-area construction for a=0.01. (E^*, β^*) is the fixed point. $\beta_c(\infty)$, E_- and E_+ are the infinite-lattice parameters at the transition.

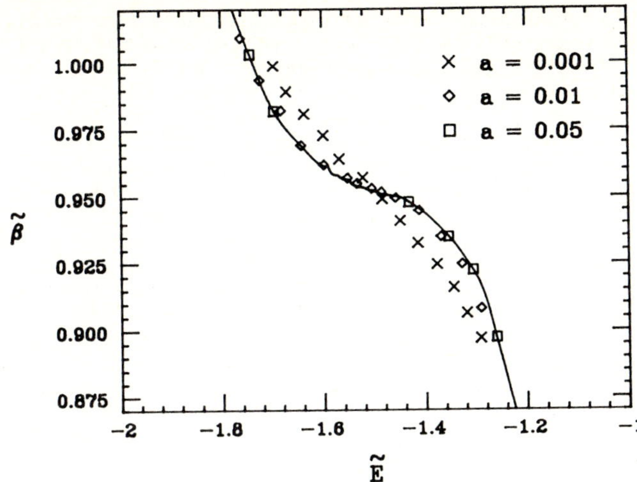

Fig.4. Estimating the microcanonical curve, $\tilde{B}(\tilde{E})$, at a second-order transition using (8) and (9). The data are for d=2, q=3, L=8. Each data point in the transition region represents averages over 10^6 MCS. Scaling breaks down for small a. The curve is from straight-line interpolations of the data for a=0.05.

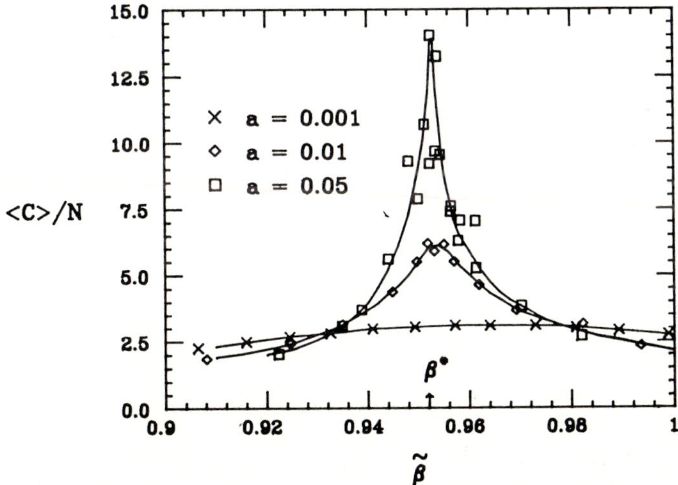

Fig.5. The specific heat data as a function of temperature corresponding to the data of Fig.4. <C> is obtained from (2). The curves are smooth guides to the eye. The second-order transition is remarkably sharpened as one moves away from the canonical ensemble. There are no more a-dependencies in <C> for a>0.05.

The sharpening of the curves in Fig. 5 for large a raises the question: is there a cusp in the specific heat as $a \to \infty$? If so, there is a possibility of sharp transitions in finite samples provided we look for them in the microcanonical ensemble. (Note that the theorem of Yang and Lee [10] which proves the analyticity of the thermodynamic potentials of finite samples in the (grand) canonical ensemble is indeed corroborated by Fig. 5 which shows that the transition becomes more smeared as $a \to 0$). One must also keep in mind the possibilities that 1) there is only a small but non-zero rounding as $a \to \infty$ and that 2) the cusp is a subtle artifact of the method itself. If there is a cusp, is it present for any value of N, say N=2? We are unable to answer any of these questions at present because of both theoretical and numerical difficulties. On the theoretical side, the problems are as follows. First, the discreteness of $\rho(E)$ for Potts models makes derivatives like $\beta = \partial(\ell n\ \rho)/\partial E$ ill-defined. Second, exact calculations of $\rho(E)$ for systems with continuous symmetry does not seem to be an easy task. Lastly, the interpretation that β is the inverse temperature loses meaning when N is too small. These theoretical difficulties are compounded by the numerical problems (apparent in Fig. 5) associated with sampling in the

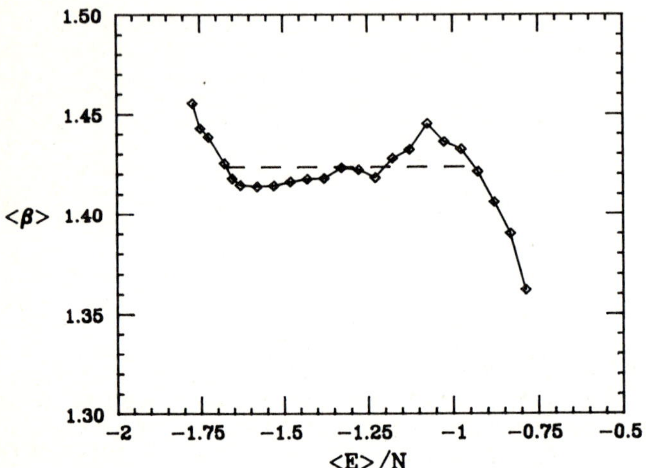

Fig.6. Data for L=50 at the first-order transition of Fig.3. The curve consists of straight-line interpolations between data points which represent averages over 5000 MCS each. The horizontal dashed-line is the equal area construction and yields $E_-=-1.672$, $E_+=-0.938$ and $\beta_c(\infty)=1.424$. The exact results are -1.6643, -0.9682 and 1.4261 respectively.

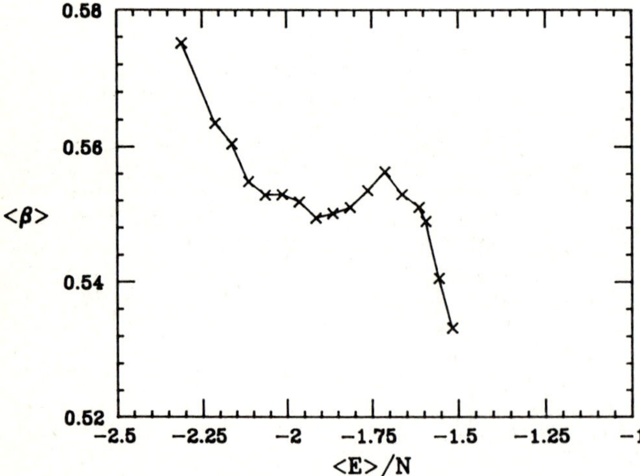

Fig.7. $\langle\beta\rangle$ vs. $\langle E\rangle/N$ curves for the d=3,q=3 model. Data are for L=4 and a=0.05 each point being an average over 50000 MCS. The loop confirms the first-order transition diagnosed by other workers.

microcanonical ensemble. These are some of the questions we wish to address in the future.

Figures 6 and 7 highlight some of the computational advantages of the method. Fig. 6 shows the first-order transition in the d=2, q=10 model using a much larger lattice than in Fig. 3. Using only 5000 MCS per data point, we are able to get good estimates of the infinite-lattice parameters at the transition. One needs over 10^6 MCS to obtain results of comparable accuracy using the canonical ensemble [5]. The reason for the efficiency of the Gaussian ensemble is that, by choosing a large enough, we can avoid the double-peaked distributions and consequent metastabilities associated with canonical ensemble sampling [5].

Figure 7 shows the results of simulating the d=3, q=3 model on a small lattice. The loop is clearly seen and confirms the first-order transition obtained through other approximate calculations [6]. We emphasize here the ease with which the order of the transition is diagnosed: the data of Fig. 7 were obtained in a single run which consumed 15 minutes of CPU time on a scalar computer! While extrapolations as $N \to \infty$ are unavoidable for confirming the nature of the transition, the fact that the N'-effects are larger for smaller N shows that this is a useful preliminary method for investigating phase diagrams.

Conclusions

We have demonstrated that the interpolating nature of the Gaussian ensemble makes it a useful tool in the study of phase transitions. The computational advantages of the method stem from the fact that it is a Monte Carlo method which can approach the microcanonical ensemble. Thus, by making N' small enough, the method provides a simple distinction between first- and second-order transitions. We have shown the way in which the N'-dependencies arise in the static properties of finite samples and confirmed the analysis thorugh simulations. The sharpening of the specific heats at second-order transitions when $N' \to 0$ raises the possibility of sharp transitions in finite samples. We hope to address this question and also finite-N scaling in the Gaussian ensemble in the near future.

REFERENCES

1. J.H.Hetherington: J. Low Temp. Phys., 66, 145 (1987); J.H.Hetherington, D.R.Stump: Phys. Rev. D, 35, 1972 (1987); D.R.Stump, J.H.Hetherington: Phys. Lett. B, 188, 359 (1987).
2. M.S.S.Challa, J.H.Hetherington: Phys. Rev. Lett., 60, 77 (1988).
3. M.S.S.Challa, J.H.Hetherington: submitted to Phys. Rev. B.
4. N.Metropolis, A.W.Rosenbluth, M.N.Rosenbluth, A.H.Teller and E.Teller, J. Chem. Phys., 21, 1087 (1953). See also "Monte Carlo Methods in Statistical Physics", ed. K.Binder (Springer-Verlag, New York, 1979) for a review of Monte Carlo methods.
5. M.S.S.Challa, D.P.Landau and K.Binder, Phys. Rev. B, 34, 1841 (1986).
6. For a review of Potts models, see F.Y.Wu: Rev. Mod. Phys., 54, 235 (1982).
7. R.K.Pathria, "Statistical Mechanics" (Pergamon, New York, 1972), pp. 376-377.
8. R.J.Baxter, J. Phys. C, 6, L445 (1973).
9. T.Kihara, Y.Midzuno and T.Shizume, J. Phys. Soc. Jpn., 9, 681 (1954).
10. C.N.Yang and T.D.Lee: Phys. Rev., 87, 404 (1952); ibid, p.410.

Classical Spin Dynamics in the Two-Dimensional Anisotropic Heisenberg Model

G.M. Wysin [1], M.E. Gouvea [1,*], A.R. Bishop [1], and F.G. Mertens [2]

[1] Los Alamos National Laboratory, Los Alamos, NM 87545, USA
[2] Physics Institute, University of Bayreuth, D-8580 Bayreuth, Fed. Rep. of Germany

ABSTRACT

Properties of vortices and dynamic correlation functions have been studied for the two-dimensional classical ferromagnetic Heisenberg model with easy-plane anisotropy, as a function of the anisotropy parameter $\lambda = J_z/J_x$. Continuum limit equations of motion exhibit two types of static vortex solutions, with and without out-of-plane spin components. We have studied numerically the stability of these solutions and have found that for $\lambda \lesssim 0.7$ only the planar vortex (zero out-of-plane spin components) is stable, and for $\lambda \gtrsim 0.8$ only the vortex with nonzero out-of-plane spin components is stable. Approximate spin configurations for non-zero-velocity vortices are presented. A vortex ideal gas phenomenology is used to calculate the dynamic correlation function $S(\vec{q},\omega)$. Above the Kosterlitz-Thouless transition temperature, the calculation predicts a Gaussian central peak for the out-of-plane correlations and a squared Lorentzian for the in-plane correlations. These results are compared with results of a Monte Carlo-Molecular Dynamics simulation.

1. INTRODUCTION

Quasi-two-dimensional (2D) spin systems provide challenging realizatons of models ideally suited for large-scale numerical simulations of dynamical behavior. Model Hamiltonians including anisotropic (easy-plane) exchange and possibly in-plane symmetry breaking are believed to describe materials of current experimental interest, such as $BaCo_2(AsO_4)_2$, Rb_2CrCl_4 [1,2] and $CoCl_2$-GIC [3]. In a classical description, the models support interacting nonlinear domain walls, vortices, and spin waves, and therefore can be used to test theories for the stability, energy dispersion, and interactions between these excitations. Continuum limit theory can be used as a guide, and can be compared with numerical simulations on finite lattices, where the effects of discreteness may modify the dynamics. These can be compared with experimental results, especially inelastic neutron scattering data. The low-frequency long-wavelength dynamic correlations are particularly relevant as signatures for topological excitations, such as vortices and domain walls.

The model Hamiltonian we consider is the easy-plane anisotropic Heisenberg ferromagnet with nearest-neighbor interactions:

$$H = -J \sum_{(i,j)} (S_i^x S_j^x + S_i^y S_j^y + \lambda S_i^z S_j^z). \qquad (1)$$

The exchange parameter is $J > 0$, and $0 \leq \lambda \leq 1$ measures the degree of anisotropy. For the XY model ($\lambda = 0$), the system is expected to undergo a Kosterlitz-Thouless (KT) topological phase transition at a finite temperature T_{KT} [4]. Static spin-

* Supported by CNPq (Brazil).

Springer Proceedings in Physics Vol. 33: Computer Simulation Studies in Condensed Matter Physics
Editors: D. P. Landau · K. K. Mon · H-B. Schüttler

spin correlations are expected to change from power law to exponential as the temperature T is raised above T_{KT}, due to a finite density of vortex-antivortex pairs becoming unbound. For $T > T_{KT}$, we therefore expect there will be a low density of free vortices interacting weakly with one another, in a background spin field of the remaining bound pairs and renormalized spin waves [5]. This is reminiscent of the soliton ideal gas picture for 1D magnets [6]. Phenomenologically, we can determine the vortex ideal gas contribution to the dynamic structure function $S(\vec{q},\omega)$, as in Mertens et al. [7], for in-plane and out-of-plane correlations. The theory includes phenomenological parameters that are temperature dependent, including an rms vortex velocity $\bar{u}$, and a correlation length ξ equal to half the mean vortex-vortex separation. We will also assume a vortex number density $n_v \simeq (1/2\xi)^2$. As in the 1D soliton models, a central peak ($\omega \simeq 0$) in $S(\vec{q},\omega)$ is expected, relating to the presence of zero-velocity coherent nonlinear structures (i.e., vortices). These results will also apply to nonzero-λ cases; T_{KT} decreases very slowly with λ until λ is very near 1 [8]. However, the out-of-plane correlations might be expected to be modified with increasing λ. This depends on details of the λ-dependence of the out-of-plane spin configurations for moving vortices, which we consider below.

The total $S(\vec{q},\omega)$, due to all excitations and their interactions, can be determined from numerical simulations using a combined Monte Carlo-Molecular Dynamics (MC-MD) approach [9]. A (microcanonical) MD integration of equations of motion is performed, using initial configurations generated by a MC simulation. Comparisons of the phenomenology with numerical experiments and laboratory experiments must allow for effects of nonlinear interactions between the interactions. In particular, higher order spin wave processes also can contribute to the central peak; we are presently considering this problem.

We have three primary goals here. First, we consider the dynamics of an isolated vortex, in the continuum limit and numerically on a lattice, to determine the energy dispersion, stability, and approximate out-of-plane spin configuration. This is done for stationary and moving vortices. Second, the vortex ideal gas phenomenology is reviewed, and the vortex spin configurations are used to predict the free vortex contributions to $S(\vec{q},\omega)$. Third, we compare the phenomenological predictions with numerical simulations on a square lattice, specifically for the central peak (CP) properties of the XY model.

2. CONTINUUM LIMIT VORTEX PROPERTIES

For the continuum classical dynamics of Eq. 1, we define the in-plane spin field $\phi(\vec{r})$ and out-of-plane spin field $\theta(\vec{r})$ by

$$\vec{S}(\vec{r}) = (\cos\theta \cos\phi, \cos\theta \sin\phi, \sin\theta), \tag{2}$$

where $\vec{r} = (r,\varphi)$ in polar coordinates. Assuming small spatial derivatives (for example, $a\, \partial\theta/\partial r \ll 1$, $(a/r)\, \partial\theta/\partial\varphi \ll 1$, a = lattice spacing), the continuum equations of motion are [10, 11]

$$\dot{\theta} = \cos\theta\, \nabla^2\phi - 2\sin\theta\, \vec{\nabla}\theta \cdot \vec{\nabla}\phi, \tag{3}$$

$$\dot{\phi}\cos\theta = [2\delta - (\delta\, |\nabla\theta|^2 + |\nabla\phi|^2)/2]\sin 2\theta - (1 - \delta\cos^2\theta)\nabla^2\theta, \tag{4}$$

where the unit of time is $\hbar/JS$, and

$$\delta = 1 - \lambda\ . \tag{5}$$

The Hamiltonian contains terms due to θ and ϕ gradients, as well as an anisotropy term (we set $a = 1$):

$$H = \frac{JS^2}{2}\int d^2r\, \{(1-\delta\cos^2\theta)\, |\nabla\theta|^2 + \cos^2\theta\, |\nabla\phi|^2 + 4\delta\sin^2\theta\}. \tag{6}$$

For static configurations, $\dot{\theta} = \dot{\phi} = 0$ and Eq. 3 admits the solution for a vortex (or antivortex) at the origin:

$$\phi = \pm \tan^{-1} (y/x) + \phi_o, \tag{7}$$

where ϕ_o is an arbitrary constant. The corresponding out-of-plane component can have two distinct configurations. First, there is a "planar" solution, with $\theta_{pl} = 0$, and energy $E_{pl} = \pi JS^2 \ln (R/r_o)$, where R is the system size and $r_o \simeq a$ is a vortex core cut-off for the radial integration. The second solution involves nonzero out-of-plane spin components, with asymptotic values

$$\theta_{out} \simeq A(r_v/r)^{1/2} e^{-r/r_v}, \quad r \to \infty \tag{8}$$

$$\theta_{out} \simeq \pi/2 - \tilde{A}\, r/r_v , \qquad r \to 0 \tag{9}$$

$$r_v = \frac{1}{2} \left(\frac{\lambda}{1-\lambda}\right)^{1/2}. \tag{10}$$

If we match the asymptotic functions at $r = r_v$, then

$$A = \pi e/5 \text{ and } \tilde{A} = 3\pi/10 . \tag{11}$$

The energy E_{out} is a function of λ and r_o. The decay of θ_{out} for large r is characterized by a vortex radius r_v. For the XY model, $r_v = 0$, and the out-of-plane solution degenerates to the planar solution. For nonzero λ, relative stability is determined by $E_{out}-E_{p\ell}$. However, this difference is very sensitive to the value of r_o. To compare with simulations on a square lattice, $r_o \simeq a$, but it may be more appropriate to use $r_o \simeq a/\sqrt{2}$ because this is the smallest r to a lattice site for a vortex centered in a unit cell. Alternatively, the stability can be examined directly on a discrete lattice by numerical means.

For the XY model, the out-of-plane vortex correlations must derive from moving "planar" vortices, which can have nonzero out-of-plane components. An approximate traveling wave solution to Eq. 4 can be found by assuming $\sin\theta \ll 1$ and $(\vec{r}-\vec{u}t)$ time dependence, giving

$$\sin\theta \simeq \frac{\vec{u} \cdot \hat{e}_\varphi r}{1-4r^2}, \tag{12}$$

where $\hat{e}_\varphi$ is the azimuthal unit vector.

The spin profile is proportional to the velocity. The angular dependence on $\hat{e}_\varphi$ is to be expected--the out-of-plane profile necessarily cannot be isotropic in φ if we want the profile to define a distinct direction for the velocity. However, we see that Eq. 12 is discontinuous and diverges at $r = 1/2$; retaining the neglected nonlinear terms presumably would suppress the divergence and force $\sin\theta$ to cross zero near $r = 1/2$. For arbitrary λ, asymptotic solutions for a moving planar vortex are consistent with Eq. 12;

$$\theta \simeq \vec{u} \cdot \hat{e}_\varphi\, r, \qquad r \to 0, \tag{13}$$

$$\theta \simeq \frac{-\vec{u} \cdot \hat{e}_\varphi}{4\delta} \frac{1}{r}, \qquad r \to \infty. \tag{14}$$

Whether the asymptotic $r \to 0$ solution is relevant for vortices on a lattice is questionable; the continuum "solution" varies too rapidly near $r = 1/2$ to justify the assumption of small spatial derivatives. We consider this question and vortex stability using numerical simulation.

3. VORTEX MOTION ON A LATTICE

For numerical simulations, the discrete equations of motion are

$$\dot{\vec{S}}_i = \vec{S}_i \times \vec{F}_i - \varepsilon\, \vec{S}_i \times (\vec{S}_i \times \vec{F}_i) , \tag{15}$$

$$\vec{F}_i = J \sum_{(i,j)} (S_j^x \hat{x} + S_j^y \hat{y} + \lambda S_j^z \hat{z}). \tag{16}$$

The sum on j is only over the nearest neighbors of i. The parameter ε is the strength of a Landau-Gilbert damping, which was included for testing vortex stability and for damping out spin waves generated from non-ideal initial conditions. Neumann boundary conditions were used for simulating single vortices; periodic boundary conditions were used for vortex-antivortex pairs. The equations for the xyz spin components were integrated using a fourth order Runge-Kutta scheme with a time step of 0.04 (using time unit $\hbar/JS$). Conservation of energy and spin length served as checks of numerical accuracy (to about 1 part in 10^5). Single vortex motion was studied on a 40 x 40 square lattice; vortex-antivortex pair motion was studied on a 100 x 100 square lattice.

The stability of static planar vortices was tested by using Eq. 7 as initial condition, with $\theta = 0$, centered in a unit cell. The time evolution was followed to $t = 400$, with damping $\varepsilon = 0.1$. For $\lambda \lesssim 0.7$, the planar vortex was found to be stable; a small decrease in energy occurred due to small changes in the boundary spin configuration, but the out-of-plane component remained zero everywhere. For $\lambda \gtrsim 0.8$, however, a nonzero out-of-plane component developed after $t \simeq 100$, and relaxed into configurations as in Fig. 1, approximately described by Eq. 8 and 9 for the out-of-plane vortex. These simulations were also repeated using the out-of-plane vortex as initial condition. Again, we found that for $\lambda \lesssim 0.7$ the planar vortex is the stable configuration and for $\lambda \gtrsim 0.8$ the out-of-plane vortex is the stable configuration.

Some insight about moving vortex profiles can be obtained by using static planar vortex-antivortex pairs as initial condition. With $\varepsilon = 0.1$, it is found that the pairs move easily toward each other for $\lambda \gtrsim 0.8$. For $\lambda \lesssim 0.7$, however, the attraction is considerably weaker and falls off strongly with distance. An instantaneous configuration is shown in Fig. 2. The azimuthal dependence of the

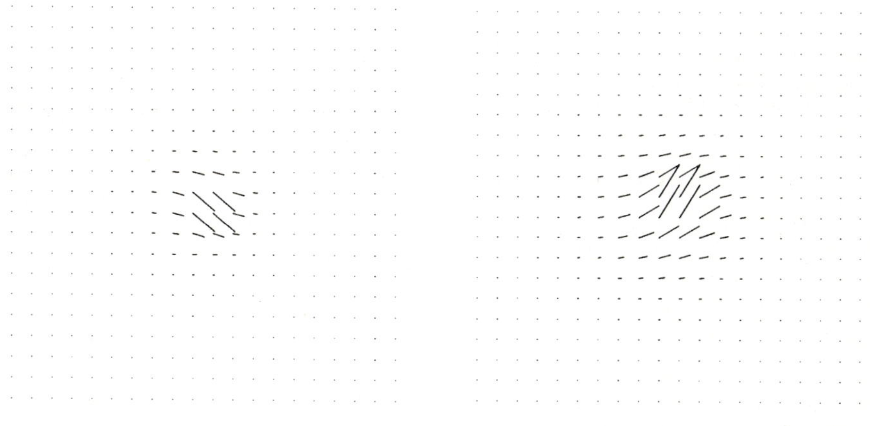

$\lambda = 0.8$ $\lambda = 0.9$

Fig. 1. Single vortex out-of-plane angles θ, after integration for 400 time units starting from a planar vortex ($\theta = 0$). The lengths of the lines are proportioned to θ; the angles from the horizontal axis are θ.

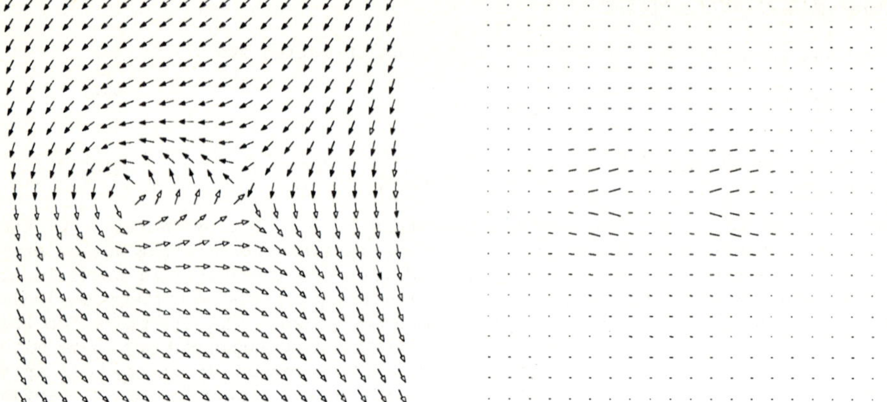

Fig. 2. Vortex-antivortex pair motion for $\lambda = 0.7$, at $t = 15$, starting from a planar pair at $t = 0$ ($\theta = 0$). A 20 x 20 segment of the 100 x 100 lattice simulated is shown. Note the dependence of θ on the azimuthal coordinate for each vortex.

out-of-plane component is clear. There is no sign change in $\sin\theta$ as a function of r. It should be noted that the vortex motion is not along the line connecting the centers, but includes a net drift in the orthogonal direction as well [5].

We have also tested profiles for isolated moving vortices without damping. The XY planar vortex, Eq. 12, is not found to be a good traveling wave solution. However, if we modify the r-dependence to $\sin\theta \sim (1-e^{-Br^2})/r$, which does not change sign, then we find that fairly stable moving profiles can be obtained. The details of the small-r dependence of θ still need to be clarified.

4. VORTEX IDEAL GAS PHENOMENOLOGY

The correlations in a dilute gas of free vortices with average separation 2ξ and with a Maxwellian velocity distribution with rms velocity $\bar{u}$ were considered in Mertens et al. [7]. Based on the single vortex stability regimes, we expect that the out-of-plane correlations will be determined by moving planar vortices for $\lambda \lesssim 0.7$ or by out-of-plane vortices for $\lambda \gtrsim 0.8$. Using the large r asymptotic solutions, we can predict the small-wavevector behavior of $S(\vec{q},\omega)$. Assuming incoherent scattering of the vortices, the out-of-plane space-time correlation function is

$$S_{zz}(\vec{r},t) = n_v S^2 \int d^2R \int d^2u \, P(\vec{u}) \sin\theta(\vec{r}-\vec{R}-\vec{u}t) \sin\theta(\vec{R}), \tag{17}$$

where n_v is the free vortex density and $P(\vec{u})$ is the Maxwell-Boltzmann velocity distribution. For $\lambda \lesssim 0.7$, and using Eq. 14 for θ for moving planar the vortices, space-time Fourier transform of Eq. 17 is

$$S_{zz}(\vec{q},\omega) = \frac{S^2}{32\sqrt{\pi}\delta^2} \frac{n_v \bar{u}}{q^3} e^{-(\omega/\bar{u}q)^2}, \quad \lambda \lesssim 0.7 . \tag{18}$$

Similarly, for $\lambda \gtrsim 0.8$, the vortex contribution to $S_{zz}(\vec{q},\omega)$ is found using Eq. 8 for θ for out-of-plane vortices, and we find

$$S_{zz}(\vec{q},\omega) = \frac{S^2}{4\pi^{5/2}} \frac{n_v}{\bar{u}} \frac{|f(q)|^2}{q} e^{-(\omega/\bar{u}q)^2}, \quad \lambda \gtrsim 0.8 \tag{19}$$

with vortex form factor for $qr_v \ll 1$,

$$f(q) \simeq \pi^{3/2} A r_v^2 [1 - \frac{15}{16} (qr_v)^2 + \frac{945}{4 \cdot 16^2} (qr_v)^4 - \cdots]. \tag{20}$$

In both cases there is a Gaussian central peak with width $\Gamma_z = \bar{u}q$. This reflects the assumption of a Maxwellian velocity distribution. The integrated intensities are

$$I_z^{pl} = \frac{S^2}{32\delta^2} \frac{n_v \bar{u}^2}{q^2} , \qquad \lambda \lesssim 0.7 , \tag{21}$$

$$I_z^{out} = \left(\frac{S}{2\pi}\right)^2 n_v |f(q)|^2 , \qquad \lambda \gtrsim 0.8 . \tag{22}$$

The small-q divergence of Eq. 21 can be removed by taking into account the finite system size or other similar cutoffs.

For the in-plane correlations, a different approach is needed because the in-plane spin components have no Fourier transform and they are globally sensitive to the presence of vortices. A vortex passing between 0 and r breaks the topological long range order in cos ϕ, diminishing the correlations and changing cos ϕ by a factor of -1. The vortices behave similarly to 2D sign functions. Carefully counting the number of vortices passing between 0 and r in time t leads to [7]

$$S_{xx}(\vec{r},t) \simeq \frac{S^2}{2} \exp \{-[(r/\xi)^2 + (\gamma t)^2]^{1/2}\} , \tag{23}$$

$$\gamma \equiv \sqrt{\pi}\, \bar{u}/2\xi . \tag{24}$$

Taking the space-time Fourier transform leads to a squared Lorentzian central peak

$$S_{xx}(\vec{q},\omega) \simeq \frac{S^2}{2\pi^2} \frac{\gamma^3 \xi^2}{\{\omega^2 + \gamma^2 [1 + (\xi q)^2]\}^2} . \tag{25}$$

The CP width and integrated intensity are

$$\Gamma_x \simeq \frac{[\pi(\sqrt{2} - 1)]^{1/2}}{2} \bar{u}\xi^{-1} [1 + (\xi q)^2]^{1/2} , \tag{26}$$

$$I_x \simeq \frac{S^2}{4\pi} \frac{\xi^2}{[1 + (\xi q)^2]^{3/2}} . \tag{27}$$

These results can be compared with the MC-MD simulation, for the XY model. We used a 100 x 100 lattice with periodic boundary conditions, allowing access to q $\simeq 0.02$ (π/a). First, an MC algorithm of 10^4 steps per spin was used to produce three equilibrium configurations at a desired temperature. These were used as initial conditions for MD using 4th order Runge-Kutta time integration with time step 0.04, sampling time $N_s \times 0.04$, and total integration time $512 \times N_s \times 0.04$. The sampling interval $N_s = 4 - 32$ depending on the wavevectors of interest. A Gaussian window function was applied to $S(\vec{q},t)$ before using an FFT algorithm for the time Fourier transform. $S(\vec{q},\omega)$ was averaged over the three initial conditions.

Typical data are shown in Fig. 3 for three temperatures, where $T_{KT} \simeq 0.8$ J. In-plane as well as out-of-plane spin wave peaks are seen to soften as T is increased towards T_{KT}. Additionally, CP intensity develops. Spin wave frequencies determined from $S(\vec{q},\omega)$ are shown in Fig. 4 .

Estimates (upper limits) of numerical CP widths and intensities are shown in Figs. 5 and 6 for T = 1.0. Agreement with the phenomenology is very good, using u and ξ as adjustable parameters. Fitting Γ_x to Eq. 26 gives $\bar{u} = 0.9$ and $\xi/a = 3.0$ (Fig. 6). Using these values in Eq. 21 gives $I_z = 7 \times 10^{-4}/q^2$; for small q this is larger than the MC-MD calculation of I_z by a factor of 2 (Fig. 5). It is not clear whether finite site corrections for small q are responsible for this difference. It is encouraging that the orders of magnitude are comparable, although the MC-MD data for I_z do not appear to vary as $1/q^2$. Multi-spinwave and other effects due to vortex-vortex and vortex-spinwave interactions may also be responsible for modifying the out-of-plane CP intensity to lower values than predicted for small q. Alternatively, it is probable that part of the in-plane width is due to multi-

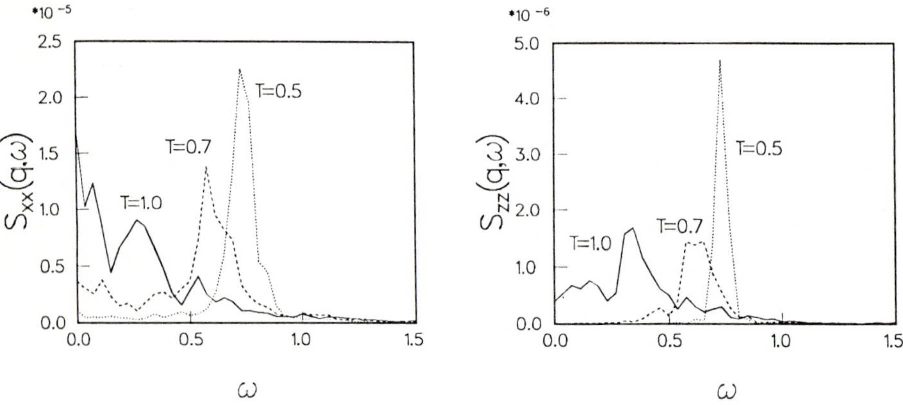

Fig. 3. Results from MC-MD simulations of the XY model on a 100 x 100 lattice, for $\vec{q} = (0.10, 0.10)\ \pi/a$, for a) in-plane correlations and b) out-of-plane correlations. Spin wave softening and development of central peak intensity are seen with increasing temperature.

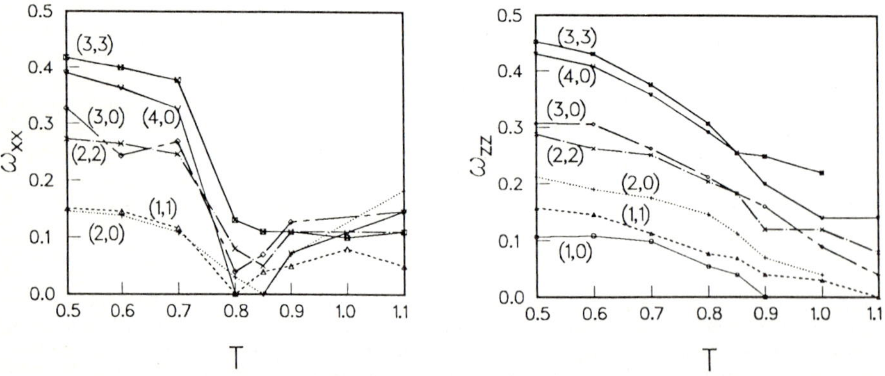

Fig. 4. Spin wave frequencies for the XY model, for q's in units of (1/50) (π/a), as functions of temperature (with J = 1), determined from a) in-plane correlations $S_{xx}(\vec{q},\omega)$, and b) out-of-plane correlations $S_{zz}(\vec{q},\omega)$. The frequencies for $T \gtrsim 0.8$ are crude estimates, especially for ω_{xx}; the softening of ω_{xx} is much stronger than for ω_{zz}.

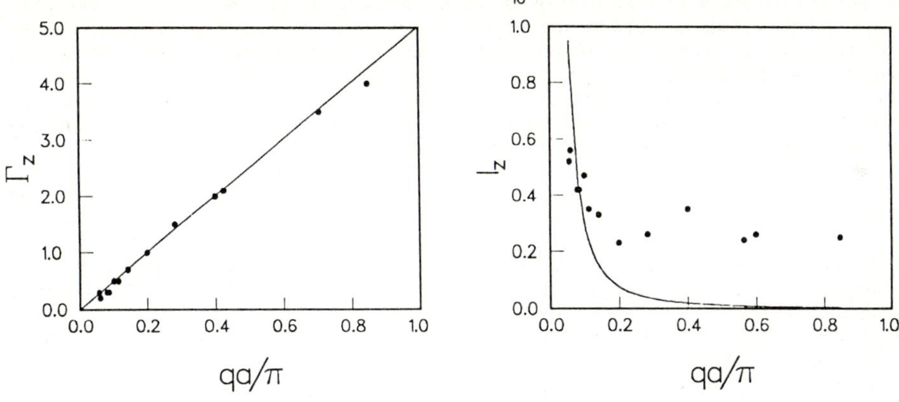

Fig. 5. Central peak properties of $S_{zz}(\vec{q},\omega)$ for the XY model at T = 1.0. In a) we show estimates of the CP width Γ_z from MC-MD data and compare with $\Gamma_z = \bar{u}q$ using $\bar{u} = 1.6$. In b) we show estimates of the CP intensity I_z from MC-MD data and compare with $I_z = 3 \times 10^{-4}/q^2$ (c.f. Eq. 21).

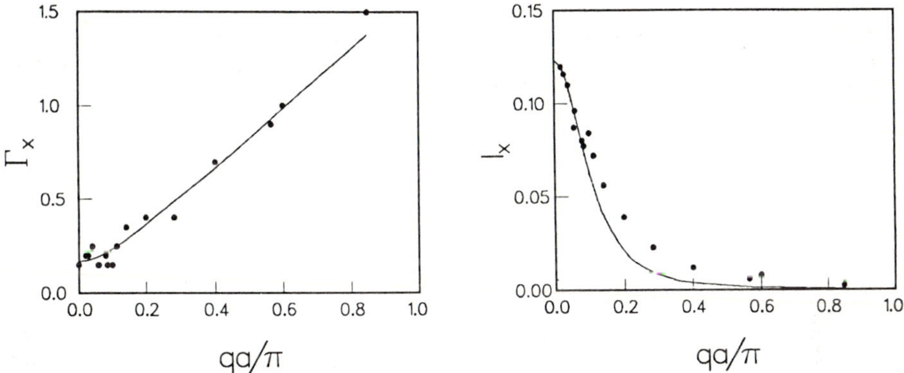

Fig. 6. Central peak properties of $S_{xx}(\vec{q},\omega)$ for the XY model at T = 1.0. In a) we show estimates of the CP width Γ_x from MC-MD data and compare with Eq. 26 using $\bar{u} = 0.9$ and $\xi = 3.0a$. In b) we show estimates of the CP intensity I_x from MC-MD data and compare with Eq. 27 using $\xi = 2.4a$.

spinwave processes, and by fitting the observed Γ_x to Eq. 26, we effectively underestimate the vortex-vortex separation 2ξ and thus overestimate the vortex density. This would then result in a predicted out-of-plane intensity due to vortices that is too large. Investigation of these effects will be important for more precise comparisons.

5. CONCLUSIONS

We have made continuum and discrete numerical calculations for classical vortex dynamics in a 2D ferromagnet, for the microscopic single vortex behavior and for the collective behavior of a finite system in equilibrium. From simulations on a square lattice, the planar ($\theta = 0$) static vortex predicted in continuum theory is stable only for $\lambda \lesssim 0.7$, and the out-of-plane static vortex is stable only for $\lambda \gtrsim 0.8$. Therefore, for $\lambda \lesssim 0.7$, out-of-plane correlations due to vortices should be determined by the velocity-dependent out-of-plane spin components of moving planar vortices. Using a large-r asymptotic expression for a moving planar vortex, we have calculated $S(\vec{q},\omega)$ within an ideal vortex gas phenomenology. The phenomenology gives a fairly good description of the central peak behavior of the XY model above T_{KT}, as found in MC-MD simulations. Corrections to the ideal gas phenomenology will involve accounting for the effects of interactions between the excitations. This may be easiest for the XY model, and will provide an instructive example for dealing with additional modifications due to in-plane anisotropy and due to other lattices, such as triangular and honeycomb.

REFERENCES

1. L. P. Regnault and J. Rossat-Mignod: In Magnetic Properties of Layered Transition Metal Compounds, eds., L. J. De Jongh and R. D. Willett (Reidel, Dordricht 1987).
2. M. T. Hutchings, J. Als-Nielsen, P. A. Lindgard, and P. J. Walker: J. Phys. C. 14, 5327 (1981).
3. M. Elahy and G. Dresselhaus: Phys. Rev. B30 7225 (1984).
4. J. M. Kosterlitz and D. J. Thouless: J. Phys. C. 6, 1181 (1973); Prog. Low Temp. Phys. Vol. VIIB, Chapter 5, ed. D. F. Brewer (North-Holland, Amsterdam 1978).
5. D. L. Huber: Phys. Lett. 76A, 406 (1980); Phys Rev. B26, 3758 (1982).
6. H. J. Mikeska: J. Phys. C 11, L29 (1978); J. Phys. C 13, 2313 (1980).
7. F. G. Mertens, A. R. Bishop, G. M. Wysin, and C. Kawabata: Phys. Rev. Lett. 59, 117 (1987).
8. C. Kawabata and A. R. Bishop: Sol. State Comm. 60, 169 (1986).
9. C. Kawabata, M. Takeuchi, and A. R. Bishop: J. Magn. Magn. Mat. 54-57., 871 (1986); J. Stat. Phys. 43, 869 (1986).
10. S. Hikami and T. Tsuneto: Prg. Theor. Phys. 63, 387 (1980).
11. Note that Eq. 3, 4, and 6 differ from the dynamic Eqs. of Ref. 10.

Simulation Study of Light Scattering from Soot Agglomerates

R.D. Mountain

Thermophysics Division, National Bureau of Standards,
Gaithersburg, MD 20899, USA

Simulation studies can be of considerable help in clarifying measurement techniques in that "complete" information on the system being investigated is available. This point is illustrated here where a Langevin dynamics model of the formation of sooty smoke via an agglomeration process is used to determine what sort of information is available in light scattering measurements. [1] It is generally accepted that soot forms by agglomeration of small, roughly spherical particles which are composed mainly of carbon. [2] The typical size of these particles, which we refer to as primary particles, is on the order of 20-40 nm. These primary particles diffuse and occasionally stick together forming agglomerates which continue to diffuse and grow. The resulting clusters have an open structure which can be characterized by a fractal dimension, d, which is a useful statistical measure for classifying geometrically irregular objects. [3]

An example of the structure of a soot agglomerate, as imaged in a transmission electron micrograph, is shown in Fig. 1. The objects produced in the simulations described here closely resemble such clusters.

A simple model of the growth process can be developed using stochastic differential equations to describe the motion of the primary particles and of the resulting agglomerates. This is because the primary particles move diffusively in the background gas in which the particles are embedded. The model is formulated in terms of equations of motion of the Langevin type

$d\mathbf{r}/dt = \mathbf{v}$, and

$m d\mathbf{v}/dt = m\Gamma\mathbf{v} + \mathbf{f}$,

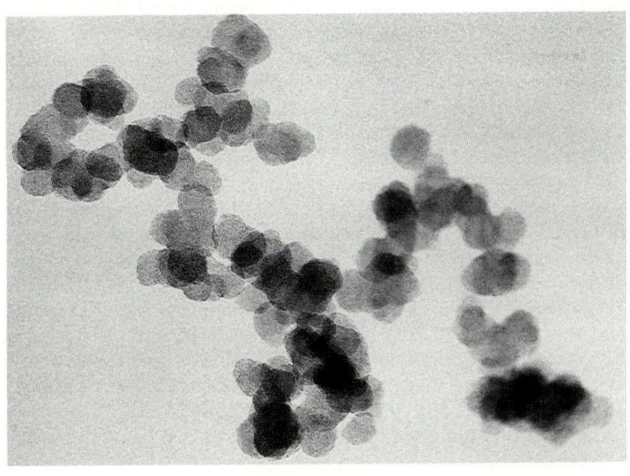

Figure 1. A TEM image of a soot agglomerate, courtesy of G.W. Mulholland.

Springer Proceedings in Physics Vol. 33: **Computer Simulation Studies in Condensed Matter Physics**
Editors: D. P. Landau • K. K. Mon • H-B. Schüttler

where $\mathbf{r}$ and $\mathbf{v}$ are the position and velocity of a cluster of mass m, Γ is a friction coefficient which is related to the fluctuating force $\mathbf{f}$ by a fluctuation-dissipation relation

$$\langle \mathbf{f}(t) \cdot \mathbf{f}(t+s) \rangle = 6m\Gamma k_B T\delta(s)$$

when the cluster is in thermal equilibrium with the background gas through which it diffuses. [4] The temperature of the gas is T and k_B is Boltzmann's constant. The agglomeration process is modeled by solving a set of such equations, one pair for each of the distinct clusters or isolated primary particles, along with a boundary condition that if any two primary particles touch, then the clusters to which the primary particles belong stick together forming a new cluster. This model is sometimes called cluster-cluster aggregation. [5]

Before discussing the details of how the model is solved, let us digress and consider just what is meant by a solution of a stochastic differential equation. Clearly the solution has statistical rather than deterministic content. Suppose one has a stochastic equation of the form

$$dX/dt = F(X) + g,$$

where the stochastic term g is characterized by expectation values such as

$$\langle g(t) \rangle = 0 \text{ and } \langle g(t)g(t+s) \rangle = A\delta(s).$$

A set of solutions would provide answers to questions such as: given X(t), what is the probability of finding X(t+s)? If we knew the conditional probability of finding X(t+s) given X(t), P(X(t+s);X(t)), then the answer to that question could be obtained by sampling from the conditional probability. In effect, a solution to the stochastic differential equation provides a sample taken from the conditional probability P, and a collection of solutions provides a realization of P.

The standard discussion of the solution of the translational Brownian motion problem for physicists is found in the elegant review article by S. CHANDRASEKHAR [4]. This should be "must" reading for students of statistical physics.

ERMAK and BUCKHOLZ [6] have used the fully worked out solution to the translational Brownian motion problem to construct numerical solutions of the stochastic equations of motion for a diffusing particle. What they have done is to use the conditional probability of finding $\mathbf{r}(t+s)$ and $\mathbf{v}(t+s)$, given $\mathbf{r}(t)$ and $\mathbf{v}(t)$ to construct trajectories for the diffusing particle. We use this approach when generating the clusters formed by agglomeration. It is also possible to generate solutions of the differential equations using numerical integration techniques provided that the statistical properties of the fluctuating term are carefully taken into account. [7]

The numerical integration approach is necessary if rotational degrees of freedom are to be included in the description of the system. Since we do not include rotational motion in our model, only a few observations on how to deal with rotations will be included here. First, the rigid objects should be described using a quaternion representation. [8-11] There are some distinct advantages in using quaternions, the first being that it provides a singularity free method for transforming between laboratory-fixed and body-fixed frames of reference. Also, the Langevin equations for rotational Brownian motion are readily integrated, along with the translational Langevin equations, so the process of sampling the conditional probability function can occur using a unified integration scheme.

The simulation of an agglomeration process requires some fairly extensive bookkeeping arrangements in order to keep track of which primary particles are in which cluster and to update cluster information when agglomeration occurs. It is also desirable to have some sort of neighbor list so that the test for agglomeration can be made efficiently. The necessary bookkeeping can be managed using linked lists which provide connections between the cluster in which a primary particle resides and the sorted order of the primary particles in the x-, y-, and z-directions. Having the primary particles in sorted order simplifies the determination of cluster properties, such as the radius of gyration, when periodic boundary conditions are used. The code used in the simulations described here makes use of these features. [12] In particular, using a neighbor table construction based on having the particles in sorted order proves to be quite efficient when large numbers of particles are used in the simulation. [13] The specific introduction of a unit of time based on physical considerations makes it possible to examine growth kinetics of agglomerates under a variety of conditions. [14]

In the simulations discussed here, the mass and diameter, σ, of the primary particles are taken as the units of mass and length. Initially, 4000 primary particles are distributed randomly in a cubic box with periodic boundary conditions. The volume of the box is chosen so that the number density of the primary particles is 0.0167. This density corresponds to the situation of free molecular motion of the gas molecules. The initial velocities of the particles are selected from an equilibrium distribution with temperature, T. The unit of time, τ, is taken to be the time required for a primary particle moving with the mean thermal speed to move a distance σ. The equations of motion are integrated using a time step of 0.05τ. Under these conditions, 600 time steps are required to generate a set of agglomerates which are then used to estimate the scattering properties.

One of the properties of interest is the fractal dimension, d, of the clusters as it relates the number of primary particles, N, in a cluster to the radius of gyration of the cluster,

$$N = AR_g^{\ d}.$$

The constant A is found empirically from our simulations to have a value between 5 and 6. This is illustrated in Fig. 2 for a set of 43 clusters generated in order to examine light scattering by agglomerates. The clusters

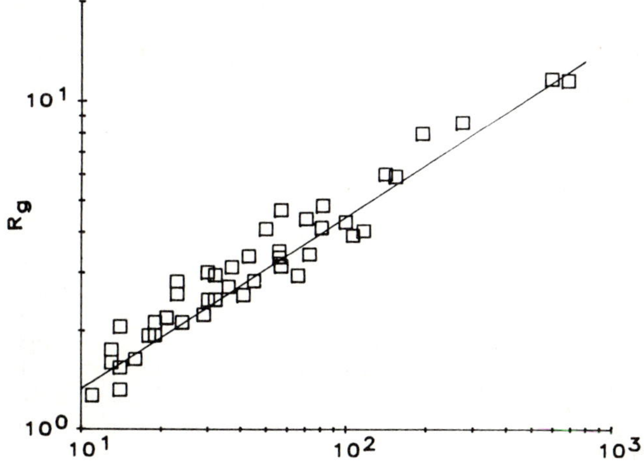

Figure 2. The radius of gyration vs N for a set of agglomerates. The solid line represents the case with fractal dimension d=1.9.

in this set contain between 11 and 700 primary particles. Our experience is that the fractal relation between N and R_g holds for N>10. The fractal dimension inferred from this plot is $1.8<d<1.9$. That means that these clusters are "transparent" in that the interior of the cluster is not geometrically screened by the particles on the surface.

The facts that (a) our simulated soot clusters are transparent, that is $d<2$, and (b) in the physical system, the primary particles are much smaller than the wavelength of visible light mean that the simple Rayleigh-Debye light scattering theory can be used to estimate the intensity of the scattered light. [15] In that theory, the scattered intensity relative to the intensity of the incident light with wavelength λ is expressed as

$$I(q) = k^4|\alpha^2|\sin^2\psi\, S(q).$$

Here $k=2\pi/\lambda$, α is the polarizability of the primary particles, and ψ is the angle between the polarization direction of the incident light and the propagation direction of the scattered light. S(q) is the orientationally averaged structure factor of the cluster

$$S(q) = \langle|\Sigma\exp(i\mathbf{q}\cdot\mathbf{r}_j)|\rangle \qquad (1)$$

and $q = 2k\sin(\vartheta/2)$ where ϑ is the scattering angle. The sum in (1) runs over the N particles in the cluster. Because of the irregular shape of the agglomerates, an average over the orientations of the cluster relative to the polarization and propagation direction of the scattered light is needed to make contact with experimental observations. This average is indicated by the angular brackets <...>. In the following discussion it should be kept in mind that lengths are expressed in units of σ, the primary particle diameter. The term "small q" means $R_g q<<1$, while "large q" means $1<R_g q$ but $q\sigma<1$.

The results obtained for the orientationally averaged structure factor for 43 clusters are displayed in Fig.3. There the structure factor divided by N^2 is shown as a function of $R_g q$. Although there is some scatter in the data for larger values of q, these results strongly suggest that $S(q)/N^2$ is a function of $R_g q$ for various size agglomerates. There is a good deal of information in

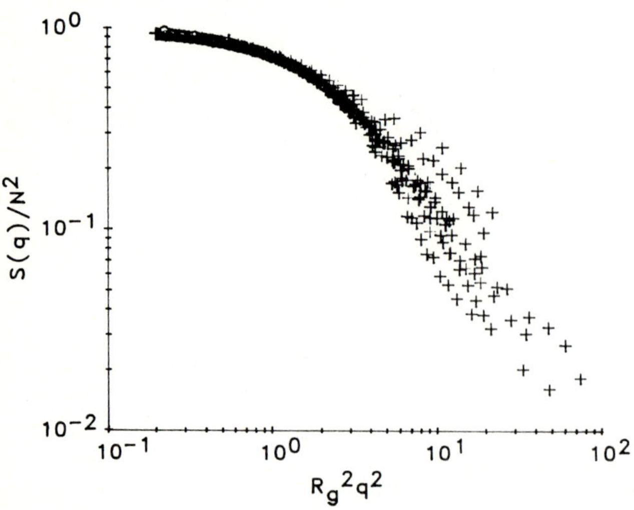

Figure 3. Scaled structure factor for the set of agglomerates.

the scattering data. First of all, the small q region contains information about R_g, as for small q

$$S(q)/N^2 = 1 - \frac{1}{3}(R_g q)^2 + .. O(q^4)$$

which is a well known result for scattering by polymers. [15] The structure factor for large q decreases as q^{-d} since the agglomerates are fractal objects. [3] It is in principle possible to infer not only R_g and d from the scattering data, but to also obtain an estimate of the primary particle size by using integrated intensity data in conjunction with the simulation results. This topic is discussed in detail elsewhere. [16]

One of the characteristics of fractals is that g(r), the distribution of pairs in the object separated by a distance r, has for a fractal of infinite extent, the form

$$g(r) = C/r^{D-d}, \qquad (2)$$

where D is the dimension of the space in which the fractal is embedded; in our case D = 3. Because g(r) and S(q) are Fourier transform pairs, the scaling property of S(q) with $R_g q$ implies that g(r) is a function of r/R_g. This is illustrated in Fig.4 where the pair distribution function of the set of clusters used to construct S(q) is shown. The solid line is the distribution of pairs constructed from the simulation results and the dashed line is (2) with C = 1 and d = 1.9. Since the agglomerates are of finite extent, the power law form is modified by a cutoff function h(r) defined by

$$g(r) = (C/r^{D-d})\, h(r). \qquad (3)$$

This function has the property that h(0)=1 and that it goes rapidly to zero for $r>R_g$. The cutoff function h(r) for these clusters has been determined using (3) and may be represented by the form $h(r) = \exp(-0.2(r^{2.5}))$. The value 2.5 should be understood to mean that it is greater than 2 and less than 3.

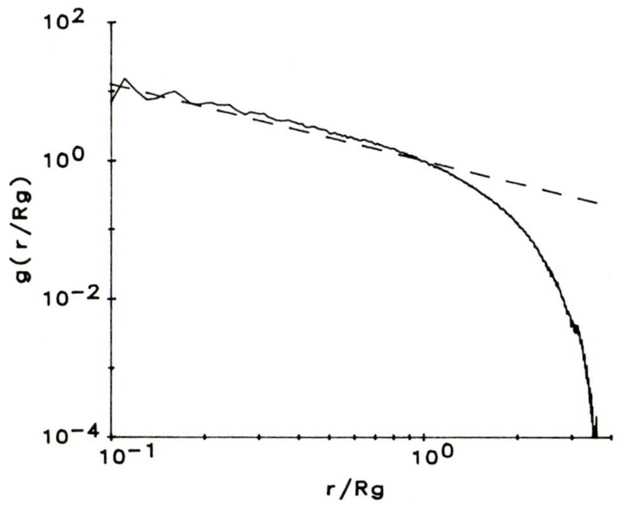

Figure 4. The distribution of pairs of primary particles in the agglomerates. The dashed line represents (2) for d=1.9.

The use of a stochastic model to examine the information contained in liqht scattering by agglomerates is an example of how the computer simulation technique can be used to guide the design of measurement methods. While the underlying theory of the model was known to CHANDRASEKHAR [4], it is only through simulation techniques that practical information can be extracted from it.

References

1. H. M. Lindsay, M. Y. Lin, D. A. Weitz, P. Sheng, R. Klein and P. Meakin: Faraday Discuss. Chem Soc. 83, 153 (1987)
2. R. J. Samson, G. W. Mulholland, and J. W. Gentry: Langmuir, 3, 272 (1987)
3. J. Teixeira: In On Growth and Form, ed. by H. E. Stanley and N. Ostrowsky (Martinus Nijhoff Publishers, Dordrecht 1986) p. 145
4. S. Chandrasekhar: Rev. Mod. Phys. 15, 3 (1943)
5. R. Jullien and R. Botet: Aggregation and Fractal Aggregates (World Scientific Publishing, Singapore 1987)
6. D. L. Ermak and H. Buckholz: J. Comput. Phys 35, 169 (1980)
7. T. Schneider and E. Stoll: Phys. Rev. B 17, 1302 (1978)
8. R. Gerling and A Hüller: Z. Phys. B 40, 209, (1980)
9. D. J. Evans and S. Murad: Mol. Phys. 34, 327 (1977)
10. R. Sonnenschein: J. Compt. Phys. 59, 347 (1985)
11. D. C. Rapaport: J. Compt. Phys. 60, 305 (1985)
12. F. Sullivan, R. D. Mountain: Comput. Phys. Comm. 42, 43 (1985)
13. F. Sullivan, R.D. Mountain and J. O'Connell: J. Comput. Phys. 61, 138 (1985)
14. R. D. Mountain, G. W. Mulholland and H. Baum: J. Colloid and Interface Sci. 114, 67 (1986)
15. B. J. Berne and R. Pecora: Dynamic Light Scattering (John Wiley and Sons, New York 1976)
16. R. D. Mountain and G. W. Mulholland, work in progress.

Simulation of Non-equilibrium Growth and Aggregation Processes

P. Meakin

Central Research and Development Department,
E.I. du Pont de Nemours and Company, Wilmington, DE 19898, USA

The introduction of the diffusion-limited aggregation (DLA) model by Witten and Sander has stimulated interest in a broad range of simple non-equilibrium growth and aggregation models. Here it is shown how the DLA model has influenced the development of a simple but quite realistic model for mechanical failure in thin film deposits.

I. INTRODUCTION

The development of the diffusion-limited-aggregation (DLA) model by Witten and Sander [1] has stimulated a still growing interest in a broad variety of non-equilibrium growth and aggregation processes. Although the development of the DLA model was initially motivated by experimental work [2] on iron particle aggregates which were found to have a fractal geometry [3], it does not provide a realistic representation of this type of aggregation process. However, the DLA model is still of considerable interest because it provides a basis for understanding a broad variety of other growth processes (random growth processes in which the growth probabilities are determined by a field obeying the Laplace equation) and because the theoretical challenge of developing a rigorous understanding of this simple (to define) process has not yet been fully met. Witten and Sander [1,4] recognized the relationship between their model and the more general process of growth in a Laplacian field and this relationship was made more explicit by the introduction of the dielectric breakdown model by Niemeyer et al. [5]. At this stage it became natural to think of extensions of the DLA and dielectric breakdown models to growth processes controlled by vector and tensor fields. The first model of this type (a mechanical breakdown model controlled by a displacement field obeying the Navier equation) was developed by Louis et al. [6,7]. Unlike the DLA model this mechanical breakdown model does not seem to be directly applicable to most real experimental processes. However, this model is of interest as a generalization of the DLA model and it has stimulated the development of more realistic mechanical failure models [8-10].

Initially, interest developed in the DLA [1,4] and the cluster-cluster aggregation models [11,12] which were developed to model colloidal aggregation processes (such as that studied by Forrest and Witten [2]) because of the fractal structures which they generated. More recently considerable interest has also developed in the kinetics of these growth and aggregation processes (particularly in the case of the cluster-cluster aggregation models). At the present time it is not possible to predict theoretically the fractal dimensionality (or other geometric scaling properties) of the structures generated by most non-equilibrium growth and aggregation models from the growth "rules". Consequently, much of our knowledge and understanding of these

Springer Proceedings in Physics Vol. 33: **Computer Simulation Studies in Condensed Matter Physics**
Editors: D. P. Landau · K. K. Mon · H-B. Schüttler

processes has come about as a result of computer simulations. In the future simulation results will be needed to discriminate between competing theoretical ideas, introduce challenging new theoretical problems and stimulate new experimental work.

2. DIFFUSION-LIMITED AGGREGATION

In the diffusion-limited aggregation (DLA) model particles are added, one at a time, to a growing cluster or aggregate of particles via random walk trajectories. Figure 1 illustrates a two-dimensional version of this model carried out on a square lattice. At the stage of growth illustrated in Figure 1 a small cluster has already formed. The particles are imagined to have come from infinity but are actually launched from a random position on a circle which just encloses the cluster (in this case this launching circle has a radius of R_{max} + 5 lattice units, where R_{max} is the maximum radius of the cluster measured from the original "seed" or growth site). Two typical random walk trajectories are shown in Figure 1. Trajectory t_1 eventually brings the random walker into an unoccupied perimeter site. At this stage this perimeter site is filled (growth occurs) and a new random walker is launched from the launching circle (whose radius increases as the cluster grows). Trajectory t_2 eventually moves the particle a long way from the cluster and this trajectory is terminated when it reaches the killing circle which has a radius of 3 R_{max} in this simulation (this is satisfactory for small scale simulations but a more typical value for the radius of the killing circle in more recent large scale simulations is 100 r_{max}). The procedures described above and illustrated in Figure 1 are repeated many times until a large cluster has grown.

The early two dimensional results of Witten and Sander and extensions to spaces and lattices with (Euclidean) dimensionalities of 3-6 by Meakin [13,14] were carried out using models which quite closely resembled that illustrated in Figure 1.

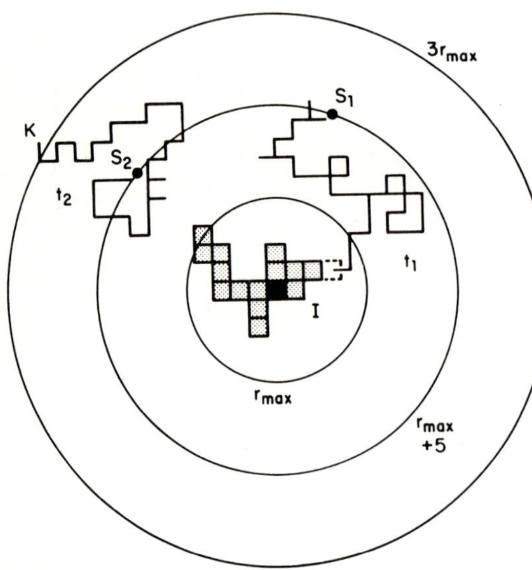

Fig. 1. An early stage in a square lattice model simulation of diffusion-limited aggregation. The original seed or growth site is shown in black and the other sites which are occupied at this stage are shaded. Two typical trajectories starting at random positions on the launching circle are shown. Trajectory t_1 reaches an unoccupied surface site (growth site), which is indicated by dashed edges, and this site is occupied. Trajectory t_2 reaches the termination circle which in this case has a radius of 3 R_{max} where R_{max} is the maximum radius of the cluster. This trajectory will be terminated and a new trajectory started at a random position on the launching circle.

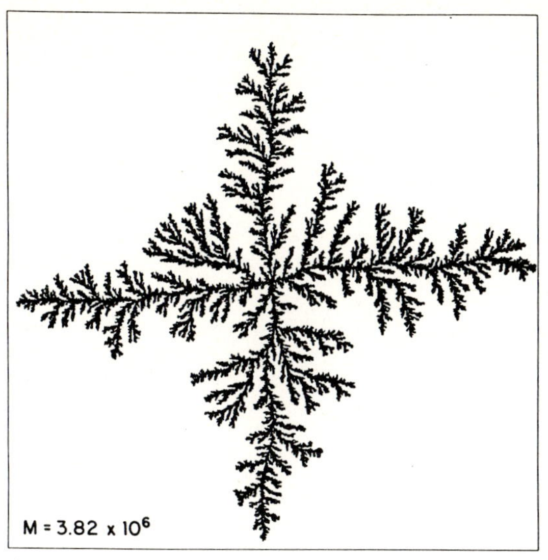

Fig. 2. A 3.82×10^6 site square lattice DLA cluster generated using the algorithms of Ball and Brady [15]. The influence of the anisotropy of the square lattice on the overall shape of the cluster is apparent in this figure.

In recent years a significant effort has gone into the development of much more efficient 2d algorithms [15-17]. The key to developing more efficient DLA algorithms is to allow the random walkers to take long steps when they are far from the cluster. In this way the speed of large scale two dimensional simulations can be increased by a factor of about 10^3 and smaller (but still significant) improvements can be obtained for $d \geqslant 3$. The "state of the art" in two dimensional DLA simulations is illustrated in Figure 2. Here a square lattice DLA cluster containing almost 4×10^6 sites [18] is shown. This cluster was generated using the algorithm of Ball and Brady [15] and requires about 16 hours of CPU time on an IBM 3081 computer. For $d \geqslant 3$ most of our knowledge concerning the structure of DLA clusters is based on quite small scale simulations (a few clusters containing up to about 10^4 sites or particles [13, 14]). However, unpublished [19] cubic lattice model simulations of about 100 50,000 site clusters give an effective fractal dimensionality of very close to 2.50. Clusters of this size are probably close to the small size (noise dominated) limit in which the effects of lattice anisotropy are unimportant. We [20] have recently developed improved algorithms for both hypercubic lattice and off-lattice DLA in dimensions 2-8. Preliminary results from the lattice models give effective fractal dimensionalities very close to the mean field limit [21] values of $(d^2+1)/(d+1)$ for $d \geqslant 3$. For d=2 the noise dominated fractal dimensionality has a value of about 1.71 (about 0.04 greater than the mean field value) [22].

3. MECHANICAL BREAKDOWN MODELS

A simple mechanical breakdown model closely related to DLA was developed by Louis et al. [6,7]. In this model a triangular network of nodes and Hookean springs is used to represent a two dimensional material (a square network cannot be used since the shear modulus would be zero). At the start of a simulation the network is stretched isotropically by a small amount (typically 0.1%) to avoid

Fig. 3a.

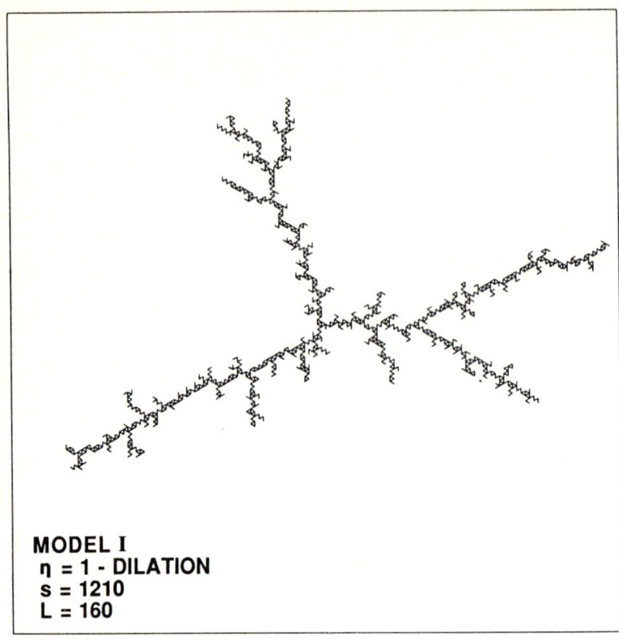

Fig. 3b.

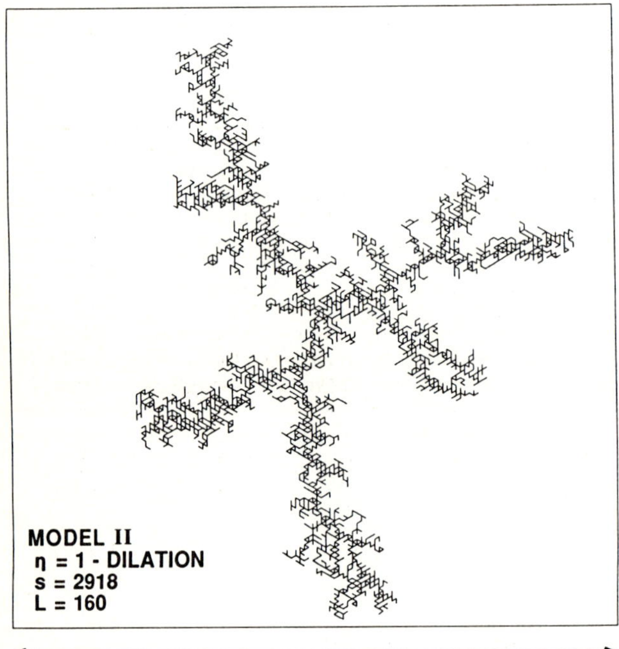

Fig. 3. Cracking patterns formed in the DLA-like mechanical breakdown model. Figures 3a, 3b and 3c show the location of the broken bonds (in the unstrained network) obtained from simulations carried out using models I, II and III with η = 1.0. The number of broken bonds (s) is 1210, 2918 and 1696 respectively.

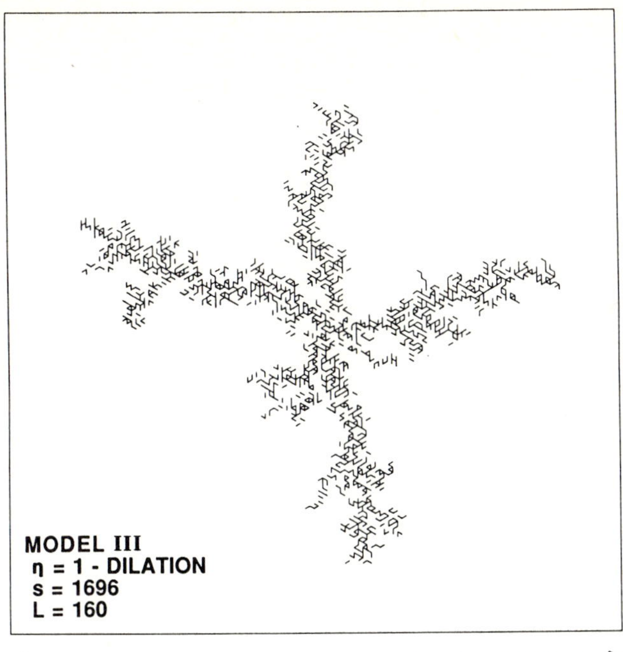

Fig. 3c.

non-linear effects. In all of the simulations described here we [23] have used an array of 160^2 nodes and $3x160^2$ bonds to represent the material. At the start of a simulation a bond at the center of the network is broken and the system is relaxed to its equilibrium configuration. At this stage the strains (δ) associated with all of the bonds at the surface of the "crack" formed by the removal of the first broken bond are found. In the next stage one of the bonds at the surface of the crack is selected with a probability P given by

$$P_i = (\delta_i)^\eta , \qquad (1)$$

where δ_i is the extension of the ith surface bond and η is an exponent analogous to the exponent η used in the dielectric breakdown model [5]. The selected bond is then broken. The network is again relaxed to equilibrium and the third broken bond is selected with a probability given by equation (1). This process of bond breaking and relaxation is repeated many times until a sufficiently large crack has grown. We have explored three versions of this model using both isotropic dilation and shear strain. In model I only those bonds which form the outline of the crack are allowed to break. In model II only those bonds associated with nodes at the end of broken bonds are allowed to break and in model III all of the bonds associated with all of the nodes at the crack perimeter are capable of being broken at the next stage in the crack growth process. After the first bond has been broken, the number of surface bonds capable of being broken is 4, 10 and 18 for models I, II and III respectively. Figure 3 shows the location of the broken bonds in simulations carried out using models I, II and III with a value of 1.0 for the bond breaking probability exponent η. Similarly, Figure 4 shows the results of a simulation carried out with η = 0.5 using model I.

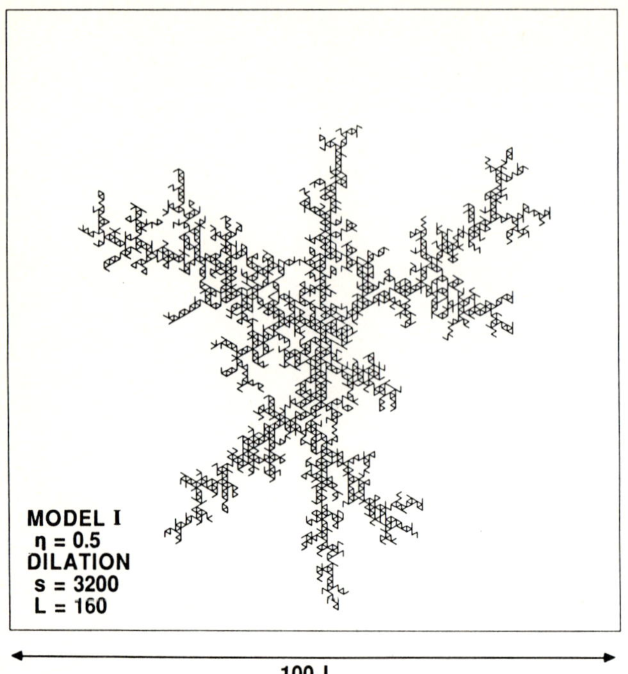

Fig. 4. A cracking pattern generated using model I with a bond breaking probability exponent (η) of 0.5.

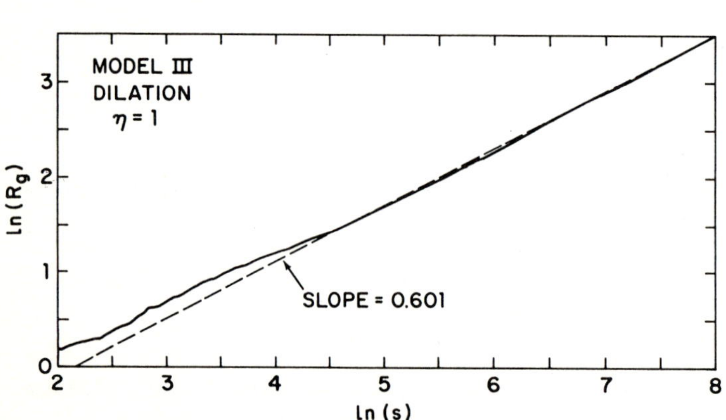

Fig. 5. Dependence of the cracking pattern radius of gyration (R_g) on the number of broken bonds (s) obtained from 8 simulations carried out using model III with η = 1.

For all of these models we have measured the dependence of the radius of gyration (R_g) of the broken bonds (using their positions in the unstrained network) as a function of the number of bonds broken (s). A typical result obtained using model III is shown in Figure 5. In all cases our results suggest a power law relationship between R_g and s

$$R_g \sim s^{\beta} \tag{2}$$

consistent with a fractal cracking pattern. In the example shown in Figure 4, the exponent β of about 0.6 corresponds to a fractal dimensionality (D_β) of about 5/3 (very close to the value associated with 2d DLA [1]). However, at this stage we cannot be sure if the results obtained using models I, II and III are universal and if they are, we cannot be sure if they belong to the same universality class as DLA. Much more extensive, larger scale simulations will be needed to resolve these questions and these will require new algorithms.

4. ELASTIC FRACTURE IN THIN FILMS

A simple model for elastic fracture in thin films is illustrated in Figure 6. As in the Louis-Guinea [6,7] model the surface layer is represented by a triangular network of nodes and springs. However, this model [10] differs from that of Louis and Guinea in several important respects. (1) The elastic network is attached to a rigid underlying substrate by bonds which have a small force constant but which do not break, (2) All of the bonds are eligible to be broken at all stages in the simulation and (3) The bond breaking probabilities are given by

$$P_i \sim e^{(k\delta_i^2/2)} . \tag{3}$$

In addition, the initial bond extension is typically about 10% of ℓ_o instead of 0.1% ℓ_o where ℓ_o is the equilibrium bond length. This model is intended to represent the mechanical failure of a high modulus film attached to a low modulus substrate or a high modulus film weakly bonded to a rigid substrate. At the start of a simulation each bond has a length of 1.0 and the network of bonds and nodes forms a perfect triangular lattice. Each node is attached to the underlying substrate by a weak bond and the force exerted by the weak bond on the ith node is given by

$$\underset{\sim}{f} = k_2(\underset{\sim}{S}_i - \underset{\sim}{S}_i^{\,o}) . \tag{4}$$

Here $\underset{\sim}{S}_i$ is the position of the ith node and $\underset{\sim}{S}_i^{\,o}$ is its position at the start of the simulation. In most of our simulations the

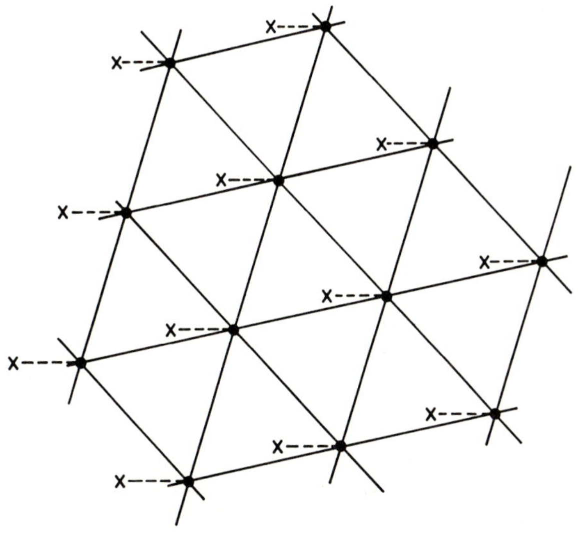

Fig. 6. A schematic representation of the model for elastic fracture in thin films. The nodes (large dots) are connected by strong bonds to form a triangular lattice. Each node is joined to the underlying substrate by a weak bond (---) at its original position at the start of the simulation. Throughout the simulation the distance from the nodes to the underlying substrate is constant (only horizontal motion is allowed). This figure shows the original configuration in which each node is associated with six bonds.

equilibrium bond length was assumed to be 0.90909... corresponding to an initial extension ratio of 1.1 (The initial value for δ is 0.090909... for all of the bonds). In all other respects the model is very similar to that of Louis and Guinea. The simulation proceeds via a sequence of bond breaking and relaxation steps.

Figure 7 shows the results of a simulation carried out using a network of 200x200 nodes and 120,000 bonds. In this simulation the value for the force constant k is quite large (k = 1200) and the bonding to the substrate is quite weak (k_2 = 12). Under these conditions a relatively slow crack initiation period is followed by the rapid growth of a few very linear cracks which reduce the strain in the surface layer. The cracks which grow at a later stage are less linear and propagate more slowly.

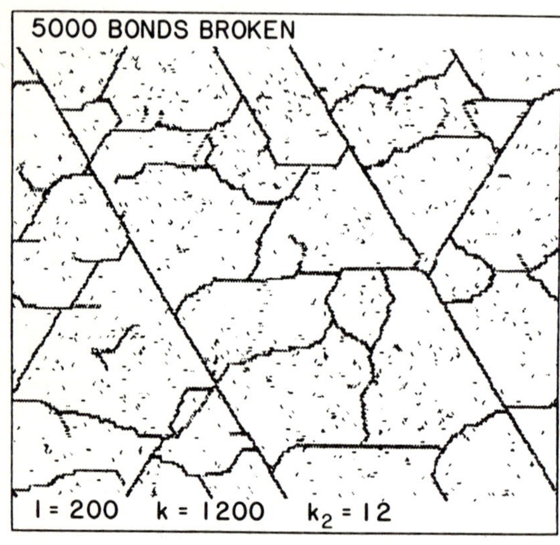

Fig. 7. The results of a simulation of elastic fracture in a surface film attached to a rigid substrate. This figure shows the locations of the broken bonds (in the unstrained network) after 5000 bonds have been broken. The long linear cracks were the first to grow and these cracks were connected at a later stage by the slower growth of the more irregular cracks.

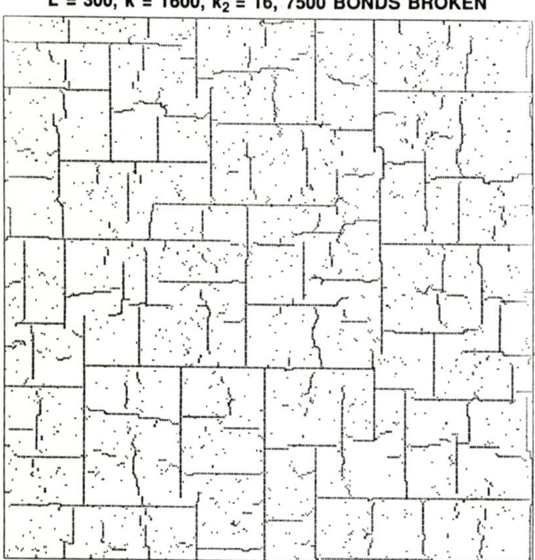

Fig. 8. The results of a simulation of elastic fracture in a surface film attached to a rigid substrate using a square network instead of a triangular network.

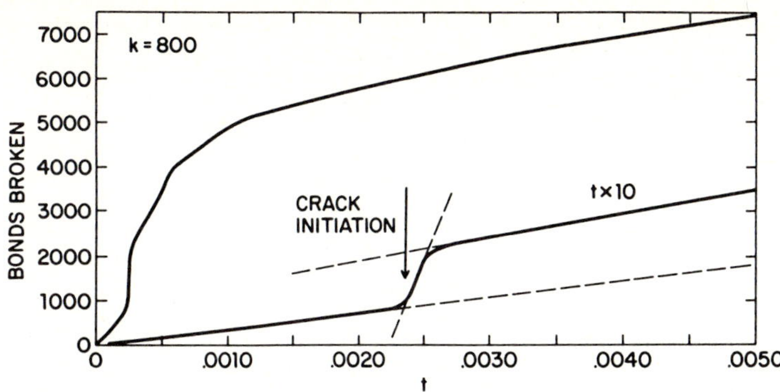

Fig. 9. Time dependence of the number of broken bond in the surface layer obtained from triangular network simulations with k = 800 and k_2 = 8. The figure indicates a slow initiation period in which little strain energy is released and only isolated defects are formed, followed by a period of rapid crack propagation in which about 30% of the strain energy is released. This rapid release of the stress in the surface layer slows down subsequent crack initiation and propagation events. As more cracking occurs, the stress becomes still smaller and further slowing down is observed.

Figure 8 shows the results of a similar simulation carried out using a network of 300x300 nodes joined by 180,000 bonds to form a square network. The cracking patterns shown in Figures 7 and 8 resemble quite closely those which occur in real systems such as ceramic coated metals, paint films, etc.

All of the models discussed in this paper can be made time dependent. For example, in this model for elastic fracture, a time scale can be introduced by incrementing the time by $1/(N\ P_{max})$ each time a bond is selected at random and broken with a probability of P/P_{max}. Here P is the bond breaking probability (from equation (3)) for the selected bond and P_{max} is the maximum bond breaking probability for any bond in the network. The quantity N is the total number of unbroken bonds. The time is incremented by $1/(N\ P_{max})$ whether or not the randomly selected bond is actually broken. Alternatively, if a bond is selected with a probability P_i/P_{TOT} and broken (where P_{TOT} is the total bond breaking probability) then the time should be incremented by $1/P_{TOT}$ after each bond has been broken where P_{TOT} is the sum of all of the bond breaking probabilities. Figure 9 shows the results obtained from a time dependent simulation of this type.

5. SUMMARY

In this short paper I have tried to indicate some recent directions in the computer simulation of non-equilibrium growth and aggregation models and point out some areas in which unproved algorithms and/or larger scale simulations are needed. I have also tried to show how work on simple models can lead to new approaches towards the development of more realistic models which might eventually find practical applications. One important aspect of recent work which was discussed only very briefly here is the kinetics of growth and aggregation processes [24,25].

6. ACKNOWLEDGMENT

The work on DLA was carried out in collaboration with R. C. Ball, P. Ramanlal, L. M. Sander and S. Tolman and the work on DLA-like cracking models was done with F. Guinea, G. Li, E. Louis and L. M. Sander.

7. REFERENCES

1. T. A. Witten and L. M. Sander, Phys. Rev. Lett. 47, 1400 (1981)
2. S. R. Forrest and T. A. Witten, J. Phys. A12, L109 (1979)
3. B. B. Mandelbrot, "The Fractal Geometry of Nature", W. H. Freeman and Company, New York (1982)
4. T. A. Witten and L. M. Sander, Phys. Rev. B27, 5686 (1983)
5. L. Niemeyer, L. Pietronero and H. J. Wiesmann, Phys. Rev. Lett. 52, 1033 (1984)
6. E. Louis, F. Guinea, and F. Flores in "Fractals in Physics" Proceedings of 6th International Symposium on Fractals in Physics ICTP, L. Pietronero and E. Tosatti, eds., Elsevier - North Holland, Amsterdam, p. 177 (1986)
7. E. Louis and F. Guinea, Europhys. Lett. 3, 871 (1987)
8. Y. Termonia and P. Meakin, Nature 320, 6061 (1986)
9. Y. Termonia, P. Meakin and P. Smith, Macromolecules, 18, 2246 (1985)
10. P. Meakin, Thin Solid Films, 151, 165 (1987)
11. P. Meakin, Phys. Rev. Lett. 51, 1119 (1983)
12. M. Kolb, R. Botet and R. Jullien, Phys. Rev. Lett. 51, 1123 (1983)
13. P. Meakin, Phys. Rev. A27, 604 (1983).
14. P. Meakin, Phys. Rev. A27, 1495 (1983).
15. R. C. Ball and R. M. Brady, J. Phys. A18, L809 (1985).
16. P. Meakin, J. Phys. A18, L661 (1985).
17. P. Meakin in "On Growth and Form: Fractal and Nonfractal Patterns in Physics", NATO ASI Series E100, H. E. Stanley and N. Ostrowski, eds., Martinus Nijhoff, Dordrecht (1986).
18. P. Meakin, R. C. Ball, P. Ramanlal and L. M. Sander, Phys. Rev. A35, 5233 (1987).
19. P. Meakin and L. M. Sander, unpublished.
20. S. Tolman and P. Meakin, unpublished.
21. M. Muthukumar, Phys. Rev. Lett. 50, 839 (1983).
22. P. Meakin and L. M. Sander, Phys. Rev. Lett. 54, 2053.
23. P. Meakin, L. M. Sander, G. Li, E. Louis and F. Guinea, unpublished.
24. F. Family and D. P. Landau, "Kinetics of Aggregation and Gelation", Elsevier-North Holland, Amsterdam (1984).
25. T. Riste and R. Pynn, "Time Dependent Effects in Disordered Materials" NATO ASI, Geilo, Norway (1987).

Growth by Gradients: Fractal Growth and Pattern Formation in a Laplacian Field

F. Family

Department of Physics, Emory University, Atlanta, GA 30322, USA

1. Introduction

The formation of a wide variety of patterns is controlled by the strength of the gradient of a field at the interface between the structure and the outside. For example, the rate of growth in aggregation processes [1,2] depends on the concentration of diffusing particles, in solidification [3] it depends on the temperature gradient, in dielectric breakdown [4] it is proportional to some power of the electric field, in electrodeposition [5,6] it depends on the gradient of the electrical voltage, and in viscous fingering [7,8] it is the pressure gradient. It has long been speculated [9] that concentration gradients of nutrients, light and energy resources are the factors controlling the development of a variety of biological patterns as well, even though much less is known about pattern formation in biological systems [9].

In addition to the fact that the controlling mechanism in the growth of these patterns is the field gradient, there exists an even deeper connection. The common denominator among many growth processes is that the controlling fields, like temperature, concentration, pressure, and electrical potential, satisfy either the diffusion equation or the Laplace equation which is a special case of the diffusion equation. The differences in morphologies are often reflections of variations in boundary conditions and external parameters. Under quasi-stationary conditions [3], *i.e.* in the limit of infinitely slow motion, the diffusion equation can be approximated by the Laplace equation. Therefore, all these processes are examples of pattern formation in a Laplacian field [1-10] and are governed by the same underlying mathematical equations. This generality has stimulated much of the current research in pattern formation, because there exists the possibility of understanding many diverse systems by studying one set of basic equations.

The formation of patterns in a Laplacian field generally occurs under conditions that are far from thermodynamic equilibrium. This implies that the analytical methods of equilibrium statistical mechanics cannot be applied to these systems. For this reason, much of our current understanding of fractal [11] and nonfractal pattern formation processes comes from recent computer simulations [12] of a variety of model systems. These efforts have been very successful in elucidating the essential features of growth in Laplacian fields and producing results that are in excellent agreement with experiments [13]. The purpose of these lectures is to present an overview of some of the recent work that we have carried out, using a combination of numerical [14,15,18,19] and analytical [16,17,19] methods, in trying to develop a better understanding of fractal and non-fractal pattern formation in Laplacian growth processes.

2. Growth in a Laplacian Field

In general, the larger the gradient of the controlling variable or the field, the faster the growth at a given point on the interface. Often the growth rate is a non-linear

Springer Proceedings in Physics Vol. 33: **Computer Simulation Studies in Condensed Matter Physics**
Editors: D. P. Landau • K. K. Mon • H-B. Schüttler

function of the gradient. If the normal gradient to the interface is ∇u then the growth rate normal to the interface, v_n, is assumed to have the form

$$v_n \propto |\nabla u|^{\eta} \tag{1}$$

where η is the degree of nonlinearity between the gradient and the velocity. This type of nonlinearity is particularly important in dielectric breakdown patterns [4]. Although the growth of patterns in different systems is controlled by different physical variables, the field $u(\mathbf{r}, t)$ for diffusion-limited growth is a solution of the diffusion equation

$$D\nabla^2 u = \frac{\partial u}{\partial t} \tag{2a}$$

where D is the diffusion coefficient. Let us consider the growth of a pattern in free space. Suppose that it is moving with constant velocity v in the z direction. In the moving frame, the diffusion equation (2a) becomes

$$D\nabla^2 u + v\frac{\partial u}{\partial z'} = \frac{\partial u}{\partial t} \tag{2a'}$$

where x'=x, $y'=y$, $z'= z-vt$ and we have used the chain rule to write (2a'). In the steady state, the right hand side of (2a') is zero. Therefore (2a') reduces to

$$\nabla'^2 u + \frac{v}{D}\frac{\partial u}{\partial z'} = 0 \quad . \tag{2a''}$$

In the quasi-stationary state [3], *i.e.* in the limit of infinitely slow motion ($v \to 0$) or infinitely fast diffusion ($D \to \infty$), the above equation reduces to the Laplace equation

$$\nabla^2 u = 0 \ . \tag{2b}$$

In addition to (1) and (2), the field u is assumed to be constant at the interface and at some distance much larger than the size of the growing pattern, *i.e.* $u(\infty)$=constant.

Simple linear stability analysis [3] shows that under conditions (1) and (2) a smooth surface is unstable against small perturbations of all wavelengths, such as noise. Therefore, if a part of the surface moves faster, then the gradient at the tip of this protrusion is increased, leading to further growth at this tip. This effect results in continuous branching of the tips. Thus, any wrinkling at the interface is amplified, leading to more wrinkling at the next length scale, up to the largest length scale. Under these conditions the resulting structures are fractal patterns. Classic examples of this type of growth are the diffusion limited aggregation (DLA) process [2] and the dielectric breakdown model (DBM) [4]. An example of a DBM aggregate of 5,000 particles grown on a square lattice is shown in Fig. 1.

If (1) and (2) were the only factors controlling the growth, all Laplacian patterns would be stringy and tenuous fractal objects. The factor that stabilizes this instability is the surface tension [3]. The effect of surface tension is to smooth out the perturbations at the surface so that the instability due to the increased gradient at sharp tips is reduced. Surface tension is introduced by specifying the value of u at the interface Γ,

$$u_{\Gamma} = u_0(1 - d_0\kappa) \tag{3}$$

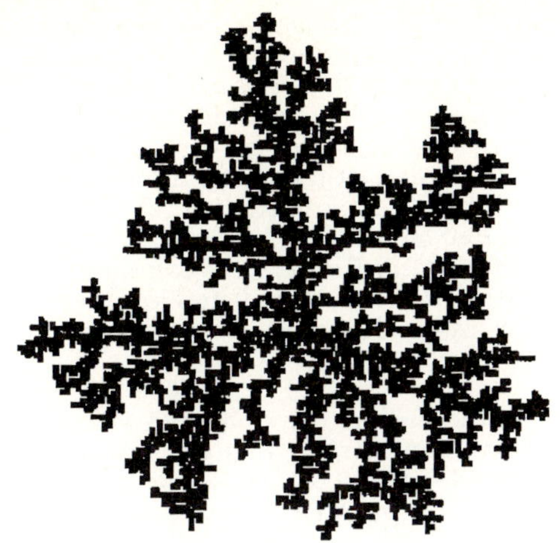

Figure 1: A Laplacian fractal generated by the dielectric breakdown model

where u_0 is the value of the field at the interface in the absence of surface tension, the capillary length d_o is proportional to the surface tension, and κ is the surface curvature. In case of dendritic growth, u is proportional to the temperature. Suppose a bump appears at the interface. Then by (3) the temperature at that point is slightly reduced because of positive curvature, and heat will flow there causing the bump to melt back. Thus the effect of surface tension is to reduce the gradient at sharp tips and stabilize smooth surfaces.

Equations (1-3) describe the essential physics of an important class of non-local growth processes with moving boundary conditions, such as aggregation [1,2] dendritic solidification [3], electrodeposition [5,6], dielectric breakdown [4] and viscous fingering[7,8]. But despite their generality and intensive efforts, little progress has been made in obtaining the growing patterns by analytically solving the equations. On the other hand, numerical methods [2,4,12] have been developed in the past several years which are very effective in producing a variety of diffusion-limited or Laplacian patterns. The most important of these methods is the random walk simulation technique developed by Witten and Sander [2] to study diffusion-limited aggregation. In DLA [2,12], particles are launched from far away and allowed to diffuse. Once a diffusing particle comes in contact with a growing aggregate it sticks to it and becomes part of it. This process of releasing random walkers continues until a large cluster is formed. Using the random walk approach for solving the Laplace equation, both fractal growth [2,12] and diffusion-limited pattern formation [20-22] have been studied.

An alternative approach based on aggregation is DBM [4], which was originally developed for studying branched pathways of charge flow during electrical breakdown of an insulator [4]. In DBM [4], the Laplace equation is solved directly on a lattice using a finite difference technique. Fig. 1 is an example of a DBM cluster for η=1. One of the advantages of this approach is that growth conditions with arbitrary values of η can be simulated, whereas in the DLA algorithm, only integer values of η can be studied [12]. The results that I will discuss in these lectures are based on direct solutions of the Laplace equation rather than the DLA approach.

3. FVT Model

Family, Vicsek and Taggett (FVT) [14] have proposed a model in which (1-3) are solved on a lattice using an aggregation type approach. Thus the motion of the interface in the FVT model is simulated by addition of particles to a growing cluster. The growth starts from a seed particle placed at the center of a lattice. After each time step, the field u is calculated by solving the lattice Laplace equation (2b) with boundary conditions (1) and (3) at the surface, and u=0 on a circle of radius R, which is taken to be much larger than the size of the cluster. After the field u is calculated, the gradients at all the perimeter sites are determined. These gradients are then normalized to the maximum gradient at the surface by dividing them by the value of the largest gradient on the interface.

In the FVT model, the normalized gradient at each site is compared with a random number and all perimeter sites with a gradient larger than the random number are filled. More recently, Family, Platt and Vicsek [18] have generalized this model by comparing the gradients with a parameter p whose choice is dictated by the physics of the problem. In this model [18] only those sites having a normalized gradient larger than the parameter p are filled. When p is a random number, we recover the FVT model. Here we discuss the type of patterns that are generated by three different choices of p.

In sec. 4 we discuss the results of the FVT model in which p is random. In this case the resulting patterns [14] are random fractal structures that are in the same universality class as DBM [4] and DLA [2]. Due to noise averaging property [23-24] of the FVT model [14], the generated clusters have an anisotropic structure. We introduce the m-spoke model [16], which is an analytic approach for studying the fractal properties of anisotropic clusters, in sec. 5. The case where p is a fixed constant during the process is discussed in sec. 6. With this choice of p, the Laplacian growth process is purely deterministic and generates exact fractal carpets. Finally, in sec. 7 we discuss the deterministic growth model for dendritic solidification recently proposed by Family, Platt and Vicsek [18]. This model corresponds to making p a time dependent variable such that the growth process would satisfy the boundary conditions appropriate for dendritic solidification [3]. This approach [18] has been found to be capable of producing dendritic patterns, including snowflakes, that are strikingly similar to real dendrites and snow crystals.

4. Random Fractals

We first consider the case in which p is a random variable, *i.e.* its value varies from one perimeter site to another perimeter site. This is equivalent to the FVT model. In the simulation, the value of the normalized gradient at a perimeter site is compared with a random number and the site is occupied if the value of the gradient is larger than the random number. The process of picking a new random number for each surface site and testing and occupying the site is repeated until all the perimeter sites have been checked. After this process is finished, the field u is determined everywhere using a relaxation method and the above steps are repeated again. Clearly, in this approach a finite number of particles are added to the growing cluster before the field u is relaxed.

The typical fractal patterns obtained from our model for various η are shown in Fig. 2. The most striking difference between these figures and Fig. 1 is the highly anisotropic shape of the clusters. This anisotropy, which is due to the symmetry of the square lattice, is often compensated by the fluctuations due to the randomness in the growth process [23,24]. For example, in the original DBM (Fig. 1) and DLA, only

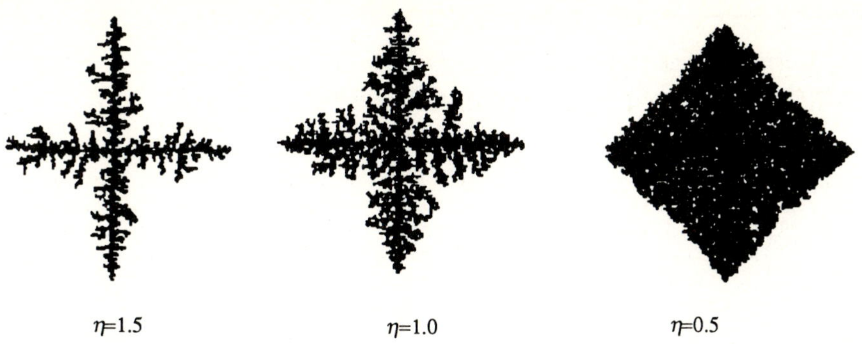

Figure 2: Random fractal patterns obtained in the FVT model for various η.

a single particle is added to the growing cluster at each time step, and the fluctuations dominate the growth. The reason we observe large anisotropy here is that in our model [14] the probability that a particle is added to the cluster does not depend on the position of the newly added particles. This results in a decrease of fluctuations at short length scales and enhancement of lattice anisotropy [14].

The anisotropic shape of the clusters in Fig. 2 suggest that perhaps the length and the width of the arms of the clusters diverge with different exponents. In order to study the scaling properties of the clusters and determine their fractal dimension, we have investigated the dependence of the radius of gyration R_g, the maximum length of the cluster arms $R_{//}$, and the width of the arms $R_{\perp}$, on the number of sites or the mass of the cluster, N, in two and three dimensions. We have determined the exponents ν, $\nu_{//}$, and $\nu_{\perp}$, which are defined by

$$R_g \sim N^{\nu} \qquad R_{//} \sim N^{\nu_{//}} \qquad R_{\perp} \sim N^{\nu_{\perp}} \ . \tag{4}$$

One possible definition of the fractal dimension is $D=1/\nu$. We find that the exponents for small cluster sizes are smaller than the corresponding values for larger scale simulations, because larger clusters become anisotropic. For example, for η=1 the exponents for both small (N~6,000) and large (N~24,000) clusters were calculated. For the smaller scale simulations, we find [19] $\nu_{//}=0.586$ and $\nu_{\perp}=0.569$. This shows that even though the clusters have an anisotropic shape, their scaling behavior is still similar to isotropic clusters, and the values of the exponents are close to the off-lattice DLA exponents obtained by Meakin [12]. For the larger clusters ($N\geq 24{,}000$), $\nu_{//}=0.654$ and $\nu_{\perp}=0.448$. These results are in close agreement with Meakin's square lattice DLA clusters with up to $4\mathrm{x}10^6$ particles [12]. However, simulations of highly-noise reduced DLA [25] clusters suggest that asymptotically $\nu \approx \nu_{//} \approx \nu_{\perp} \approx 0.66$ [25]. This implies that $\nu_{//}$ reaches its asymptotic value much faster than $\nu_{\perp}$. The values of the fractal dimension $D=1/\nu$ and the exponent $\nu_{//}$ for various η obtained from square lattice simulations are shown in Table 1. For η=1, our result [19] is in excellent agreement with the noise reduced DBM result of Nittmann and Stanley [24], and the DLA result of Meakin [12,25].

We have also studied the FVT model in three dimensions. Again, for small cluster sizes our data [19] for various η are in good agreement with the results of small scale simulations of the dielectric breakdown model in three dimensions [26]. However, for larger cluster sizes (N~60,000 for η=1) the effect of anisotropy is more

TABLE 1

The fractal dimension $D=1/\nu$ and the exponent $\nu_{//}$, for different η, for the FVT model in two dimensions

η	D	$\nu_{//}$
0.5	1.89	0.548
0.6	1.82	0.564
0.7	1.73	0.591
0.8	1.65	0.614
0.9	1.58	0.643
1.0	1.55	0.653
1.5	1.33	0.746
2.0	1.28	0.784

pronounced and the values of the exponents are increased. We find that within the statistical errors, the exponents $\nu_{//}$ and $\nu_{\perp}$ are equal. A summary of our three dimensional results are listed in Table 2. The values of the exponents for $\eta>1$ are probably not indicative of the asymptotic behavior, because the data were obtained on clusters having arm lengths and widths of no more than 10 or 20 lattice spacings.

TABLE 2

The fractal dimension $D=1/\nu$ and the exponent $\nu_{//}$ for the random fractal model in three dimensions

η	D	$\nu_{//}$	$\nu_{\perp}$
0.5	2.56	0.391	0.402
0.6	2.50	0.400	0.433
0.7	2.39	0.420	0.452
0.8	2.31	0.434	0.471
0.9	2.24	0.447	0.485
1.0	2.14	0.467	0.474
2.0	1.81	0.553	0.645
3.0	1.70	0.587	0.819

5. The m-spoke Model

The geometry of the clusters in our model [14], and in the noise reduced DLA [23-25] on various lattices, is very anisotropic. In the presence of an m-fold anisotropy, arising either from the symmetry of the underlying lattice or the anisotropy of the growth mechanism [27], a large cluster resembles a star with m-spokes, where each arm of the star is a "cigar" of length $R_{//}$ and width $R_{\perp}$. As a simplified representation of this type of clusters, Family and Hentschel [16] have introduced the m-spoke model, which is an idealized model of asymptotically large clusters. As shown in Fig. 3, in this model the geometry of an anisotropic cluster is approximated by a star consisting of m arms of length $R_{//}$. The field u is determined exactly [16] by solving the Laplace equation for the m-spoke model using conformal mapping. Knowing the field at the tips of the cluster, we then use the approach of Turkevich and Scher [28] and Ball *et al.* [27] to determine the exponent $\nu_{//}$ exactly in d dimensions. The general case of nonlinear growth, *i.e.* $\eta>1$, has been studied by Matsushita, Family and Honda [17], and more recently Platt and Family have extended these results to d dimensions [19].

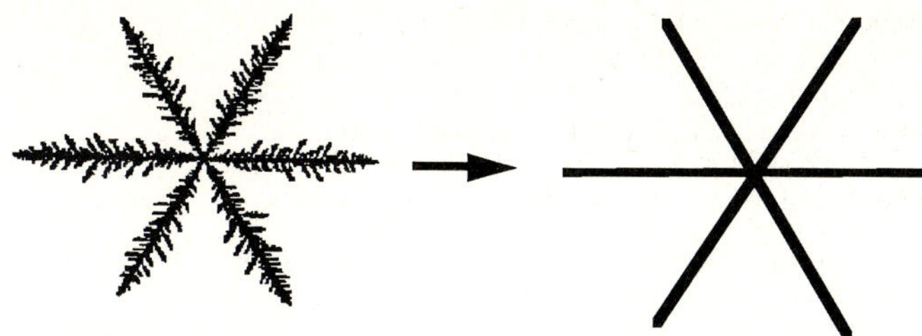

Figure 3: In the m-spoke model, anisotropic clusters like the one on the left are approximated by the star-shaped cluster shown on the right .

The Laplace equation is first solved [16,17] for the distribution of the field u around the m-spokes using conformal mapping techniques. In order to determine u, we assume that the potential on the spokes is zero and u on a large circle concentric with the spokes is constant. Knowing u everywhere, we use (1) to determine the flux at a distance r from the tip of the arm. The maximum growth probability at the tip is given by [28]

$$P_{max} \sim \frac{\int_0^a |\nabla u(r)|^\eta dr}{\int_0^{R_{||}} |\nabla u(r)|^\eta \, dr} \tag{5}$$

where a is a short length scale cutoff (of the order of the lattice spacing). Extending the results of Family and Hentschel [16] and Matsushita *et al.* [17] to d-dimensions, we find [19]

$$P_{max} \sim \left(\frac{a}{R_{||}}\right)^x \tag{6a}$$

where

$$x = \left(\frac{1-d}{2}\right)\eta + d - 1 \quad . \tag{6b}$$

The maximum growth probability is equal to the rate of growth of $R_{||}$, *i.e.*

$$P_{max} \sim \frac{dR_{||}}{dN} \quad . \tag{7}$$

Using (6) in (7) yields

$$\nu_{||} = \frac{2}{(1-d)\eta + 2d} \quad . \tag{8}$$

This result implies that for $\eta \geq 2$, $\nu_{||}=1$. For η near 1, where we expect the m-spoke model to be most appropriate, the values of the exponents predicted by (8) are in good agreement with the two- and three-dimensional results given in Tables 1 and 2, respectively. For small η, the tip angle is clearly much larger than zero, which is the value assumed in the m-spoke model. Therefore, (8) is not expected to be appropriate for small η. For η>>1, the denominator in (5) has singularities and this approach might not be suitable [19]. However, at the present time accurate numerical data are not available to test the predictions of (8) in this regime.

6. Exact Fractals: Laplace Carpets

Instead of choosing a different random number for each perimeter site we follow the approach of ref. [18] and let p be a fixed constant throughout the growth process. Unlike all other Laplacian growth models, there is no noise in the growth process here, *i.e.* this is a *deterministic* growth model [18]. Consequently, the patterns generated in this process are *exact* fractals [18]. The resulting patterns on a square lattice for p=0.20, 0.30, and 0.40 and the values of the respective fractal dimensions are shown in Fig. 4. In the limit p=0, all the perimeter sites are filled and the result is a dense polygon having the symmetry of the underlying lattice. As shown in Fig. 4, for finite p, the patterns are regular fractals with a fractal dimension which varies from 2 to 1 as p is increased from 0 to 1.

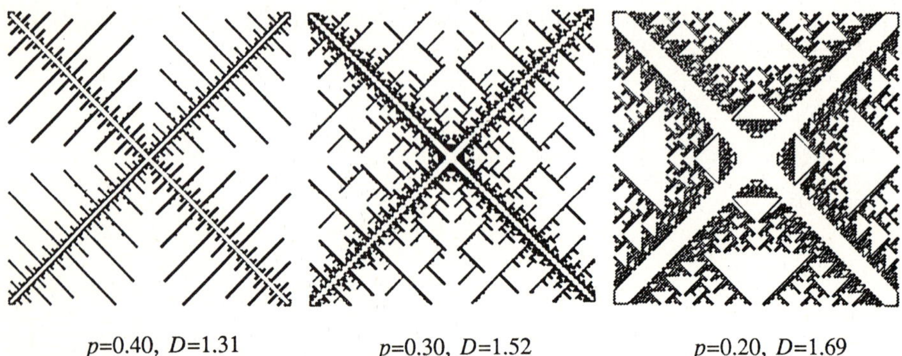

p=0.40, D=1.31 p=0.30, D=1.52 p=0.20, D=1.69

Figure 4: Laplace carpets obtained in the deterministic growth model by allowing the interface to advance with a threshold p which is fixed throughout the growth process.

The Laplace carpets closely resemble the type of patterns that are generally found in cellular automatan [29]. It would be instructive to investigate the possibility of a deeper connection between these two processes than visual similarity. Such a result would provide the first direct connection between a local growth model — cellular automaton — and a nonlocal Laplacian growth process.

7. Dendritic Solidification: Snowflakes

The growth of dendritic crystals is a profound example among a wide range of spontaneous pattern forming phenomena in physics, chemistry, biology and engineering. The formation of snowflakes is perhaps the most fascinating and puzzling example of these processes. Although subject to intensive efforts [3], previous methods had not produced such regular and intricate dendritic structures as are found in nature. Therefore, the development of the deterministic growth model [18] which for the first time was capable of producing realistic dendritic patterns, including snowflakes, has been an important step in developing effective methods for understanding the formation of snowflakes and dendritic crystals.

The deterministic growth model [18] is based on direct solution of (1-3) on a regular lattice. The field u is numerically determined by a relaxation technique such as the Gauss-Seidl over-relaxation method. For fixed d_0, *i.e.* fixed value of the surface tension, the new values of u at the surface are calculated from the Gibbs-Thomson condition (3), by numerically determining the surface curvature at each perimeter site. The interface is advanced by occupation of the perimeter sites, in analogy with the FVT [14] model. Since random growth processes [2,4,12] cannot produce such symmetric patterns as snowflakes, our model is based on a deterministic growth rule: all perimeter sites having a normalized gradient larger than p are occupied.

In dendritic solidification [3], by heat conservation, the normal velocity at a point on the interface is proportional to the temperature gradient, *i.e.* η=1 in (1). When p is a constant, v_n is constant at surface sites having a normalized gradient larger than p and zero otherwise. In contrast, dendritic growth is governed by (1), and the interface growth velocity must be proportional to the local gradient [3]. This implies that within a time interval Δt, p must vary linearly with the time so that sites having the maximum gradient are always filled, while those with smaller gradients are filled less frequently, depending on the local gradient. In order to implement this boundary condition in our model, we discretize the time interval into c steps and assume that during this time interval p is given by [18]

$$p = a + b\,(t \bmod c)\ , \tag{9}$$

which is a piecewise function approximating a straight line of slope b. Since in every time step some particles are added to the cluster, this characteristic time interval introduces a length scale equal to c lattice spacings. The effect of this length scale is most pronounced for small size clusters with zero surface tension. This can be clearly seen in the dendritic crystal pattern shown in Fig. 5a, where because of the small size of the arms (~100 lattice spacings) the periodic modulation on the surface of the four needles is equal to c. On the other hand, in the needle crystals shown in Fig. 5b the capillary length d_0 is finite and this effect is not observable.

The above method can produce practically all types of observed two-dimensional dendritic patterns by changing the surface tension in (3) and the parameters a and b in (9). The various limiting cases include faceted growth, needle crystals and regular fractal structures. For intermediate values of the parameters, combinations of these patterns are obtained. The parameter a in (9) can

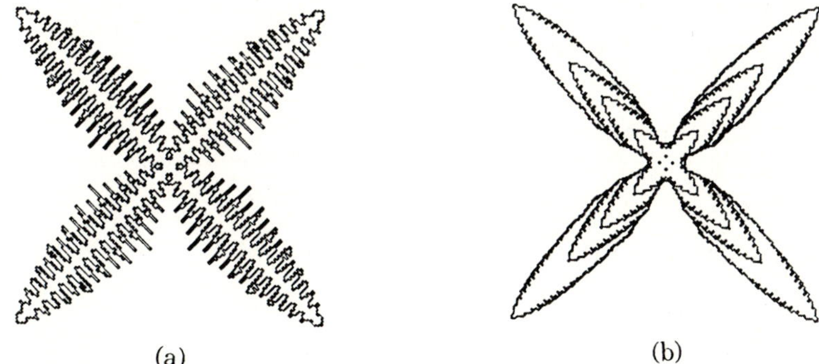

Figure 5: Examples of dendritic needle crystals obtained from the deterministic growth model. (a) Dendritic crystal with zero surface tension. (b) The time evolution of the pattern with d_0=0.4 shows how stable parabolic tips appear as a result of finite surface tension and anisotropy.

be varied in order to simulate the effects of changing the environmental conditions during the growth. The effectiveness of the deterministic growth method is best demonstrated by the great variety of possible morphologies it generates. In Fig. 6, we show only a few selected examples of six-fold symmetric snowflakes generated by varying the environmental conditions during the growth. All of these figures were generated by the stepwise function (9), with c=5 and varying a. To obtain a faceted, near equilibrium interface pattern, a was made negative ($0>a>-0.1$), and to obtain a boundary with sharp, dendritic shape, a was made greater than zero ($0.5>a>0$). As a was changed, b was adjusted to account for the fact that at the end of each time interval the straight line must pass through the point p=1.

Figure 6: A few examples of six-fold symmetric snowflakes obtained from the deterministic growth model by varying the environmental conditions during the growth.

8. Conclusions

The formation of complex patterns and structures in a Laplacian field is common to a wide range of processes of scientific and industrial applications [1-10]. We have shown that simulations of growth in a Laplacian field based on aggregation processes can generate a variety of the type of patterns observed in nature. This includes random fractal structures [14], exact fractal carpets [18] and dendritic crystal shapes and snowflakes [18]. This approach appears to be much more effective in producing complex structures such as dendritic patterns than the previous methods based on numerical solutions of the solidification equations. These types of models are useful tools for sorting out various long-standing and puzzling aspects of fractal and nonfractal pattern formation in a Laplacian field.

Acknowledgements

I would like to thank Dan Platt, Becky Taggett, and Tamás Vicsek for their collaborations on parts of the work discussed here. I would also like to thank Daniel Hong and Scott Anderson for useful comments on the manuscript. This work was partially supported by the Office of Naval Research. Acknowledgment is also made to the donors of the Petroleum Research Fund, administered by the ACS, for partial support of this research.

References

1. F. Family and D. P. Landau, (eds), *Kinetics of Aggregation and Gelation* (North -Holland, Amsterdam, 1984).
2. T. A. Witten, Jr., and L. M. Sander, *Phys. Rev. Lett.* **47**, 1400 (1981).
3. J. S. Langer, *Rev. Mod. Phys.* **52**, 1 (1980).

4. L. Niemeyer, L. Pietronero, and H.J. Wiesmann, *Phys. Rev. Lett.* **52**, 1033 (1984).
5. R. M. Brady and R. C. Ball, *Nature* **309**, 225 (1982).
6. M. Matsushita, M. Sano, Y. Hayakawa, H. Honjo and Y. Sawada, *Phys. Rev. Lett.* **53**, 286 (1984).
7. L. Patterson, *Phys. Rev. Lett.* **52**, 1621 (1984).
8. J. Nittmann, G. Daccord, and H. E. Stanley, *Nature* **314**, 141 (1985).
9. H. Meinhardt, *Models of Biological Pattern Formation*, (Academic Press, New York, 1982), and references therein.
10. T. Vicsek, *Europhysics News*, (to be published, 1988).
11. B. B. Mandelbrot, *The Fractal Geometry of Nature* (Freeman, San Francisco, 1982).
12. P. Meakin, in *Critical Phenomena and Phase Transitions*, C. Domb and J. L. Lebowitz (eds.), (Academic Press, New York, 1988), Vol 8, p.335.
13. J. Feder, *Fractals*, (Plenum Press, New York, 1988).
14. F. Family, T. Vicsek, and B. Taggett, *J. Phys. A* **19**, L727 (1986).
15. F. Family, Y. C. Zhang and T. Vicsek, *J. Phys. A* **19**, L733 (1986).
16. F. Family and H.G.E. Hentschel, *Faraday Discuss. Chem. Soc.* **83**, 139 (1987).
17. M. Matsushita, F. Family, and K. Honda, *Phys. Rev. A* **36**, 3518 (1987).
18. F. Family, D. E. Platt and T. Vicsek, *J. Phys. A* **20**, L1177 (1987).
19. D. E. Platt and F. Family, *to be published* (1988).
20. T. Vicsek, Phys. *Rev. Lett.* **53**, 2281 (1984).
21. L. P. Kadanoff, *J. Stat. Phys.* **39**, 267 (1985).
22. P. Meakin, F. Family, and T. Vicsek, *J. Colloid and Int. Sci.* **117**, 394 (1987).
23. T. Vicsek and J. Kertész, *J. Phys.* **A19**, L257 (1986).
24. J. Nittmann and H. E. Stanley, *Nature* **321**, 663 (1986).
25. P. Meakin, *Faraday Discuss. Chem. Soc.* **83**, 113 (1987).
26. S. Satpathy, *Phys. Rev. B* **33**, 5093 (1986).
27. R. C. Ball, R. M. Brady, G. Rossi and B. R. Thompson, *Phys. Rev. Lett.* **55**, 1406 (1985).
28. L. Turkevich and H. Scher, *Phys. Rev. Lett.* **55**, 1026 (1985).
29. S. Wolfram, *Theory and Applications of Cellular Automata*, (World Scientific, Singapore, 1986).

Dynamics of Dense Polymers: A Molecular Dynamics Approach

G.S. Grest[1] *and K. Kremer*[2]

[1]Corporate Research Science Laboratory, Exxon Research and Engineering Company, Annandale, NJ 08801, USA
[2]Institut für Physik, Universität Mainz, D-6500 Mainz, Fed. Rep. of Germany

1. INTRODUCTION

The physics of polymeric materials[1,2] is one of the most challenging problems in condensed matter physics today. It is a problem of great interest both from a fundamental viewpoint and for their various technical applications. In addition to theortical and experimental approaches, computer simulations[3-11] have played an important role in our present understanding of polymers. For static properties Monte Carlo methods have been widely used and give excellent results for static critical exponents. To investigate dynamic properties three different methods--Monte Carlo (MC)[3-7], molecular dynamics (MD)[8,9] and Brownian dynamics methods[10]--have been used. Detailed microscopic dynamics of a specific polymer model has also been studied[11]. We show that while MD simulations have not been widely used for studies of polymeric systems, the method is in fact very general and efficient. Here we describe a continuum approach[12] which can be effectively used in a wide variety of systems, such as linear, ring or star polymers, from the dilute, single chain limit to a dense melt consisting of many long, entangled chains. For a dense polymer melt, we present for the first time simulations which cover the entire regime from Rouse to reptation dynamics and give strong evidence for the latter.

For a single, dilute polymer chain, the MC and Brownian dynamics simulation methods give Rouse relaxation. The Rouse model[13] for the dynamics of a single polymer describes the motion of an ideal chain immersed in a viscous solvent. This model neglects the self-repelling of chain monomers as well as any hydrodynamic effects[14]. Although this so-called "free-draining" limit is somewhat artifical, it has been very useful for understanding polymer dynamics. The Rouse model describes the motion of a single monomer by a Langevin equation:

$$\frac{\partial \vec{r}_i}{\partial t} = \mu\left(\vec{F}_i + \frac{3k_BT}{\sigma^2} \frac{\partial^2 \vec{r}_i}{\partial n^2} + \vec{\phi}_i \right). \tag{1}$$

$\vec{r}_i$ is the position of the ith monomer, μ the mobility, and $\vec{F}_i$ the force from the surroundings other than the chain itself. The last two terms describe the elastic force along the chain and the inner-chain interactions $\vec{\phi}_i$. The diffusion constant $D=\mu k_BT/N$ for a chain of N monomers using the Einstein relation. For an ideal chain (random walk) one finds a longest relaxation time (Rouse time) $\tau_N \sim N^2$. For the motion of a single monomer this leads to the following relation for the monomer-monomer diffusion:

$$g_1(t) = \langle[\vec{r}_i(t)-\vec{r}_i(0)]^2\rangle \propto \begin{cases} t^{1/2}, & \tau_o \ll t \ll \tau_N \\ t, & t \gg \tau_N \end{cases} \tag{2}$$

where τ_o is a microscopic time. At the Rouse time, $g_1(\tau_N) \approx \langle R_G^2\rangle$, the mean square radius of gyration. Qualitatively, the monomer motion is governed by fluctuations around the center of gravity until the average distance reaches approximately R_G;

Springer Proceedings in Physics Vol. 33: **Computer Simulation Studies in Condensed Matter Physics**
Editors: D. P. Landau · K. K. Mon · H-B. Schüttler

then the diffusion of the chain as a single object dominates. For self-avoiding chains[15] one find that $<R_G^2>\sim N^{2\nu}$, with $\nu=0.59$ (d=3) instead of 1/2 and $g_1(t) \sim t^{0.54}$.

In a MC simulation the chain moves stochastically through random moves of a small number of monomers at a time. The number of bonds involved typically varies from two to four. This technique reproduces the Rouse behavior and is especially suited for scalar computers. It suffers from two disadvantages. First, it is very difficult to write an efficient parralel MC algorithm for polymers[3,6]. The second problem is that for dense systems, the motion is almost reduced to cooperative fluctuations. In order to have a high acceptance of the moves, the density ρ has to be relatively small (ρ=0.5 in Ref. 8). Thus the chains have to be very long in order to mimic a melt of long polymers[6,8]. This method has however been very successful and will remain very important at low and moderate densities. In a quasi-Brownian dynamics simulation[10] the Smoluchowski equation is solved for the position of the monomers by a MC algorithm. Due to the mainly stochastic nature of the motion, only moderate densities ($\rho\leq 0.5$) have been used. In both methods the Rouse dynamics is built in due to the stochastic nature of the chain motions. The third alternative is to use a standard MD method[8,9]. Here Newton's equations of motion are solved directly for each monomer. Since both momentum and energy are conserved, one simulates a microcanonical ensemble. However, in order to sample the entire phase space for a single chain, there must be an exchange of energy and momentum with the surroundings. This has been done in the past by taking the solvent molecules explicitly into account. Typically, one needs at least 20 times as many solvent particles as monomers of the chain. This makes the algorithm so inefficient that it has only been used for short chains.

Clearly it is important to find a simulation technique which works at high density since many of the unresolved problems in polymers dynamics envolve dense melts or semi-dilute solutions. To be precise we want to have a method which is (1) almost as efficient as MC for dilute chains and reproduces the Rouse model and (2) should be effective at high densities. In addition, the algorthim should be general enough so that one can incorporate constant-pressure methods[16] to calculate moduli for an entangled melt or polymer glass. To acheive this aim, we[12] propose a MD algorithm where each particle is coupled to a heat bath. Schneider and Stoll[17] used a similar method to simulate a system with a distortive phase transition. However, they made the coupling so weak that they could neglect the effect of the heat bath on the dynamics of the system. We want just the opposite, since the coupling of the polymer to the surrounding is not weak. Thus, we solve the equation of motion:

$$\ddot{\vec{r}}_i = -\vec{\nabla} U_i - \Gamma \dot{\vec{r}}_i + \vec{W}_i(t), \tag{3}$$

where Γ is the bead friction and $\vec{W}_i(t)$ describes the random force of the heat bath acting on each monomer. $\vec{W}_i(t)$ is a Gaussian white noise with

$$<\vec{W}_i(t)\cdot\vec{W}_j(t')> = \delta_{ij}\delta(t-t')6k_BT\Gamma. \tag{4}$$

Using the Einstein relation this gives $D_o=k_BT/\Gamma N$. The potential[9,10,12,18] $U_i=\Sigma_j U_{ij}$ has two parts U^o and U^{ch}. U^o is a shifted, purely repulsive Lennard-Jones potential, truncated at a cutoff $r_c=2^{1/6}\sigma$, between any two monomers. The potential $U^{ch}(r)$ is an attractive potential between neighboring monomers along the chain. The parameters for this potential are given in Ref. 12 and differ from those used in Refs. 9 and 10 in order to avoid bond cutting. Typically the time step $\Delta t=0.006\tau$, where $\tau=\sigma(m/\epsilon)^{1/2}$ in terms of Lennard-Jones parameters. The equations were solved using a fifth-order predictor corrector loop, though one may use any of the standard algorithms for second order differential equations. We tested several algorithms and provided that Γ is not so large that the inertia term in Eq. (3) becomes irrelvant, any of the standard algorthims should be appropriate. Note that Δt cannot be mapped directly onto a microscopic time, because a single monomer corresponds to a number of real bonds, which depend on the chemistry. Most of our simulations have been carried out for $k_BT/\epsilon=1.0$. The average bond length was found to be 0.97σ and a persistent length 1.34σ for a single chain. In order to compare our chain lengths with experimental ones, we note that the persistence length[19] for PDMS is approximately 2.45 and for polystyrene 3.15. Using a molecular weight of 72 (PDMS) and 104 (polystyrene), we find that a chain of N=100 in our model corresponds to chains of approximately 13000 (PDMS) and 24500 (polystyrene). However, using the persistence length to map our model to experiment is not unique. A better estimate, particularly for dense melts, can be made using dynamic data, where the entanglement length of our model can be mapped onto experimental values. Since the entanglement length depends strongly on both temperature and density, while the persistence length does not, mapping of dynamic lengths, which we present in Sec. 3, should be more accurate.

In our studies, both for single chains[12] as well as dense melts[20], we used a value of the bead friction Γ between 0.5 to $2.0\tau^{-1}$. For a given bead friction, the motion of a monomer is undamped (with the chain constraints) for $t<\Gamma^{-1}$, while for $t>>\Gamma^{-1}$ the motion is Rouse like. We found that for $\Gamma^{-1}=2\tau$, the ballistic motion of a monomer on a single, dilute chain is confined to about two to three bond lengths. For larger distances the motion is diffusive. This range of values for Γ is a convenient compromise. For smaller Γ, the coupling to the heat bath is too weak and a single chain[12] or star polymer[21] samples phase space very ineffectively. It does not show the desired Rouse behavior until very late times. For larger values of Γ, the viscosity damping and random force term dominate over the inertia term in Eq. (3). This means that the motion is dominated by Langevin dynamics even for very early times and there could be little cooperative motion of the monomer which is important for movement to occur particularly in dense melts[20] or in the dense interior of a star[21]. This limit would make the algorthim very ineffective. This short time ballistic regime will show up in any autocorrelation or dynamic function, e. g. Eq. (1), for short time $t\leq 2\text{-}3\Gamma^{-1}$ or 4-6τ for $\Gamma=0.5\tau^{-1}$. However, since the total length of the runs are typically between 10^3 to $10^5\tau$ and sometimes longer, this does not cause any difficulty.

To test the method, we[12] first carried out detailed studies of linear and cyclic chains of 50 to 200 monomers. We confirmed that the model gave the predicted $t^{0.54}$ power law for $g_1(t)$ for a self-avoiding chain undergoing Rouse relaxation. Since the advantage of the method over traditional simulation techniques is at high density, we have studied two such systems in detail. The first is the structure and relaxation of many-arm star polymers[21,22] and the second is a dense melt of long, entangled linear polymers[20].

2. STAR POLYMERS

A special class of branched polymers[23] is a star polymer, which may be thought of as linear polymers joined at one of their ends to a common center. Well-characterized star polymers have been synthesized with up to 18 arms connected to a single center[24,25]. The static properties of star polymers have been analyzed using a scaling picture[26,27]. In this picture, the blob model is used to describe the overall structure of a single star polymer. The idea is that the density of monomers from the center decays as a power law. At a given distance from the center, a sphere of radius r is cut by f arms. The star should look like a semi-dilute solution with a screening length $\xi(r)$, where $\xi(r)$ is a function of r and f. At a given distance r from the center, there are f blobs, one for each polymer chain. Since f blobs cover a sphere of radius r, the blob radius, $\xi(r)\sim rf^{-1/2}$. As in any semi-dilute solution[1], each blob contains $\xi^{-1/\nu}$ monomers. From this picture, one can easily verify that density of monomers from the center $\rho(r)$ falls off as[26,27]

$$\rho(r)\ \alpha\ f^{(3\nu-1)/2\nu}\ r^{(1-3\nu)/\nu} = f^{0.65}r^{-1.30}. \tag{5}$$

Qualitatively, this scaling behavior is illustrated in Fig. 1, in which we show a projection of a typical configuration of a star polymer with f=10, 30 and 50 arms with N=50 monomers per arm. In Fig. 2, we show data[21] for $\rho(r)$ scaled as in Eq.

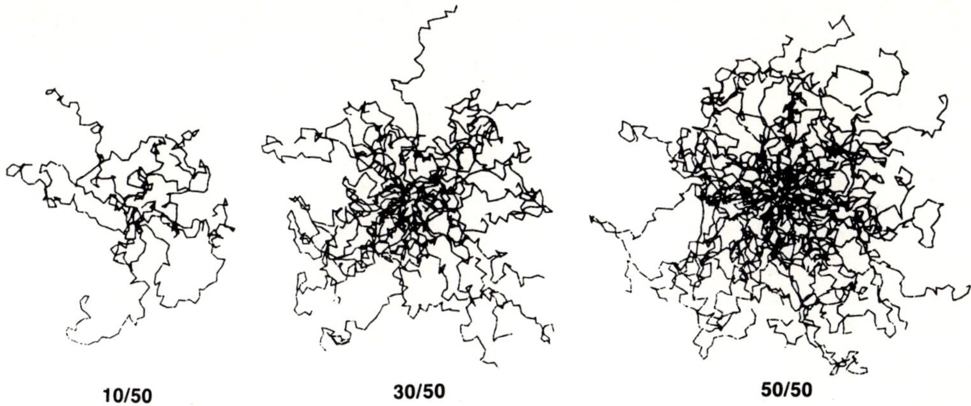

Figure 1 Projections of a typical configuration of a star of f=10, 30 and 50 arms with N=50 monomers per arm. The pictures give an impression of the increasingly homogenous density in the f=30 and 50 systems while the f=10 star obviously is governed much more by single-chain properties.

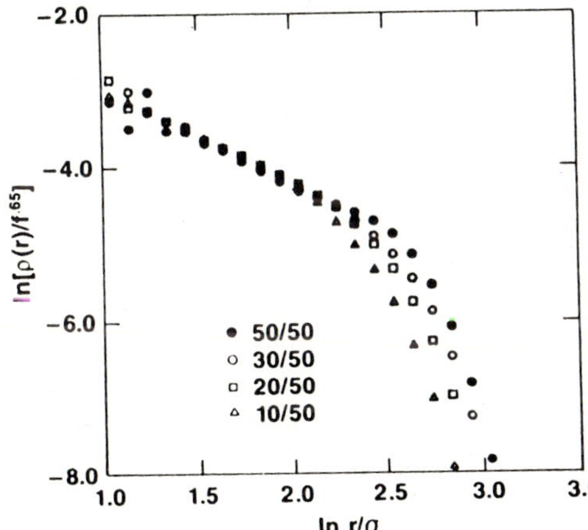

Figure 2 Scaling plot of density $\rho(r)/f^{0.65}$ vs. r/σ for four stars with N=50 monomers per arm.

(5). The slope of ln $\rho(r)$ vs. ln r is approximately 1.25, in reasonably good agreement with the expected value of 1.30.

This scaling also should be exhibited by the radius of gyration $\langle R_G^2 \rangle$ and average center-end distance $\langle R^2 \rangle$, which are predicted to have the form

$$\langle R^2 \rangle \; \alpha \; \langle R_G^2 \rangle \; \alpha \; N^{2\nu} f^{1-\nu} = N^{1.18} f^{0.41} \quad . \tag{6}$$

Figure 3 gives our results[21] for $\langle R_G^2 \rangle / \langle R_{G1}^2 \rangle$ and $\langle R^2 \rangle / \langle R_{G1}^2 \rangle$ vs. f, where $\langle R_{G1}^2 \rangle$ is the radius of gyration for a single polymer chain of N monomers. Also shown are experimental results for $\langle R_G^2 \rangle / \langle R_{G1}^2 \rangle$ for polystyrene and polyisoprene[25]. $\langle R_G^2 \rangle$ and $\langle R^2 \rangle$ clearly follow a power law with a slope of 0.39±0.03, within the error bars of the expected value of 0.41. From these results we can conclude that the scaling theory as well as our relatively simple coarsed grained bead spring model give a very good description of the static properties of star polymers.

In collobration with Witten and Milner, we[22] have generalized the scaling picture to describe the dynamic relaxation processes in stars. We have identified three typical relaxation mechanisms. The first describes elastic deformation of

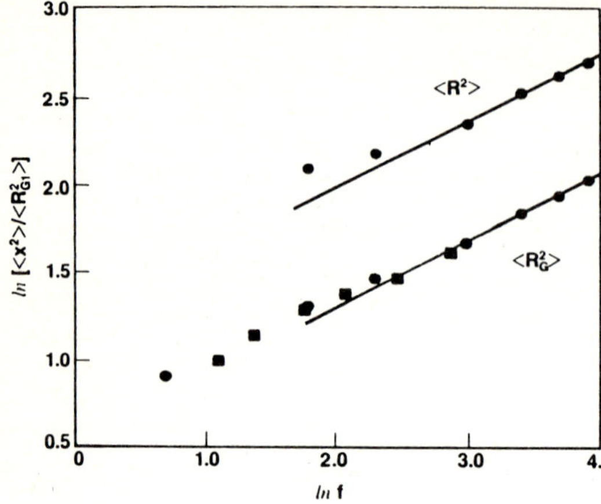

Figure 3 Radius of gyration $<R_G^2>/<R_{G1}^2>$ and average square end-center distance $<R^2>/<R_{G1}^2>$ vs. f for stars with N=50, where $<R_{G1}^2>$ is the radius of gyration for a single polymer chain. The solid squares are data for polystyrene and polyisoprene from Ref. 25. The solid lines indicate the predicted asymptotic power, 0.41.

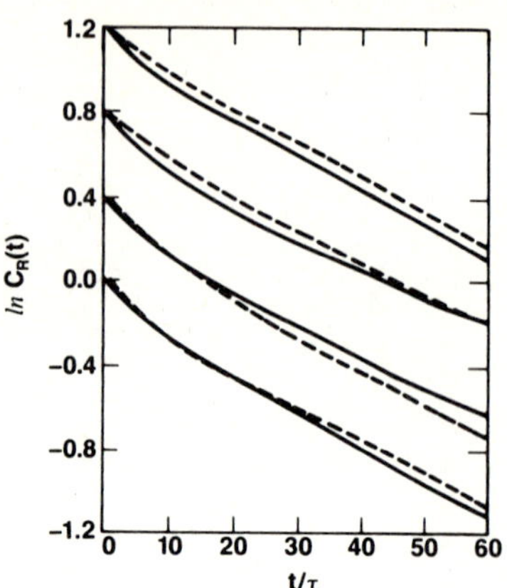

Figure 4 Autocorrelation function for the end-center distance (solid lines) and single arm radius of gyration (dashed) for four stars. Note that the curves have been displaced for clarity.

the overall shape and its relaxation is nearly independent of f. A second type of relaxation occurs via rotational diffusion and is expected to have a relaxation time which scales with $N^{2\nu+1}f^{2-\nu}$. A third relaxation process is the disentanglement of two or more arms. Here the longest relaxation time should increase exponentially with $f^{1/2}$. We have analyzed our molecular dynamics data to look for these three relaxation processes. Relaxation of the shape fluctuations can be studied by measuring the autocorrelation function $C_R(t)$ for R_G, the radius of gyration of a single arm R_{Ga} or the scalar center-end distance R for each arm. In Ref. 22, we showed that the longest relaxation time τ_{BR}, with BR for breathing, should scale as $N^{2.18}f^{-0.09}$ for Rouse dynamics. In Fig. 4, we present results for $C_R(t)$ for R and R_{Ga}. The relaxation is exponential with a characteristic time $\tau_{BR} \approx 60\ \tau$. The final relaxation time depends imperceptibly on f, consistent with the prediction. We were unable to measure the rotational time for these large objects since the diffusion is very slow. This would require significantly longer runs than presently accessible. Finally, we also investigated the disentanglement time τ_e for stars with between 10 to 50 arms. While we[22] observed a stong dependence of the relaxation on the number of arms f, the decay of the appropriate autocorrelation function was not sufficient to determine τ_e accurately enough to test our scaling predictions.

3. POLYMER MELTS

One of the long standing problems in polymer physics is to understand the dynamical properites of dense polymer melts. Due to their topological interactions, melts display a rich and unusual viscoelastic behavior[2,18]. Though experiments have been very important in elucadating many of the properties of these complex systems, they are unable to probe the microscopic origin of the interactions directly. For this reason, computer simulations of the motion of highly entangled polymers can play an important role in understanding these complex systems. In collobration with Carmesin, we[20] have begun a large scale molecular dynamics simulation study of a melt of linear polymers which covers the whole range from the short chain (Rouse) to the highly entangled (reptation) regime.

The dynamics of polymer melts is typically discussed in terms of the Rouse and reptation model[1,2]. For short chains, the surrounding monomers cause a stochastic motion of an individual monomer. This motion is constrained by the connections to monomers along the chain, leading to standard Rouse behavior. The longest relaxation time is just the Rouse time $\tau_N \sim N^2$. The diffusion constant $D \sim N^{-1}$ and viscosity $\eta \sim N$. Experimentally for $N > N_e$, the entanglement length, this dependence changes to $D \sim N^{-2}$ and $\eta \sim N^{3.4}$[28]. This large N regime is usually explained by the reptation model of EDWARDS[29] and DeGENNES[30]. In this model, the chains on a length scale larger than $d_T \sim N_e^{1/2}$ move predominatly along their own contour. The chain has a Rouse like relaxation up to a time $\tau_e \sim N_e^2$, after which one has a Rouse relaxation along the coarse grained path of a chain with bonds having N_e monomers each. Since this is a one dimensional diffusion along a random walk path, the chain needs a time $\tau_{rep} \sim (N/N_e)N^2$ to leave the orginal tube. Consequently, one gets $\eta \sim N^3$[2]. This simple model has been very successful in describing the qualitative aspects of the dynamics of a melt. The difference in the measured and predicted exponent for the viscosity remains an unresolved puzzle, though there have been numerous suggestions for the difference, none of them particularly satisfactory. However, the microscopic implications of the model are unambigious, particularly for the mean square displacement $g_1(t)$. The reptation model predicts that $g_1(t) \sim t^{1/2}$ for $t < \tau_e$, $\sim t^{1/4}$ for $\tau_e < t < \tau_N$, $\sim t^{1/2}$ for $\tau_N < t < \tau_{Rep}$ and finally $\sim t$ for $t >> \tau_{Rep}$. These four distinct time regimes for the mean square displacement, particularly the $t^{1/4}$ regime, are a direct and distinct consequence of the reptation model. Several attempts have been made to find evidence for the reptation model using computer simulations[3,6,7] by measuring $g_1(t)$, however to date none have been successful, mostly because the chains were too short for the density studied. Here we[20] present the first unambigious evidence for reptation from a simulation. The difference from earlier studies is that we work at much higher densities (ρ=0.85) where the tube diameter is small and concentrate on analyzing the motion of monomers near the center of the chain to reduce end effects.

Before one begins an extended study and test for reptation, it is important to determine whether the longest chains one can handle effective in the simulation are in fact long enough to be in the reptation regime. Since the relaxation time $\tau_N \sim N^3$, very long runs are necessary to ensure that a melt is in equilibrium. Ideally, the center of mass of each chain should move several times its radius of gyration. In addition, the number of chains M must be greater than $N^{1/2}$ so that each chain has enough room so that it does not interact with itself artifically as a result of the periodic boundary conditions. These conditions limited our intial study to chains up to size N=150 and M=20, in which case we were able to run up to $15 \cdot 10^6$ time steps[31]. Using the conversions discussed in Sec. 1 for the persistence length, our longest chain thus corresponds to approximately 37000 for polystyrene and 20000 for PDMS. From experiments on the plateau modulus fitted to the reptation theory, $N_e \approx 18000$ for polystryene and 9000 for PDMS[28], suggesting that our chains just cover the crossover from the Rouse to reptation regimes. However, our simulated chains turn out to be longer than a simple mapping from the persistence length indicates, because we work at rather low temperatue and high density. In fact, a detailed analysis of the normal modes of the chain shows that our chains are actually a factor of two longer than indicated from the mapping of the persistence length[20]. This coupled with the fact that the innermost monomers must feel the topological constraints much earlier that the whole chain, made us feel that we should be able to find reptation-like behavior, if it exists. To do this, we calculated $g_1(t)$, but averaged only over the innermost monomers. We also subtracted the overall diffusion of the whole system due to the random force $\vec{W}_i(t)$.

Figure 5 presents our results for $g_1(t)$ and the diffusion constant D. The data for D show a strong deviation from the Rouse behavior (6DN~constant) for N>30 and scales $\sim N^{-2}$ for large N. We found that for short chains (N<30), no deviation from

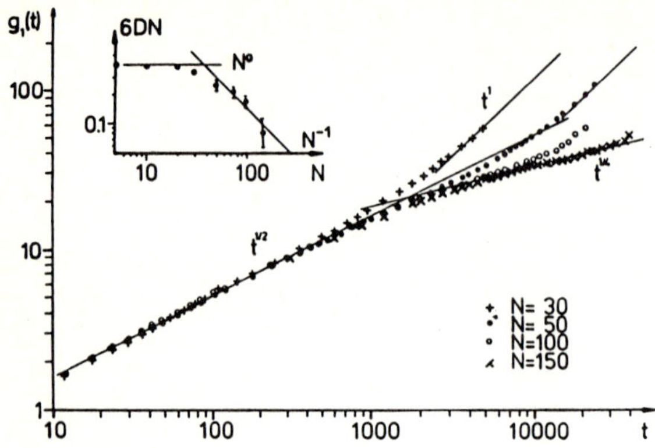

Figure 5 Mean square displacement $g_1(t)$ vs. t/τ averaged over the innermost monomers and the diffusion constant 6DN vs. N (inset). D was obtained from the center of mass diffusion of the chain.

Rouse behavior is seen in $g_1(t)$. However, for $N \geq 50$ at a time t/τ and distance $d_T^2 \approx 22$ independent of chain length the data deviate from Rouse, displaying a small dip (N=50) up to a clear $t^{1/4}$ (N=150) behavior, as expected from reptation. Using the relation[2] that $d_T^2 = 0.8 \langle R^2(N_e) \rangle$[2], we find $N_e \approx 15$, which is slightly more than 10 persistence lengths. Using $d_T^2 = 2 \langle R_G^2 \rangle$[2] would yield $N_e \approx 30$. To check the validity of the data as well as the equilibration, we calculated the Rouse or eigenmodes of the chains[20]. Here too we saw a distinct crossover from Rouse like relaxation to reptation for $N_e \approx 30$, consistent with the $g_1(t)$ data. Now that we have seen reptation, the next step is to develop a better microscopic understanding of the motion of entangled chains from our simulation data. This phase of the analysis is now underway.

4. DISCUSSION and CONCLUSIONS

In this paper, we have tried to demonstrate that molecular dynamics simulations can be very useful in understanding the dynamics of polymers. By coupling to a heat bath and viscous background, one can eliminate the need to simulate the solvent molecules explicitly. By changing Γ, local configurational fluctuations are governed by a ballistic dynamics up to times of order Γ^{-1}. This is faster than purely Brownian dynamics[10]. One strength of the technique over standard MC or Brownian dynamics is that it can be used effectively at quite high density. It is straightforward to generalize to a constant pressure algorithm[16] in order to study dense systems under stress. To demonstrate the effectiveness of the technique, we presented results for two different systems. The first was for the structure and dynamics of stars, whose density is quite high near the center yet dilute far away. We showed that our simulation results agree both with scaling theory[26,27] and with experiment[25]. This calculation required a moderate amount of computer time, roughly 100 hours on a Cray XMP. Our second example was much more cpu extensive[31] and examined the dynamics of dense polymer melts. Because the algorithm can be applied at quite high density, we were able to see clear evidence for reptation[29,30] from simulations for the first time. Future work will include a detailed analysis of the motion of the chain in the tube created by the other chains, a study of the dynamics of a melt of ring polymers as well as mixtures of linear and ring polymers and mixtures of polymers of different length chain.

We would like to thank our collobrators T. A. Witten, S. T. Milner and I. Carmesin for their important contributions to this work. We wish to acknowledge support from NATO, under travel grant 86/680 and the German Supercomputer Center HLRZ for a generous grant of time on the Cray XMP. GSG thanks SFB41 for support during his visits to Universität Mainz and KK thanks Exxon for their support during his visits to their Laboratory.

REFERENCES

1. P. G. deGennes: Scaling Concepts in Polymer Physics (Cornell Univ. Press, Ithaca 1979)
2. M. Doi, S. F. Edwards: The Theory of Polymer Dynamics (Clarendon Press, Oxford 1986)
3. A. Baumgärtner: Annul. Rev. Phys. Chem. 35, 419 (1984)
4. A. Baumgärtner, K. Kremer, K. Binder: Faraday Symp. Chem. Soc. 18, 37 (1984)
5. C. C. Crabb, J. Kovac: Macromolecules 18, 1430 (1984)
6. For a recent review of Monte Carlo studies of polymers see K. Kremer, K. Binder: Comp. Phys. Repts. in press (1988)
7. A. Kolinski, J. Skolnick, R. Yaris: J. Chem. Phys. 86, 7164 (1987)
8. Yu. Ya. Gotlib et al.: Macromolecules 13, 602 (1980)
9. M. Bishop, M. H. Kalos, H. L. Frisch: J. Chem. Phys. 70, 1299 (1982)
10. D. Ceperly, M. H. Kalos, J. L. Lebowitz: Macromolecules 14, 1472 (1981); M. Bishop et al.: J. Chem. Phys. 76, 1557 (1982)
11. E. Helfand: J. Chem. Phys. 69, 1016 (1978); 70, 2016(1979)
12. G. S. Grest, K. Kremer: Phys. Rev. A33, 3628 (1986)
13. P. E. Rouse: J. Chem. Phys. 21, 1273 (1953)
14. B. E. Zimm: J. Chem. Phys. 24, 265 (1956)
15. P. G. de Gennes: Macromolecules 9, 594 (1976)
16. M. Parinello, A. Rahman: Phys. Rev. Lett. 45, 1196 (1980)
17. T. Schneider, E. Stoll: Phys. Rev. B17, 1302 (1978)
18. R. B. Bird et al: Dynamics of Polymeric Liquids (Wiley, New York 1977), Vol. 2
19. P. J. Flory: Statistical Mechanics of Chain Molecules
20. K. Kremer, G. S. Grest, I. Carmesin: to be published (1988)
21. G. S. Grest, K. Kremer, T. A. Witten: Macromolecules 20, 1376 (1988)
22. G. S. Grest, K. Kremer, S. T. Milner, T. A. Witten: to be published (1988)
23. W. Burchard: Adv. Poly. Sci. 48, 1 (1983)
24. W. Dozier, J. Huang, L. J. Fetters: to be published (1988)
25. N. Khasat, R. W. Pennisi, N. Hadjichristidis, L. J. Fetters: to be published (1988)
26. M. Daoud, J. P. Cotton: J. Phys. 43, 531 (1982)
27. T. M. Birshtein, E. B. Zhulina, O. V. Borisov: Polymer 27, 1078 (1986)
28. J. D. Ferry: Viscoelastic Properties of Polymers (Wiley, New York 1980)
29. S. F. Edwards: Proc. Phys. Soc. 91, 513 (1967)
30. P. G. deGennes: J. Chem. Phys. 55, 572 (1971); 72, 4756 (1980)
31. The study reported here on dense melts took approximately 1500 cpu hours on a Cray XMP.

Monte Carlo Simulations of Polymer Systems

*K. Binder**

Institut für Physik, Universität Mainz,
D-6500 Mainz, Fed. Rep. of Germany, and
Department of Mathematics, Rutgers University,
New Brunswick, NJ 08903, USA

Abstract: The impact of Monte Carlo "computer experiments" in polymer physics is described, emphasizing three examples taken from the author's research group. The first example is a test of the classical Flory–Huggins theory for polymer mixtures, including a discussion of cricital phenomena. Also "technical aspects" of such simulations ("grand–canonical" ensemble, finite–size scaling, etc.) are explained briefly. The second example refers to configurational statistics and dynamics of chains confined to cylindrical tubes; the third example deals with the adsorption of polymers at walls. These simulations check scaling concepts developed along the lines of de Gennes.

1. Introduction

While one of the main motivations for Monte Carlo simulations is to test theories, they are often also relevant to experiment [1,2]. In this lecture, this latter point will be emphasized only for phase separation of fluid mixtures of linear flexible macromolecules, e.g., the mixture of polystyrene and polymethylvinylether, with molecular weights of the order of 10^5 {see e.g. [3]}. In the temperature–concentration plane, the one–phase region is separated from the two–phase region by the coexistence curve which ends in a critical point.

In the first part, we shall be concerned with critical phenomena near this point and test the "classical theory" for these phenomena, due to Flory and Huggins [4–6]. Phase separation occurs since the entropy of mixing is overruled by energetic effects driving the two partners apart; this theory [4–6] introduces one effective energy parameter, χ, reponsible for phase separation. χ plays a great practical role – e.g., it enters the expression [7] for the scattering intensity $I(\vec{q})$ from concentration fluctuations (described by the "collective structure factor")

$$[I(\vec{q})]^{-1} \sim 4/S_T^{coll}(\vec{q}) = (\phi_A N_A)^{-1} + (\phi_B N_B)^{-1} - 2\chi + 2q^2\left[\frac{\langle (R^A_{gyr})^2\rangle}{\phi_A N_A} + \frac{\langle (R^B_{gyr})^2\rangle}{\phi_B N_B}\right]. \quad (1)$$

Here $\vec{q}$ is the scattering vector, ϕ_A, ϕ_B are the volume fractions taken by the effective segments of the two polymers A,B. N_A, N_B are the number of segments per chain, and R^A_{gyr}, R^R_{gyr} the chain gyration radii. The brackets $\langle\ \rangle$ denote a configurational and thermal average.

Since all these quantities can be found from independent measurements, a fit of inverse scattering intensities to Eq.(1) yields χ [8]. However, typically [8] one finds that for each temperature T and concentration ϕ_A a different fitting parameter $\overline{\chi}(T,\phi_A)$ results: rather fitting a theory containing one adjustable parameter only, one works with a function of two variables! Of course, the feasibility of the fit hence is <u>not</u> a sensitive test for the quality of the

---

*Rutgers Supercomputer Visiting Scientist

Springer Proceedings in Physics Vol. 33: **Computer Simulation Studies in Condensed Matter Physics**
Editors: D. P. Landau • K. K. Mon • H-B. Schüttler

basic theoretical model. A more stringent test can be performed by Monte Carlo methods, as described next.

2. Monte Carlo Simulation of the Flory–Huggins lattice model [4,5]

In this model, every effective segment of a chain takes a lattice site. Every chain is represented by a self–avoiding walk (SAW) on the three–dimensional lattice. Since each site can host at most one segment, no walk can intersect itself or other chains.

Here the original model is slightly generalized, by including a third component, e.g., a low molecular weight solvent or "free volume" ("vacancies" V), see Fig. 1 [9].

This is needed in order that the chain configurations can relax via the motions shown in Fig. 1 and also is of physical interest [10].

Unmixing of the blends is caused by an energy ϵ_{AA}, ϵ_{BB} being won if two nearest neighbor sites are taken by effective segments of the same kind, or if a repulsive energy ϵ_{AB} occurs between unlike segments. This model is the main basis for discussing thermodynamic properties of polymer mixtures since 47 years.

From this model, we wish to obtain the critical temperature T_c, the coexistence curve, $S_T^{coll}(\vec{q})$, etc., via statistical thermodynamics. In principle, all configurations of the two types

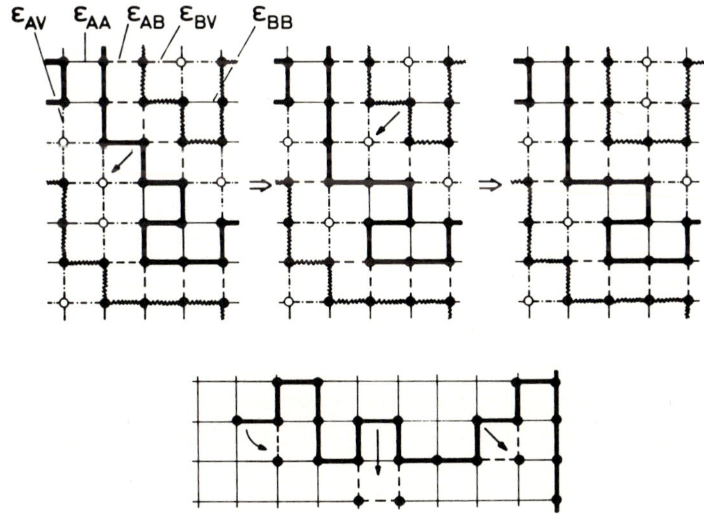

Fig. 1: Schematic illustration of the dynamic Flory–Huggins lattice model. Sites taken by effective units of a polymer chain are indicated by full dots, vacancies by empty circles. A–chains are denoted by thick bonds between effective units, B–chains by wavy bonds. Nearest–neighbor interactions between units of the same kind (ϵ_{AA},ϵ_{BB}) are denoted by thin straight lines, interactions between effective units of different kind (ϵ_{AB}) by broken lines. Interactions between chain units and solvent molecules (ϵ_{AV},ϵ_{BV}; dash–dotted lines) will be set equal to zero. Lower part illustrates the dynamics of the model–random motions occur by moving the chain ends, by exchange of two neighboring bonds which form an angle of 90° , and by the 90° "crankshaft" rotation of three neighboring bonds on the (simple–cubic!) lattice. From Sariban and Binder [9]

of chains on the lattice need to be considered, and their energy; from the Boltzmann weight for each configuration, we obtain the partition function, etc.

Of course this problem cannot be solved exactly; Flory and Huggins instead proposed a mean–field approximation, writing the free energy of the mixture as the sum of the entropy of mixing and an enthalpy term involving the χ parameter which relates to the nearest neighbor energy $\epsilon \equiv \epsilon_{AB}-(\epsilon_{AA}+\epsilon_{BB})/2$ via

$$\chi = z\epsilon/k_BT; \tag{2}$$

here z is the coordination number of the lattice. Then this free energy excess is [4,5,9]

$$F/k_BT = (\phi_A \ln\phi_A)/N_A + (\phi_B \ln\phi_B)/N_B + \phi_V \ln\phi_V + \chi\phi_A\phi_B . \tag{3}$$

If we had three low molecular weight components ($N_A = N_B = 1$), the entropy term would be exact since then every lattice site can be occupied independently of its neighbors. For polymers, however, this is no longer true, since the chains must not intersect themselves. Thus the entropy expression is only approximate.

Also the enthalpy term involves approximations: we need the energy ϵ/k_BT times the probability that a site is taken by an A–segment and its neighbor by a B–segment. This probability is not simply ϕ_A times ϕ_B: the site occupation probabilities are correlated. Also the count for the number of neighbors is inaccurate: an inner segment of an A–chain can have at most z–2 B–neighbors.

Accepting for the moment these approximations, the resulting predictions for T_c for the symmetric mixtures, $N_A = N_B = N$, is

$$N(1-\phi_V)\chi_c = N(1-\phi_V)z\epsilon/k_BT_c = 2 \quad . \tag{4}$$

We also ask: how accurate is this approximation? Another feature is that T_c is predicted to depend on the three energy parameters $\epsilon_{AA}, \epsilon_{AB}, \epsilon_{BB}$ only through $\epsilon = \epsilon_{AB} - (\epsilon_{AA} + \epsilon_{BB})/2$. These predictions now can be stringently tested by Monte Carlo: no adjustable parameter whatsoever enter the comparison between the theory and the simulation !

Let me anticipate the main result, Fig. 2 [11]: the Flory Huggins approximation for T_c is not good at all, the predicted values are far too high. For the dense limit (no vacancies) T_c is too high by about a factor of two (or about 1.3 for the Guggenheim [12] approximation) while the situation gets much worse with increasing vacancy content. Also the "one–energy–parameter approximation" breaks down, when the polymer concentration is less than about 40 %. The discrepancies between simulation and the theory get larger with increasing chain length.

Before we discuss why mean field approximations fail so badly here, we describe our Monte Carlo simulation techniques [9,11].

First, chain configurations are relaxed by local motions: end bond jumps, 90° crankshaft rotations, and the "kink–jump" motion (Fig. 1). In addition, we apply a grand–canonical simulation technique: after a chain has experienced $k \approx 100$ trial steps by which one attempts to relax its configuration, a move is attempted where the chain changes its identity: an A–chain becomes a B–chain, and vice versa. Thus the chemical potential difference $\Delta\mu$ between A and B segments is held fixed as an independent variable, rather than the concentration.

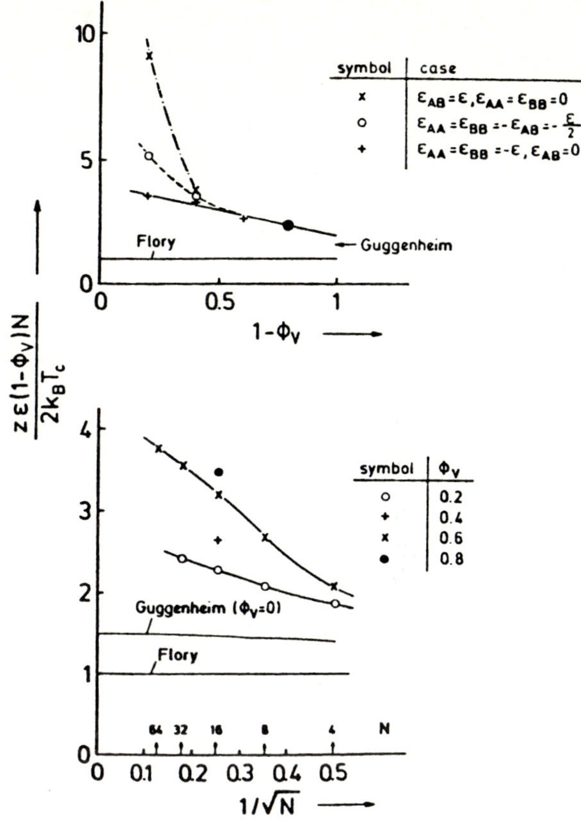

Fig. 2: Ratio between Flory–Huggins critical temperature and actual critical temperature T_c plotted vs polymer concentration (upper part) and versus the inverse square root of the chain length (lower part). Upper part refers to N = 16 (for $1-\phi_V = 0.8$ all three cases coincide, as indicated by a solid dot). Curves are only drawn to guide the eye. From Sariban and Binder [11]

We used chain lengths N = 4,8,16,32 and 64, and vacancy concentrations $\phi_V = 0.2,0.4,0.6$ and 0.8. Without vacancies no chain relaxation is possible.

Also, one cannot simulate infinitely large systems: rather finite simple cubic LxLxL boxes with periodic boundary conditions are treated, with linear dimensions from L = 12 to L = 30. In order to infer the coexistence curve and the critical temperature T_c, a finite size scaling [13–15] analysis is needed, as discussed below.

Since in the grand canonical simulation at constant $\Delta\mu$ A–chains stochastically transform into B–chains and vice versa, we record the relative excess m of the number of A–chains n_A over the number of B–chains n_B, see Fig. 3,

$$m = \langle |M| \rangle, \quad M = (n_A - n_B)/(n_A + n_B). \tag{5}$$

In the thermodynamic limit $L \to \infty$, this order parameter m of the unmixing transition would be nonzero for all temperatures $T < T_c$, but would vanish at T_c (for $\Delta\mu = 0$) and stay zero for all

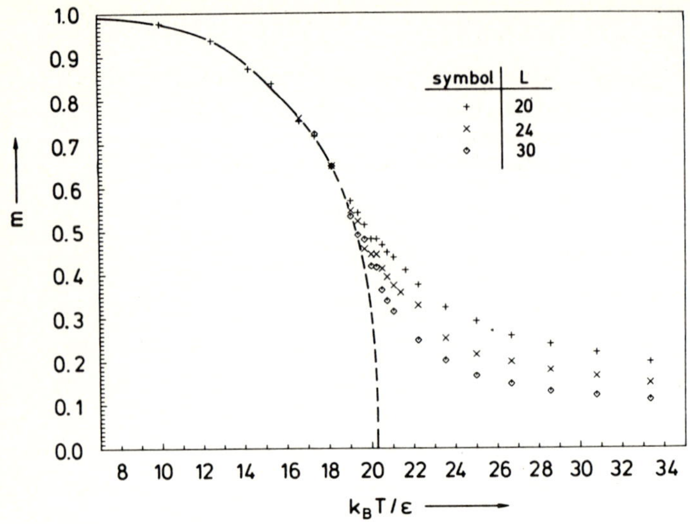

Fig. 3: Order parameter m plotted vs k_BT/ϵ, for $N = 64$, $\phi_V = 0.6$, $\Delta\mu = 0$, $\epsilon_{AB} = 0$, $\epsilon_{AA} = \epsilon_{BB} = -\epsilon/2$. Broken curve is the result of a finite size scaling analysis (Fig. 4), $m = \hat{B}(1-T/T_c)^\beta$, with $\beta = 0.32$, $\hat{B} = 1.3$, $\epsilon/k_BT_c = 0.049$. From Sariban and Binder [11]

$T > T_c$. In finite lattices this singularity at T_c is rounded off, m as defined in Eq.(5) has a "finite–size–tail" above T_c.

Similar finite size effects occur in the collective structure factor at zero wavevector, $S_{coll}(\vec{q}=0)$, with [16]

$$S_{coll}(\vec{q}) \equiv \left\langle \left[\sum_j \exp(i\vec{q}\cdot\vec{R}_j)(\delta\phi_B^j - \delta\phi_A^j) \right]^2 \right\rangle / L^3 \tag{6}$$

where $\delta\phi_B^j - \delta\phi_A^j = \phi_B^j - \phi_A^j - <|\phi_B^j - \phi_A^j|>$, which can be sampled from order parameter fluctuations

$$S_{coll}(o) = (n_A+n_B)N(1-\phi_V)\ [<M^2>-<|M|>^2]. \tag{7}$$

While for $L \to \infty$ $S_{coll}(o)$ diverges at T_c ("critical opalescence"), in finite lattices this divergence is rounded off. But the extrapolation of the position $T_c^{eff}(L)$ where the peak $S_{coll}(o)|_{max}$ occurs towards $L \to \infty$ yields an estimate of T_c. An even more accurate estimate results from the cumulant intersection method [14]. The normalized fourth order cumulant of the order parameter distribution $P_L(M)$, defined as

$$U_L \equiv 1 - <M^4> / (3<M^2>^2), \tag{8}$$

above T_c tends to zero as $L \to \infty$ {the distribution $P_L(M)$ for $\Delta\mu \to 0$ tends to a gaussian around zero}. For $T < T_c$, U_L tends towards 2/3 as $L \to \infty$ {the distribution $P_L(M)$ now is characaterized by two peaks off–set from zero, with $<m^4> \approx <m^2>^2\}$. Finite size scaling implies that U_L has an L–independent value between zero and 2/3 for $T = T_c$, and hence at T_c all cumulants for the different values of L should have a common intersection point. Within reasonable error limits this is in fact the case [9,11].

Having found T_c we use this in finite size scaling analyses: the family of curves m(T,L) {Fig. 3} should collapse on a single function $\tilde{m}$ if $mL^{\beta/\nu}$ is plotted vs. $|1-T/T_c|L^{1/\nu}$, since [13,14]

$$m(T,L) \xrightarrow[T \to T_c, L \to \infty]{} L^{-\beta/\nu} \tilde{m}\{|1-T/T_c|L^{1/\nu}\}. \tag{9}$$

Universality [17] implies that for data close enough to T_c the Ising exponents [18] apply,

$$\beta \approx 0.32, \; \gamma \approx 1.24, \; \nu \approx 0.63. \tag{10}$$

Using these exponents, a resonable "data collapsing" is in fact obtained (Fig. 4).

From the straight line with slope $\beta = 0.32$ we get the "critical amplitude" $\hat{B}$ in the power law

$$m = \hat{B}(1-T/T_c)^\beta, \; T \to T_c. \tag{11}$$

In this manner, the dashed curve in Fig. 3 was derived.

Since the "Ginzburg criterion" [19] implies that the critical behavior of phase separation in polymer mixtures should be mean–field like as $N \to \infty$ [7,20,21], i.e. have the exponents [17]

$$\beta = 1/2, \; \gamma = 1, \; \nu = 1/2, \tag{12}$$

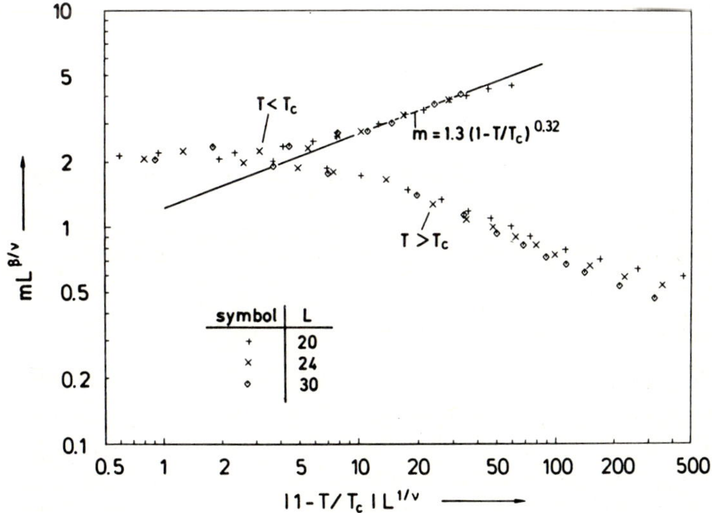

Fig. 4: Log–log plot of $mL^{\beta/\nu}$ vs $|1-T/T_c|L^{1/\nu}$ for $N = 32$, $\phi_V = 0.2$, $\Delta\mu = 0$, $\epsilon_{AB} = 0$, $\epsilon_{AA} = \epsilon_{BB}$. Ising exponents $\beta = 0.32$, $\nu = 0.63$ [18] are used. Straight line indicates the power law Eq.(11). From Sariban and Binder [9]

we have also tried a version of finite size scaling appropriate [15] for such mean field exponents. The "data collapsing" is distinctly worse [9,11]; hence for our rather short chains a crossover between Ising and mean field critical behavior is not seen. The same story comes from the collective structure factor $S_{coll}(0)$ [9,11]: Finite size scaling [13,14]

$$S_{coll}(0) = L^{-\gamma/\nu} \mathcal{S}\{(T/T_c-1)L^{1/\nu}\}, \; T \to T_c, \; L \to \infty \tag{13}$$

works rather well; again we obtain the critical amplitudes $\hat{\Gamma}_+(T > T_c)$, $\hat{\Gamma}_-(T < T_c)$ in the relation

$$S_{coll}(0)/(1-\phi_V)^2 = \hat{\Gamma}_\pm |1-T_c/T|^{-\gamma}, \; T \to T_c \; . \tag{14}$$

The mean field finite size scaling [15],

$$S_{coll}(0) = L^{-3/2} \mathcal{S}'\{(T/T_c-1)L^{3/2}\}, \; T \to T_c, \tag{15}$$

again does not yield a satisfactory "data collapsing" [9,11].

Of course, it is no surprise that the analytic approximations due to Flory–Huggins [4–6] and Guggenheim [12] yield mean field critical exponents instead of Ising exponents; but why is T_c so badly predicted (Fig. 2)? Several effects need to be considered.

While the approximations assume chain configurations independent of temperature and composition, the simulation shows that with increasing ϵ/k_BT the chain radii contract [11]. This effect is more pronounced for the minority component. Since this contraction typically is a few percent only, however, it should not have a strong effect on T_c.

Where the theory fails more dramatically is the count for the number n_c^{tot} of contacts that a chain makes with monomers of <u>other</u> chains. According to Flory $\{n_c^{tot} = n_c(AA) + n_c(AB)$ for an A–chain, intrachain AA–contacts being neglected$\}$ it should be $n_c^{tot}(F) = zN(1-\phi_V)$, according to Guggenheim [12,22] $n_c^{tot}(G) = \{(z-2)N+2\} \; (1-\phi_V)$. For the case $N = 64$, $\phi_V = 0.6$ we have $n_c^{tot}(F) = 153.6$, $n_c^{tot}(G) = 103.2$; however, the actual number n_c^{tot} is about 56 only, and there are about 17 self–contacts, which hence are not negligible. Since the Flory–Huggins approximation overestimates the number of pairs contributing to the unmixing energy by nearly a factor of three in this case, it is less of a surprise then that T_c is overestimated by a factor of about 3.8 (Fig. 2)!

Now even if we add the actual n_c^{tot} and twice the number of self–contacts (counting both "ends" of the self-contact separately), there are less contacts than expected from $n_c^{tot}(G)$. However, since each monomer must have some neighbor, there are relatively more neighbors of a monomer taken by vacancies than expected from the factor $(1-\phi_V)$. This correlation effect in the occupation probability of lattice sites is due to excluded–volume interactions, which are still present on short length scales in the mixture. These excluded volume effects are seen in the chain radii [9,11]: the radii steadily increase with increasing vacancy concentration, and stay distinctly above the non–reversal–random walk (NRRW) values [23] even in the dense limit [24],

$$\langle R_{gyr}^2 \rangle_{NRRW} = \frac{1}{4}\frac{(N-1)(N+1)}{N} - \frac{5}{16}\frac{N-1}{N} + \frac{5}{32}\frac{N-1}{N} - \frac{5}{128}\frac{1}{N^2}(1-1/5^{N-1}) \; . \tag{16}$$

The good agreement between our data at $\epsilon/k_BT=0$ and the results of other workers (e.g. [24,25]) shows that our chain configurations are indeed well equilibrated.

Now these excluded volume effects also have a pronounced consequence for the critical behavior. Disregarding them, a simple crossover scaling analysis [9] implies that the family of curves m(T,N) {resulting in the limit $L \to \infty$ from our data} should be collapsed on a crossover scaling function by plotting $mN^{1/2}$ versus $(1-T/T_c)N$; however, this does not work [9,11]: rather than obtaining a function $\tilde{\tilde{m}}(x)$

$$mN^{1/2} = \tilde{\tilde{m}}\{(1-T/T_c)N\}, \tag{17}$$

which behaves as $\tilde{\tilde{m}} \propto x^{0.32}$ for x small and as $\tilde{\tilde{m}} \propto x^{0.5}$ for x large, we obtain a set of parallel straight lines on the log–log plot. This crossover scaling analysis yields predictions for chain length dependence of the critical amplitudes $\hat{B}, \hat{\Gamma}$, and $\hat{D}$ $\{m(T=T_c,\Delta\mu)=\hat{D}(\Delta\mu/k_BT_c)^{1/\delta}$ describes the critical isotherm}, namely [9]

$$\hat{B} \propto N^{-(1/2-\beta)}=N^{-0.18}, \hat{\Gamma} \propto N^{2-\gamma}=N^{0.76}, \hat{D} \propto N^{1/\delta-(\delta-3)/2\delta} \approx N^{0.01}. \tag{18}$$

Indeed the chain length variation of $\hat{B}, \hat{\Gamma}$, and $\hat{D}$ is consistent with simple power laws (Fig. 5): however, not a single exponent prediction of Eq.(18) works out!

However, what sounds like a failure really is a major success of the scaling theory – but excluded volume effects must be incorporated properly. It has been shown by Joanny et al. [26] that the χ parameter gets "renormalized" in a power law fashion, involving a new exponent $x \approx 0.22$ which enters the crossover scaling. The physical interpretation is that the number of (unfavorable) contacts vanishes in the "semidilute" limit, the chains becoming more and more compatible because they locally don't "see" each other although they are strongly interpenetrating. The resulting crossover scaling law

$$mN^{x/[2(1+x)]} = f\left\{(1-\frac{T}{T_c})N^{x/(1+x)}\right\} \tag{19}$$

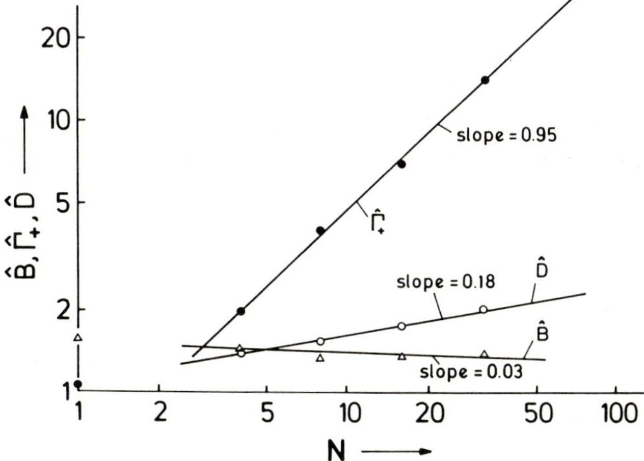

Fig. 5: Log–log plot of critical amplitudes $\hat{B}, \hat{\Gamma}_+$, and $\hat{D}$ vs chain length N, for $\phi_V = 0.2$ (Points for N = 1 refer to the standard Ising model without vacancies). From Sariban and Binder [9]

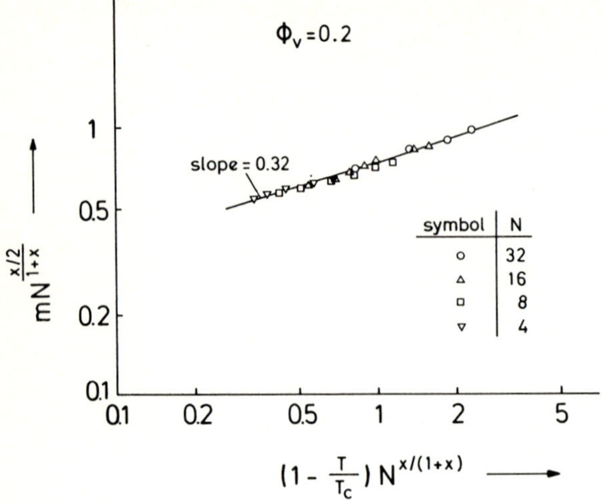

Fig. 6: Log–log plot of $mN^{x/[2(1+x)]}$ vs. $(1-T/T_c)N^{x/(1+x)}$ for $\phi_V = 0.2$ and several N, with $x = 0.22$ [26]. From Sariban and Binder [27]

is nicely consistent with the numerical results [27], see Fig. 6. Moreover, the predictions of Ref. 26 for the critical amplitudes, namely [27]

$$\hat{B} \propto N^{x(\beta-1/2)/(1+x)}, \hat{\Gamma}_+ \propto N^{1+x(1-\gamma)/(1+x)}, \tag{20a}$$

$$\hat{D} \propto N^{1/3 + [1/3+(x/2)/(1+x)](3/\delta-1)}, \tag{20b}$$

are in striking agreement with the results seen in Fig. 5 {note $\delta \approx 4.8$ [18]}.

Is there any evidence for this interesting critical behavior in real polymer mixtures? So far, there is a single observation only [28], referring to the exponent γ describing the critical behavior of the collective structure factor. The behavior near T_c can in fact be fitted well with the Ising exponent.

3. Statics and Dynamics of Polymers Confined to Cylindrical Tubes

Now the second example will be discussed briefly: polymers in confined geometries. Flexible polymers moving inside porous structures are believed to be relevant for filtration, gel permeation chromatography, heterogeneous catalysis, oil recuperation, etc. [29]. Aside from this eventual practical application, the configurational statistics of polymers in confined geometries is a challenging theoretical problem. Scaling ideas have been proposed [29–31], which we want to test.

If we consider a chain consisting of N effective units of size ℓ in dilute solution in a good solvent, its unrestricted linear dimensions are [18,32]

$$\langle R_G^2 \rangle_0 \propto \ell^2 N^{2\nu}, \nu \approx 0.59 \text{ (d=3)}, \nu = 3/4 \text{ (d=2)}. \tag{21}$$

The chain partition function Z_N involves a second exponent [18,32]

$$Z_N \propto N^{\gamma-1}(q_{eff})^N, \gamma \approx 1.16 \text{ (d=3)}, \gamma = 43/32 \text{ (d=2)}, \tag{22}$$

q_{eff} being a nonuniversal constant.

For a chain in a tube with a very large diameter, $d_T > R \equiv \sqrt{<R_G^2>_o}$, the tube has little effect. But if $d_T < R$, the chains are deformed into long "cigars" of length $R_{\parallel}$ and diameter d_T. The structure can be understood as a one–dimensional succession of "blobs" of size d_T containing about g segments each, with $d_T \ \alpha \ \ell g^{\nu}$. Since there must be N/g such blobs, the cigar length $R_{\parallel}$ is

$$R_{\parallel} \ \alpha \ d_T(N/g) \ \alpha \ d_T^{1-1/\nu} \ell^{1/\nu} N. \tag{23}$$

The crossover between the unperturbed linear dimension and this cigar length can be described by a crossover scaling function $\tilde{R}(x)$,

$$R_{\parallel} = \ell N^{\nu} \tilde{R}(\ell N^{\nu}/d_T), \ \tilde{R}(x \to 0) \to \text{const}, \ \tilde{R}(x >> 1) \ \alpha \ x^{1/\nu-1}. \tag{24}$$

This scaling description for $R_{\parallel}$ has been tested by Monte Carlo simulations for chains on the tetrahedral lattice [33], applying a dynamic relaxation technique similar to Fig. 1. Fig. 7 shows that the scaling picture works out very well.

Next we consider the dynamics of chains confined into tubes. Scaling ideas have been developed both for chains with hydrodynamic interactions [30] (the Zimm [34] model) and without them [33], describing a chain in a heat bath which induces stochastic reorientations of links in the chain with a rate W. One predicts for the mean–square displacement of inner monomers with time t an anomalous diffusion law,

$$<\vec{r}^2(t)> \ \alpha \ \ell^2(Wt)^{1/(1+1/2\nu)}, \text{ for } 1 \lesssim Wt \lesssim (d_T/\ell)^{2(1+1/2\nu)},$$
$$\text{i.e. } <\vec{r}^2(t)> \ < d_T^2. \tag{25}$$

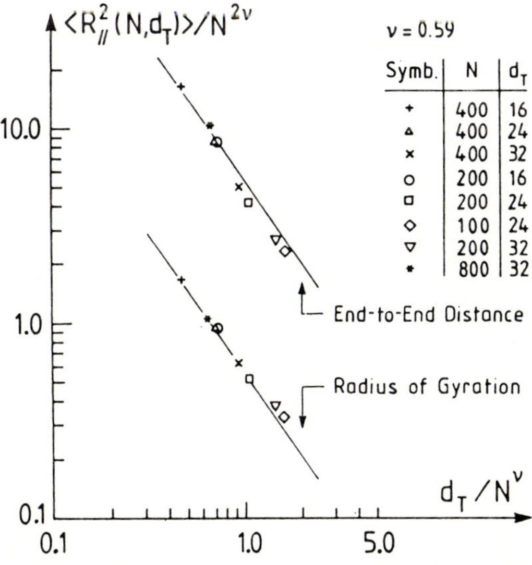

Fig. 7: Log–log plot of mean square radii in the direction parallel to the tube axis, normalized by $N^{2\nu}$ with $\nu = 0.59$, vs d_T/N^{ν}, for self–avoiding walks on the tetrahedral lattice, N = 200 to 800, and various tube diameters d_T (in units of the lattice constant). Both gyration radius and end–to–end distance are shown. From Kremer and Binder [33]

This is the standard Rouse regime, where the monomers do not yet feel the tube constraints. For larger times the perpendicular components saturates, $<\vec{r}_{\perp}^{2}> \;\alpha\; d_T^2$ independent of time t, while the parallel component can be written as

$$<r_{\shortparallel}^2(t)> \;\alpha\; d_T^2(Wt)^{1/2}\,(\ell/d_T)^{1+1/2\nu},\ \text{for}\ (d_T/\ell)^{2(1+1/2\nu)} \lesssim Wt \lesssim Wt_N, \tag{26}$$

where t_N is the largest relaxation time of the chain,

$$t_N^{-1} = W(\ell/d_T)^{2-1/\nu}/N^2. \tag{27}$$

The final regime is the diffusion of the chain as a whole,

$$<r_{\shortparallel}^2(t)> \approx 2\,D_N t,\ t > t_N,\ D_N = W\ell^2/N. \tag{28}$$

Note that the chain diffusion constant is independent of the tube diameter.

These predictions were also tested by simulations [33], see Fig. 8. The first crossover {from Eq.(25) to Eq.(26)} is seen in the perpendicular component only, since the change from $t^{1/(1+1/2\nu)} \approx t^{0.54}$ to $t^{1/2}$ is not resolved. This first crossover occurs at a time which depends on the tube diameter, but not on the chain length, consistent with the theory. Conversely, the diffusion constant is independent of the tube diameter as predicted. These results hence nicely confirm the predictions described in Eqs.(25)–(28).

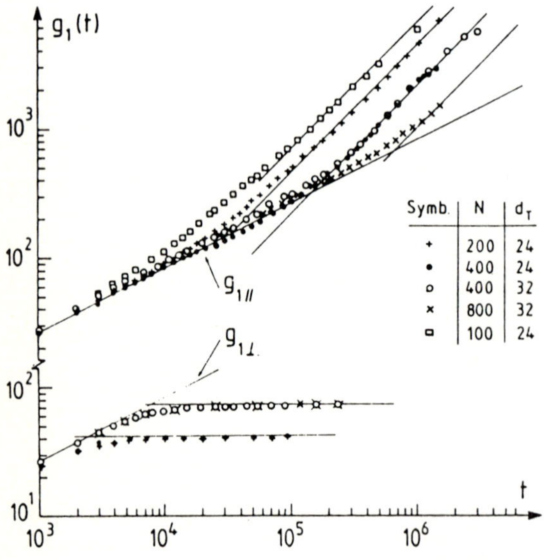

Fig. 8: Log–log plot of the mean square displacement of an inner monomer parallel ($g_{1\shortparallel}$) and perpendicular ($g_{1\perp}$) to the tube vs. time (in units of attempted moves per bead), for SAW's on the tetrahedral lattice. Straight lines have slope 1/2 (short times) or $1(g_{1\shortparallel})$, $0(g_{1\perp})$ at late times, respectively. From Kremer and Binder [33]

4. Polymer Chains Attached to Surfaces

If a polymer chain is attached to a hard wall with one or both ends, but otherwise there is no interaction between the chain and the wall, new exponents enter in the partition functions Z_N^1, Z_N^{11} for a chain with one (both) ends attached to the surface

$$Z_N^1 \;\alpha\; N^{\gamma_1 - 1} q_{eff}^N, \; Z_N^{11} \;\alpha\; N^{\gamma_{11} - 1} q_{eff}^N \; . \qquad (29)$$

These exponents have been estimated from renormalization group expansions to second order in $\epsilon = 4-d$, and from simulations. The ϵ–expansion predictions [36] $\gamma_{11} \approx 0.377$, $\gamma_1 \approx 0.695$ are in fair agreement with the Monte Carlo estimates [37], see Fig. 9.

If we allow an attractive energy ϵ for monomers next to the wall, a transition occurs at some adsorption temperature T_a where the chain configuration starts to change from three–dimensional to effectively two–dimensional [37]. At this multicritical point the exponents γ_1, γ_{11} take different values, which also have been estimated [37], see Fig. 9. There occurs a rich and complicated behavior, which is outside of consideration here, however.

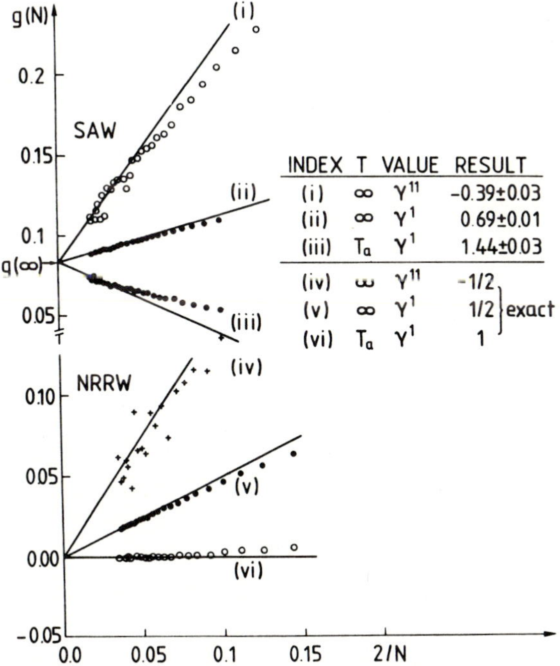

Fig. 9: Plot of $g(N) \equiv Z(N+2)/[Z(N)(z-1)^2]$ of the partition functions $Z(N) = Z_N^1, Z_N^{11}$ vs 2/N, for SAW's (upper part) and NRRW's (lower part; there the asymptotic behavior is known exactly). g(N) behaves as $[q_{eff}/(z-1)]^2 \{1+(\gamma-1)(2/N)\}$ as $N \to \infty$. Slopes of the straight lines yield estimates for γ. For Z_N^1, also included is the case where monomers at the wall experience an attractive energy $\epsilon=1$, which leads to an adsorption temperature T_a where the chain configuration changes from three dimensional to effectively two dimensional. From Eisenriegler et al. [37]

5. Concluding remarks

This lecture has attempted to show that Monte Carlo simulations play an important role in polymer physics, advancing our understanding of both theory and experiment. Of course, only a few examples of recent work could be described – for more complete discussions of various applications we refer to other reviews [38,39], where also detailed information on the "know–how" about the algorithms used for these simulations is given.

Acknowledgements: This lecture reviews research performed together with A. Sariban, K. Kremer and E. Eisenriegler; it is a pleasure to thank them for a fruitful collaboration. Support by the Supercomputer Visiting Scientist Program at Rutgers University is also acknowledged.

References

1. K. Binder (ed.), Monte Carlo Methods in Statistical Physics, 2nd edition (Springer, Berlin 1986)
2. K. Binder (ed.), Applications of the Monte Carlo Method in Statistical Physics, 2nd edition (Springer, Berlin 1987)
3. S. Reich. H. Snyder, and P. Meakin, Macromol. 16, 753 (1983)
4. P. J. Flory, Principles of Polymer Chemistry (Cornell University Press, Ithaca, N.Y. 1953)
5. M. J. Huggins, J. Am. Chem. Soc. 64, 1712 (1942); J. Chem. Phys. 9, 440 (1941)
6. R. Koningsveld, L. A. Kleintjens, and E. Nies: Croat. Chim. Acta. 60, 53 (1987)
7. P. G. de Gennes, Scaling Concepts in Polymer Physics (Cornell University, Ithaca, N.A. 1979)
8. C. Herkt–Maetzky and J. Schelten, Phys. Rev. Lett. 51, 896 (1983)
9. A. Sariban and K. Binder, J. Chem. Phys. 86, 5859 (1987)
10. I. C. Sanchez and R. H. Lacombe, J. Chem. Phys. 80, 2352 (1976); Macromolecules 11, 1145 (1978)
11. A. Sariban and K. Binder, Macromolecules 21 (1988) in press
12. E. A. Guggenheim, Proc. Roy. Soc. (London) A183, 203, 231 (1945)
13. M. E. Fisher and M. N. Barber, Phys. Rev. Lett. 28, 1516 (1972)
14. K. Binder, Z. Physik B 43, 119 (1981); Ferroelectrics 73, 43 (1987)
15. K. Binder, M. Nauenberg, V. Privman and A. P. Young, Phys. Rev. B 31, 1498 (1985)
16. Note that $S_{coll}(\vec{q})$ in Eq.(6) becomes the usual collective structure factor $S_T^{coll}(\vec{q})$ {as referred to, for instance, in Eq.(1)} if one redefines $\delta\phi_B^j - \delta\phi_A^j$ as $\delta\phi_B^j - \delta\phi_A^j = \phi_B^j - \phi_A^j - <(\phi_B^j - \phi_A^j)>$, i.e., avoiding taking the absolute value. For $T > T_c$ and $\Delta\mu = 0$, as shown in Fig. 6, $S_{coll}(o)$ is smaller than $S_T^{coll}(o)$ by a factor of $1-2/\pi$.
17. M. E. Fisher, Rev. Mod. Phys. 46, 597 (1974)
18. J. C. Le Guillou and J. Zinn–Justin, Phys. Rev. B 21, 3976 (1980)
19. V. L. Ginzburg, Sov. Phys. Solid State 2, 1824 (1960)
20. P. G. de Gennes, J. Phys. Lett. (Paris) 38, L441 (1977); J. F. Joanny, J. Phys. A 11, L117 (1978)
21. K. Binder, J. Chem. Phys. 79, 6387 (1983); Phys. Rev. A 29, 341 (1984)
22. Actually Ref. 12 considers $\phi_V = 0$ only. We thus approximately introduce a factor $1-\phi_V$ to account for vacancies in the same way as Flory–Huggins theory does.
23. C. Domb and M. E. Fisher, Proc. Cambridge Philos. Soc. 54, 48 (1958)
24. O. F. Olaj and W. Lantschbauer, Makromol. Chem., Rapid Commun. 3, 847 (1982)
25. A. Kolinski, J. Skolnick, and R. Yaris, J. Chem. Phys. 86, 1567, 7164 (1987)
26. J. F. Joanny, L. Leibler, and R. BAll, J. Chem. Phys. 81, 4640 (1984); L. Schäfer and Ch. Kappeler, J. Phys. (Paris) 46, 1853 (1985); D. Broseta, L. Leibler and J. F. Joanny, Macromolecules 20, (1987)
27. A. Sariban and K. Binder, to be published
28. D. Schwahn, K. Mortensen and H. Yee–Madeira, Phys. Rev. Lett. 58, 1544 (1987)
29. M. Daoud and P. G. de Gennes, J. Phys. (Paris) 38, 85 (1977)

30. F. Brochard and P. G. de Gennes, J. Chem. Phys. 67, 52 (1977)
31. P. G. de Gennes, J. Chem. Phys. 55, 572 (1971)
32. B. Nienhuis, Phys. Rev. Lett. 49, 1062 (1982)
33. K. Kremer and K. Binder, J. Chem. Phys. 81, 6381 (1984)
34. B. H. Zimm, J. Chem. Phys. 24, 265 (1956)
35. P. E. Rouse, J. Chem. Phys. 21, 1272 (1953)
36. H. W. Diehl and S. Dietrich, Phys. Lett. A 80, 408 (1980); Z. Phys. B 42, 65 (1981); Phys. Rev. B 24, 2878 (1981);J. S. Reeve and A. Guttmann, Phys. Rev. Lett. 45, 1581 (1980); J: Phys. A 14, 3358 (1981)
37. E. Eisenriegler, K. Kremer and K. Binder, J. Chem. Phys. 77, 6296 (1982)
38. A. Baumgärtner, in Ref. 2, Chapter 5
39. K. Kremer and K. Binder, Computer Physics Repts., in press

Molecular Dynamics: A New Approach to Hydrodynamics?

D.C. Rapaport

Physics Department, Bar-Ilan University, Ramat-Gan, Israel

1. Introduction

Molecular dynamics simulation came into being early in the computer age. Initially practiced by only the fortunate few with access to what were then considered "large" computers, the technique has become increasingly widely used /1/. At the same time the growth in computer power and storage capacity has expanded the horizon of feasibility; some of the original simulations of small equilibrium systems can now be carried out on microcomputers, whereas the latest studies of time-dependent behavior consume significant resources on the latest of supercomputers.

The intent of this lecture is to describe a relatively new application of the molecular dynamics (MD) approach -- to the study of hydrodynamic instability. Two situations involving flow instabilities, each a well-studied system from the experimental point of view, have been modeled at the detailed atomistic level with the goal of reproducing the observed macroscopic behavior on the rather limited length and time scales that are accessible to MD simulation. One of the simulations is an attempt to model eddy development and time-dependent vortex shedding observed in the flow of a fluid past an obstacle /2/. The other is a study of the Rayleigh-Bénard problem -- the spontaneous formation of convective rolls in a fluid subjected to a suitable temperature gradient /3/.

As with real flow experiments, the principal results obtained from these simulations take the form of pictures of flow patterns. Because of the complexity, characterization of time-dependent flow phenomena by quantitative measures is not a particularly well-developed aspect of fluid dynamics. At the level of comparing pictures of flow patterns however, it turns out that there is a strong similarity between experimental and MD flows, despite the enormous disparity between the length and time scales of real and simulated systems. This leads to the question of whether the MD approach will, some day, become an accepted method in the study of those hydrodynamic problems for which techniques based on numerical integration of the fluid equations (e.g., Navier-Stokes) prove unsatisfactory, as well as for studying the microscopic basis for the assumptions and phenomenological constitutive relations often required to supplement these equations. The question mark of the title is an indication that the answer is not yet available, although growing computer power suggests that it may be an issue of *when* rather than *whether*.

2. Methodology

For reasons having little to do with the processes being modeled, two very different simulation schemes were used for the two studies. The obstructed flow problem was simulated using a fluid of soft disks, and calculations for systems containing

Springer Proceedings in Physics Vol. 33: **Computer Simulation Studies in Condensed Matter Physics**
Editors: D. P. Landau • K. K. Mon • H-B. Schüttler

as many as 170,000 particles were carried out on a set of coupled array processors; more recent work on up to 200,000 particle systems has involved the use of a vector processor. The convection study made use of a much smaller system -- a mere 14,000 particles; this simulation used hard elastic disks and was carried out on a conventional, scalar computer. Both simulations addressed two-dimensional fluids; the corresponding experiments deal with hydrodynamic flows in which the third dimension appears to have negligible effect on the problem -- either flow past a long cylindrical obstacle, or convective cells in which the rolls that develop have the form of long straight cylinders. The ability to limit the simulation to two dimensions, rather than having to treat the full three-dimensional problem, facilitates the study of much larger systems (in terms of linear size) than would otherwise be possible. Computer availability dictated the choice between hard and soft particles; for reasons which will become apparent subsequently, hard disk simulations -- which prove to be the more efficient approach on a scalar processor -- apparently cannot be effectively implemented on either vector or parallel machines.

Only a comparatively brief coverage of the technical aspects of the simulations will be given here; more detailed descriptions of the methods for dealing with soft and hard disk fluids, as well as the specialized techniques for harnessing the power of parallel and vector supercomputers for MD work are to be found elsewhere /4,5/.

Soft disk MD simulation is based on the *time-driven* approach; the evolution of the system is by means of a series of fixed time steps, and the equations of motion of the particles -- each in the continually changing force field produced by its neighbors -- are integrated by some suitable numerical technique (leapfrog, predictor-corrector, etc) /6/. The numerical solution is subject to the usual errors of finite-difference schemes, and the time step is chosen in order to reduce these to a minimum. An *event-driven* approach is used for hard-disk simulation. The evolution of the system is described by a series of events, each of which consists of a binary collision between a pair of disks. The collisions are instantaneous, and particles move independently between collisions. The time intervals between collisions are of course variable, as are the steps by which the system advances; the task of maintaining a time-ordered sequence of future collision events requires special consideration /4/. In this approach there are no numerical integration errors to cause concern, only the usual errors of precision inherent in the processor itself.

The major portion of any MD simulation is the evaluation of the forces on the individual particles. In an N-particle system each particle can potentially interact with N-1 others; given the continual rearrangement that occurs in the fluid state, there is no way of determining in advance which pairs of particles lie within interaction range. In principle, the interaction computation is an $0(N^2)$ process, and for long-range interactions that are treated in full detail (and not by resorting to so-called particle-in-cell type methods /7/) there is clearly no shortcut that avoids considering all pairs of particles. For interactions of restricted range, and, in particular, for interactions that extend only slightly beyond the actual core of the particles, the amount of computation needed for a complete force evaluation should only be of order N.

Two schemes, not necessarily mutually exclusive, exist for reducing the $0(N^2)$ computation to a manageable 0(N) undertaking. The first is the *cell-list* method: divide the volume (area in two dimensions) occupied by the system into a grid of cells, and construct lists of particles belonging to individual cells. The cell size is chosen so that only particles in immediately adjacent (or the same) cells can lie within interaction range; this in turn sets a fixed upper bound to the number of particles that are potential interaction partners of any given particle. Both the initial cell subdivision and the ensuing force computation involve amounts of work proportional to N. The second approach to reducing computational effort is based on the concept of a *neighbor-list*: all particles in a shell surrounding a given particle are recorded in a list associated with the particle in question, and only these are candidates for interaction. The shell size is chosen to exceed the interaction range by an amount sufficient to permit regeneration of the neighbor list at intervals of several time steps. Cell lists can of course be used as a preparatory step in order to reduce the work needed for generating the neighbor lists.

While the use of cell lists is an essential ingredient for the efficient MD simulation of systems involving more than a minimal number of particles, use of neighbor lists is incompatible with the event-driven approach, while in time-driven simulations the added complexity of neighbor lists must be traded against the gain in computational efficiency. Scalar processor implementations of time-driven MD may stand to gain from the use of neighbor lists in addition to cell lists, but because of the additional data manipulation required it is unlikely that MD on a vector processor would benefit, since the extra data handling needed to reduce the count of potentially interacting pairs would more than likely outweigh any gains resulting from the reduced number of particle pairs actually considered.

Vector and parallel processing -- the directions future computer architectures seem destined to follow -- introduce new kinds of problems. The event-driven approach does not seem a likely candidate to benefit in terms of performance improvement due to the fundamentally sequential nature of the event handling; special purpose processors of a very different kind (based on associative or content addressable memory) would be needed for improved event-driven MD. The performance of time-driven MD on the other hand can be enhanced by resorting to both vector and parallel architectures.

Adaptation of the basic MD algorithm to run in parallel on a set of coupled processors is conceptually a very simple modification. The total region occupied by the system is sliced into a number of subregions, and each is assigned to the care of a separate processor. As the system advances in time particles will move from one subregion to another, and this entails processors transferring responsibility for these particles among themselves. This creates a requirement for efficient interprocessor communication. The other reason that processors need to be in relatively frequent contact is to enable sharing of information about particles that are close to the subregion boundaries; in order for the interaction computations to proceed concurrently in all the processors, it is essential that each be aware of the coordinates of particles lying close to, but on the opposite sides of, the subregion boundaries. While other kinds of information transfer occur during the course of an MD simulation, the two uses of communication just referred to need to be carried out at each time step, and their efficiency is thus crucial to the effective utilization of the processor network.

Vectorization of the MD computation when only short-range forces are involved calls for a less than obvious solution. A key requirement for effective vector processing is that identical arithmetic operations can be carried out on comparatively long sequences of data (the vectors) as indivisible machine instructions. The original $0(N^2)$ approach to MD obviously is capable of producing such data sequences, but in reformulating the problem in terms of cell lists to reduce the complexity to $0(N)$, the vector lengths are also reduced to very small values because they correspond to the numbers of particles belonging to individual cells, and the lower the cell occupancy the smaller the numerical factor associated with the $0(N)$. The conflicting demands of long vector length and low cell occupancy are all too apparent.

The solution is not to abandon the cell list structure, but to rearrange the data in a manner which will again produce sufficiently long vectors to make effective use of a vector processor. The major characteristic of this rearrangement is the replacement of the linked lists that associate particles belonging to common cells by a series of *layers*; each layer is able to represent a single particle belonging to each of the cells into which the region has been subdivided, while the number of layers is equal to the maximum cell occupancy. The interaction calculations are then based on pairing the contents of layers rather than cells; since the sizes of the former are considerably larger than the latter, effective use of vector hardware results. Clearly not all the layers will be fully occupied, and in fact some layers will be only very sparsely occupied; a requirement of the vector hardware is that it have the capability for processing sparse vectors, a feature that seems to be standard to many vector processors.

3. Obstructed flow

The flow of a fluid past a long cylindrical obstacle in three dimensions, and the corresponding two-dimensional case involving flow past a circular obstacle, are characterized by a single dimensionless quantity known as the Reynolds number whose value is given by $Re = UL/\nu$, where U is the flow velocity, L the diameter of the obstacle, and ν the kinematic viscosity. At very low Re the flow is Stokesian in nature, whereas at very high Re the flow is turbulent. Over a range of intermendiate Re values different kinds of steady and time-dependent flows are observed to occur /2,8/, beginning with a pair of counter-rotating eddies attached to the trailing edge of the obstacle that first appear at $Re \simeq 5$, and grow in size as Re increases; at $Re = 34$ an oscillatory wake develops, and somewhere between $Re = 55$ and 70 the von Kármán vortex street makes its initial appearance. The challenge is to reproduce these different flows, and to learn something of the microscopic origins of the instabilities that regulate the transition from one kind of flow to another, by using MD simply to follow the motions of the individual particles that constitute the fluid /9/.

The effect of the obstacle is reproduced by introducing a circular barrier from which the particles rebound elastically. This in fact corresponds to a slip boundary. In continuum hydrodynamics it is usual to assume that boundaries in contact with the fluid are of the non-slip type /10/; however, since vorticity is generated by the mere presence of the obstacle, the choice of slip *vs* non-slip may not be crucial to the outcome of the simulations. Incorporating a non-slip boundary into the microscopic simulation requires the construction of a model for surface roughness, which is a significant problem in its own right.

The boundaries of the region containing the fluid are periodic, and the flow is maintained by forcing particles into the system at the upstream end. The reentrant particles are actually the ones that have exited downstream, but the velocities of these particles are reset to the nominal flow speed on which is superimposed a random thermal velocity. In this manner three goals are accomplished: particle number is conserved, the flow is maintained, and any flow patterns that do appear are not reintroduced with the recycled fluid particles. In early work on this problem an external gravitational field was also used to help maintain the flow, although this was later found to be unnecessary. The computations were carried out on a set of four coupled array processors communicating by means of a shared memory /11/.

Typical flow patterns for an MD run of 120,000 steps for a system of 160,000 particles are shown in Figs 1-4. The actual obstacle diameter is 250 Å, while the total duration of the simulation is 1.1 nsec. Examples of the eddy pair that develops during the initial phase of the run and the time-dependent oscillatory wake that characterizes the state into which the system eventually evolves are shown in Fig 1. The arrows show locally averaged flows obtained by partitioning the region into a 60×60 array of cells and computing the mean velocity of the particles in each cell, averaged over a time interval of 50 psec; arrow length varies linearly with local velocity, but the minimum length is set to half the maximum for convenience in observing the flow direction in regions of slower flow (the reverse flow rate in the eddies is only about 10% of the downstream flow). Fig 2 shows a sequence of flow patterns (directions only, averaged over 20 psec, and restricted to a portion of the region just downstream of the obstacle) covering the initial eddy development, while Fig 3 shows the oscillatory wake travelling down the system at a later stage in the simulation. That this wake corresponds to the von Kármán vortex street is made apparent in Fig 4, which shows the same set of patterns, but in a reference frame moving at the speed of the wake itself. The recirculatory regions are the vortices, and these maintain their structure as they travel downstream.

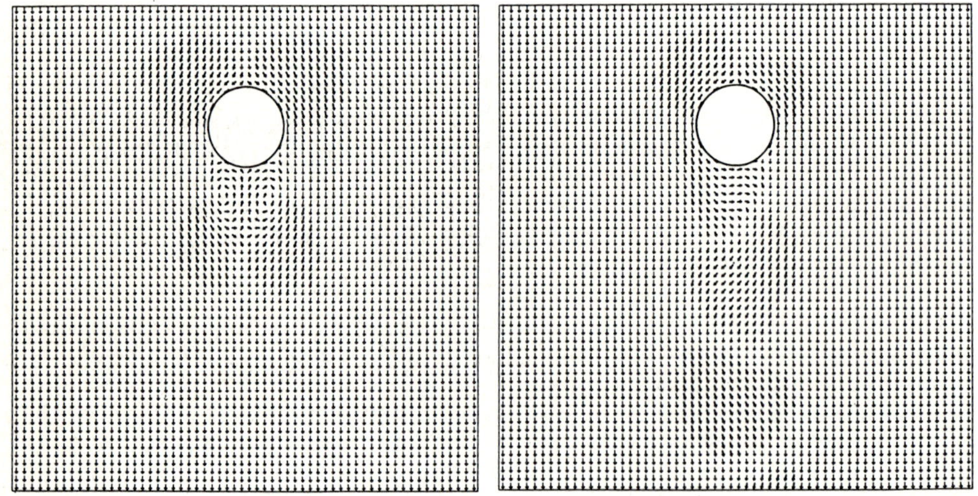

Fig 1. Eddy formation and wake oscillation in obstructed flow

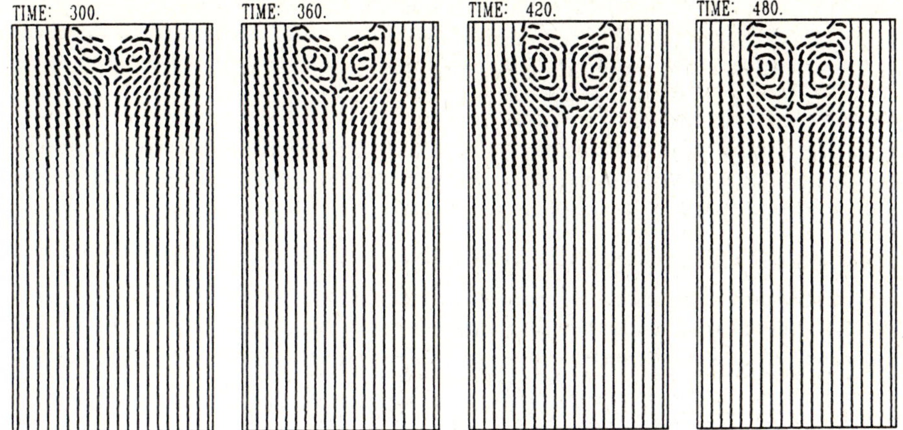

Fig 2. Sequence of snapshots showing growth of eddy pair

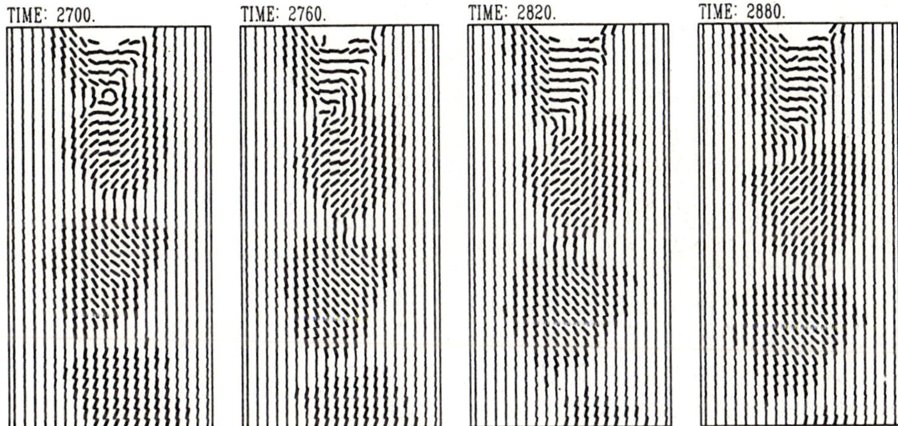

Fig 3. Snapshots showing propagation of oscillatory wake

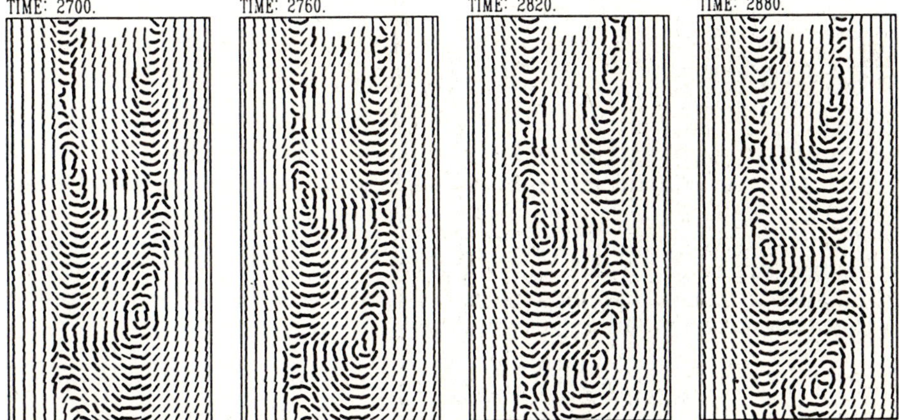

Fig 4. The vortex street of Fig 3 as seen by a moving observer

Only a very rough estimate can be made of the Reynolds number that is applicable to the conditions of this run. The reason for this is that the kinematic viscosity ν is a function of quantities that can vary considerably across the flow field -- density, temperature, shear rate -- and, unlike the case of relatively uniform macroscopic flows, it is far from obvious at what point in the flow field one ought to estimate this quantity, or alternatively, how an appropriate mean value should be computed. An approximate value of Re for the run is 25, although the flow patterns observed are more typical of a Re value above 100.

One quantity which can be estimated directly from the observed flow patterns, and which does have a macroscopic analog, is the Strouhal number $Sr = fL/U$, where f is the frequency of wake oscillation (or vortex shedding). The value measured from a series of flow patterns such as those of Fig 3 is $Sr = 0.2$; the experimental value is 0.15 at low Re, gradually increasing to 0.2 at $Re = 300$ /12/. Given the many orders of magnitude that separate MD length and time scales from those of real fluids, this agreement is satisfying, and also serves as an indication that using MD to model fluid flow is a meaningful enterprise.

A final remark addresses the need for large systems, particularly since this work has involved the largest MD computations carried out to date. A series of runs using increasing obstacle size made it apparent that a certain minimum size was needed to obtain the well-formed eddies. Sufficient space is required at the sides of the obstacle to avoid the creation of excessive velocity gradients as the fluid is forced through the narrowed region, and room downstream is needed for flow pattern development. Higher Re values cannot be achieved by increasing the flow velocity significantly, both because of a significant density drop that appears in the lee of the obstacle, and because of a desire to avoid entering the supersonic regime. Reducing the overall density of the system to produce a larger region, while keeping the same number of particles, tends to destroy the very effects one is attempting to observe, so this too is not a solution. Thus it would appear that the minimal size of systems for this kind of simulation is of the order of 100,000 particles. The quest for a stationary pair of eddies -- in the simulations so far eddy formation is only a transient phenomenon -- may well require even larger systems; recent work on systems containing 200,000 particles has still not been able to produce stabilized eddies /13/.

4. Thermal convection

The Rayleigh-Bénard phenomenon is an example of the spontaneous formation of dynamical order -- structured flow patterns -- in a dissipative system. The experiment involves a layer of fluid between two horizontal plates that is heated from below; under certain conditions a set of periodic convective rolls emerge /3/. The actual roll pattern is influenced by the lateral boundaries of the container; the most familiar patterns are made up of either cylindrical or hexagonal rolls. Two-dimensional simulation is an appropriate means of studying a vertical cross section through the cylindrical roll pattern.

Event-driven rather than time-driven simulation was used in this MD study for two reasons. Since the computations were carried out on a scalar processor the simulations stood to benefit from the more efficient method, and the fact that a relatively large temperature gradient might be needed to initiate convection meant that the time-driven approach would need a smaller integration time step than is normally used in homogeneous systems, resulting in a further drop in efficiency. The system was therefore modeled by a fluid of hard disk particles confined to a rectangular region in which the lateral boundaries were periodic, and with the bottom and top walls acting as heat sources and sinks that produced the desired vertical temperature gradient. Opposing the tendency toward upward flow produced by this temperature difference was a downward gravitational field. The system was started in a random state with a vertical velocity gradient, and the flow patterns were obtained by the same kind of coarse graining procedure described earlier.

The dimensionless quantity that characterizes the state of the Rayleigh-Bénard problem is the Rayleigh number, $Ra = d^3 \Delta T g \alpha / \kappa \nu$, where d is the thickness of the fluid layer, ΔT the temperature difference, and α and κ the thermal expansion coefficient and thermal diffusivity. Stability analysis reveals a critical value of Ra above which convective heat transport dominates over the diffusive mode; for rigid non-slip top and bottom walls (as in experiment) the value is $Ra_c = 1708$, while for slip walls (as used in MD) the value is $Ra_c = 657$ /14/. As was the case for obstructed flow, there are problems in evaluating the transport coefficients under the inhomogeneous conditions of the simulation, and thus no estimate of Ra has been made; a series of simulations over a range of imposed conditions is therefore necessary in order to learn what, if anything, occurs. Furthermore, while the onset of the instability is comparatively abrupt in the real macroscopic world, the appearance of convection at the microscopic level may turn out to be a more gradual effect; the analogy is with second order phase transitions which, though mathematically singular for infinite systems, become smeared out when the system size is finite.

In simulations carried out so far using a system of 14,000 particles in a rectangular region whose length is four times the height, two kinds of transient behavior have been observed /15/. If the relative vertical temperature change is not too large (i.e., $\Delta T < 17$ in units in which the cold wall is at $T = 1$) the system evolves into a state containing four square convective rolls. For larger ΔT, the system has been observed to develop into a state containing six narrower rolls, and after spending time in this state, to undergo a further transition to the more familiar four-roll state. The development of the six-roll pattern may be complete or only partial -- this appears to be an unpredictable occurrence, depending on subtle details of the initial state.

Figure 5 shows examples of the convective flow patterns observed in the course of one of the runs; for this run $\Delta T = 17$, and the field strength g was chosen to balance potential and thermal energy changes for particles crossing the system. Spectral analysis of the flow field reveals that both the four- and six-roll modes seen in Fig 5 are devoid of harmonics. One point that should be noted is the relatively large value of ΔT; a typical experimental value is more like $\Delta T / T = 10^{-2}$. An even more highly exaggerated value is that of the convective flow velocity which, in the simulations, reaches about one third the magnitude of the thermal velocity; experimentally the ratio of the two velocities is typically 10^{-7}. Such extreme situations are not unexpected in the comparatively small systems that can be studied at the discrete particle level -- here the fluid layer is only 300 Å in height -- but, as was the case for obstructed flow, much (if not all) of the familiar qualitative behavior is reproduced.

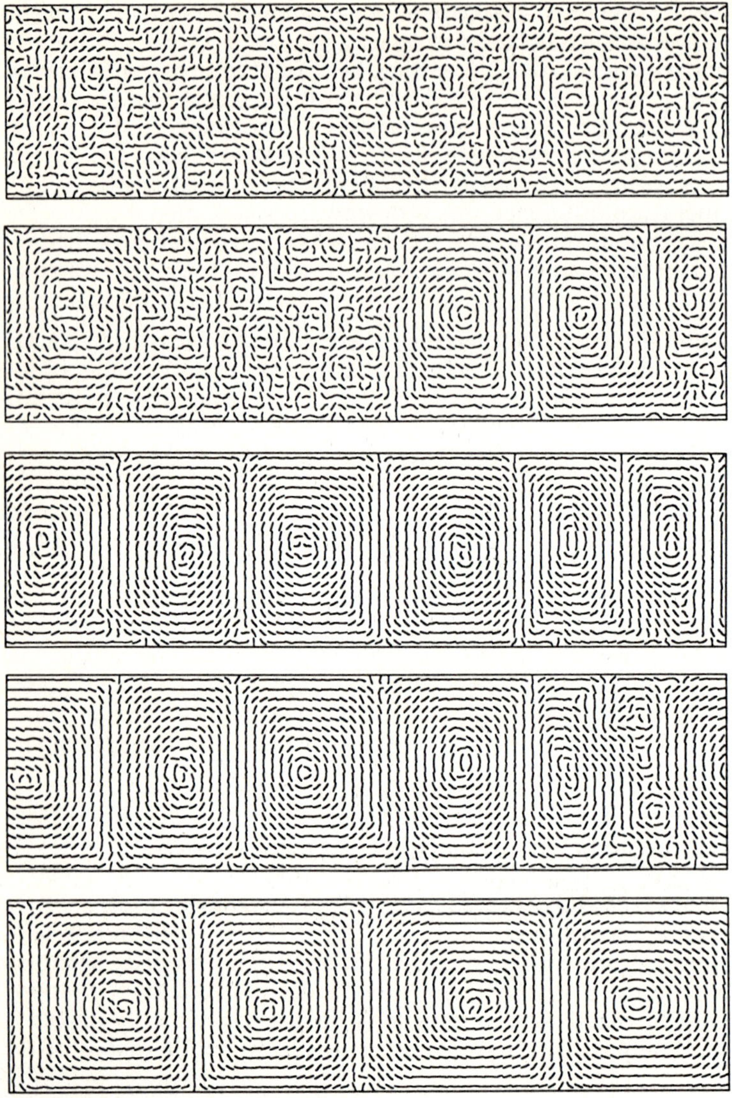

Fig 5. Development of six- and four-roll patterns in Rayleigh-Bénard convection (at times 100, 300, 1000, 1200, 2000)

5. Conclusions

Although the verdict is not yet in, the exploratory studies reported here certainly suggest that MD simulation may become a significant contributor to theoretical hydrodynamics. In addition to broadening the scope of nonequilibrium situations typified by those described in this lecture, there are many other flow problems, in both two and three dimensions, that could benefit from this approach. The significant improvements in performance -- both absolute and price-relative -- achieved by recently developed computers, and those which should become available in the foreseeable future, ought to make even more ambitious simulations (larger systems, longer time intervals, three dimensions, more complex constituents and interactions) into almost routine computations. It is entirely plausible that the molecular dynamics approach might eventually become a key tool in the numerical study of hydrodynamic phenomena.

Acknowledgement: Portions of this work were carried out during visits to the Center for Simulational Physics at the University of Georgia.

1. *Molecular Dynamic Simulation of Statistical Mechanical Systems*, Proceedings of the International School of Physics "Enrico Fermi", Course XCVII, Varenna, 1985, edited by G. Ciccotti and W.G. Hoover (North-Holland, Amsterdam, 1986).
2. G.K. Batchelor, *An Introduction to Fluid Dynamics* (Cambridge University, Cambridge, 1967).
3. P. Bergé and M. Dubois, Contemp. Phys. **25**, 535 (1984).
4. D.C. Rapaport, J. Comput. Phys. **34**, 184 (1980).
5. D.C. Rapaport, in preparation (1988).
6. H.J.C. Berendsen and W.F. van Gunsteren, in Ref. /1/, p. 43.
7. R.W. Hockney and J.W. Eastwood, *Computer Simulation Using Particles* (McGraw-Hill, New York, 1981).
8. M. van Dyke, *An Album of Fluid Motion* (Parabolic, Stanford, CA, 1982).
9. D.C. Rapaport, Phys. Rev. A **36**, 3288 (1987); D.C. Rapaport and E. Clementi, Phys. Rev. Lett. **57**, 695 (1986).
10. H. Schlichting, *Boundary-Layer Theory*, 6th ed. (McGraw-Hill, New York, 1968).
11. E. Clementi and D. Logan, IBM Kingston Report No. KGN-43, 1985 (unpublished).
12. J.H. Gerrard, Philos. Trans. R. Soc. London Ser. A **289**, 351 (1978).
13. D.C. Rapaport and D.P. Landau, work in progress (1988).
14. S. Chandrasekhar, *Hydrodynamic and Hydromagnetic Stability* (Oxford University, Oxford, 1961).
15. D.C. Rapaport, submitted (1988).

Molecular Dynamics Simulations in Material Science and Condensed Matter Physics

U. Landman *

School of Physics, Georgia Institute of Technology,
Atlanta, GA 30332, USA

1. Introduction

Basic understanding of the structure and dynamics of materials and their properties often requires knowledge on a microscopic level of the underlying energetics and interaction mechanisms, whose consequences we observe and measure. Answers to material science problems are in principle possible, embodied in solutions to the Schrodinger equation subject to the appropriate boundary conditions. However, the full implementation of such a program is impossible for most (one may dare say all) material science and condensed matter systems and we must resort to various approximations and simplification. The degree of microscopic detail with which we probe physical phenomena is determined mainly by the resolution of our experimental tools, by the ability to found the theoretical analysis on microscopic principles and by the complexity, hence solubility, of the model. In constructing theoretical models of complex material systems we must apply rigorous standards of testing and assessment as to the validity and accuracy of the proposed model. Frequently we rely at this stage on the accessibility and availability of experimental data with sufficient resolution (either as a source of values for some of the model parameters or for comparative purposes) and by the level of simplifications, dictated by considerations of feasibility and practicality, which render the model soluble. In many situations the level of complexity of the model, which is necessary in order to describe faithfully the physical phenomena, is such that analytical approaches fail to provide a solution. In these situations, which include the majority of material systems and phenomena, the use of computer-based methods [1-6] is essential.

Computer simulations [1-6] where the evolution of a physical system is simulated, with refined temporal and spatial resolution, via a direct numerical solution to the model equations are in a sense computer experiments which open new avenues in investigations of the microscopic origins of material phenomena. These methods alleviate certain of the major difficulties which hamper other theoretical approaches, particularly for complex systems such as those characterized by a large number of degrees of freedom, lack of symmetry, non-linearities and complicated interactions. In addition to comparisons with experimental data, computer simulations can be used as a source of physical information which is not accessible to laboratory experiments, and in some instances the computer experiment itself serves as a testing ground for theoretical concepts. The symbiotic relationship between the development of large-scale computer based methods for the solution of physical problems and the significant progress in the area of computers opens an era where computers are transformed from merely tools for solving theoretical models to become an integral part of the formulation of models and means of making scientific discoveries.

In this short review (and in a later one) we focus on computer simulations which employ realistic interaction potentials and other system conditions. Our aim is to demonstrate the level of reality which can be achieved via computer simulations of materials systems and the wealth of information which they provide about the fundamental mechanisms and processes which govern the properties and response of material systems. The examples which we describe are drawn from studies performed in our group and are by no means an exhaustive sample of the large body of literature on this subject (in fact it is a testimony to the rapid progress in this area that no single review covers exhaustively all that has been achieved, and that the rate at

Springer Proceedings in Physics Vol. 33: **Computer Simulation Studies in Condensed Matter Physics**
Editors: D. P. Landau · K. K. Mon · H-B. Schüttler

which summary reviews become outdated is rather fast, see references [1-6]). In this review we limit ourself to classical molecular dynamics simulations while in the later one [7] simulations of systems which contain coupled quantum and classical degrees of freedom are described. Throughout we focus on sample results rather than on detailed descriptions of the theoretical model and simulation methods (for comprehensive descriptions of these aspects references are provided).

2. Classical Molecular Dynamics Simulations [6-10]

(a) Method

The classical molecular-dynamics (MD) method [2,4,5] consists of a numerical generation of the phase-space trajectories for a system of N particles, interacting via a potential function $V(\vec{r}_1,\ldots,\vec{r}_N)$, where $\vec{r}_i$ is the coordinate of particle i. In the case of an extended system periodic boundary conditions (pbc's) are imposed. In addition, formulations have been developed which allow simulations of various equilibrium ensembles, such as constant volume microcanonical (E,Ω,N), constant pressure (HPN), or constant temperature macrocanonical (T,Ω,N) ensembles.

The starting point of a MD simulation is a well-defined microscopic description of the physical system, in terms of a Hamiltonian or a Lagrangian from which the equations of motions are derived. Thus, given a Hamiltonian

$$\mathcal{H} = \frac{1}{2}\sum_i \vec{p}_i \bullet \vec{p}_i/m_i + V(\vec{r}_1,\ldots,\vec{r}_N) \quad , \tag{1}$$

the equations of motion are

$$\frac{d\vec{r}_i}{dt} = \frac{\partial\mathcal{H}}{\partial\vec{P}_i} - \frac{\vec{p}_i}{m} \tag{2}$$

$$\frac{d\vec{p}_i}{dt} = -\frac{\partial\mathcal{H}}{\partial\vec{r}_i} = -\frac{\partial V}{\partial\vec{r}_i} \,, \quad (i = 1,\ldots,N) \,. \tag{3}$$

Numerical integration of the equations of motions allows investigations of the dynamics of equilibrium as well as non-equilibrium properties.

In equilibrium studies the properties of the system are calculated as averages along the dynamical trajectory of the system. In studies of non-equilibrium phenomena (such as crystal growth, melting and structural transformations, the transient response of a system to an impulse and steady state flow) the phase-space information obtained via the numerical solution of the dynamical equations of motion can be used to investigate the time evolution of the system, on refined temporal and spatial scales.

(b) Interaction Potentials

The potential energy of a system of N atoms is a function of the atomic coordinates and may also contain a contribution from a term which does not depend explicitly on the atomic coordinates but on the density (such as in metallic systems). Thus in general the potential energy may be expanded in a series of n-body interatomic potentials

$$V = \phi_o + \Sigma\phi_1(i) + \Sigma\phi_2(ij) + \Sigma\phi_3(ijk) + \ldots + \phi_N(ijkl,\ldots), \tag{4}$$

where ϕ_n is the n-body interaction potential which is a function of the positions of n atoms ijk...... . The sums in Eq. (4) are over all combinations (excluding redundant contributions) of n atoms in the system.

The potentials, ϕ_n, and the number of terms which are retained in a practical application depend on the nature of the system under investigation. Thus, for example, rare gases and ionic systems can be adequately described in terms of pair potentials [8], which depend only upon the distance between interacting pairs (such potentials include the 6-12 Lennard-Jones potentials for rare gases and potentials which include electrostatic interactions and Born-Meyer repulsive terms for ionic systems [9]).

For systems characterized by covalent bonding (such as semiconductors) one must go beyond the pair interactions due to the directionality of the localization of charge in the bonding regions. Potentials which include 3-body terms have been used for some time [10] for the description of equilibrium structures of crystalline semiconductors and recently generalizations for arbitrary atomic locations have been developed. In our work on silicon we have employed the potential developed by Stillinger and Weber (SW) [11] which contains 2- and 3-body interactions. The parameters of the potential have been determined [11] via empirical fits to properties of silicon in the crystalline and liquid phases.

For metallic systems it is well known [12] that the cohesive energy contains density (volume) dependent contributions. Clearly, due to the explicit and implicit dependence on the density of the system, simulations of metal system phenomena which may involve changes in volume or structural transformations require a method in which the volume of the system varies dynamically [13,14].

Alternatively, based on the philosophy of density-functional theory, a description of metallic systems which is amenable to molecular dynamics simulations is provided by effective medium theory (EMT) [15] or the related embedded atom method (EAM) [16]. The basic feature of the EMT is that the effect of the surroundings on each atom in the system can be described in terms of the average electron density which other atoms in the system provide around the atom in question. The electronic structure problem is then converted to that of the embedment of an atom in a homogeneous electron gas, which can be described in terms of a universal density dependent energy function. Thus the density dependent term gives rise to many-body interactions. Invoking the adiabatic (Born-Oppenheimer) approximation we have used the EMT method in MD simulations of premelting phenomena in aluminum described below [17].

3. Sample Results

Solid-Melt Interfaces of Silicon and Crystal Growth

Understanding the equilibrium properties of interfaces and of the non-equilibrium mechanisms, kinetics, and dynamics of crystal-growth processes and identification of the microscopic material properties and of the macroscopic control parameters related to the method of growth, which govern the growth processes and quality of the grown crystals, are of fundamental importance from both basic and applied perspectives.

It has been suggested that the structure of the solid-melt interface during zone-melting recrystallization of silicon critically determines the generation of the observed network of low-angle grain boundaries [18]. In addition ample evidence exists that surfaces of crystalline silicon become faceted upon melting and the solid-melt interface of growing Si establishes itself on the (111) crystal plane [18]. Motivated by these observations comprehensive MD studies of the solid-melt interfaces of Si [19,20] and simulations of liquid-phase epitaxial growth of Si [21] have been performed, employing the SW potentials. It was discovered that the nature and morphology of the interphase-interface exhibits crystalline face anisotropy

[19,20]. Thus while the Si(111) equilibrium solid-melt interface was found to be atomically smooth, the Si(100) interface is structured exhibiting facets established along (111) directions [19], in qualitative agreement with experiment [19] (see Fig. 1). Furthermore, it was found that faceting initiates upon melting and further refines upon achieving equilibrium at a temperature of 1665K (compared to the experimental value of 1683K).

Liquid-phase epitaxial growth [21] was initiated by driving the crystal-melt systems out of equilibrium via allowing heat conduction to the underlying substrate. It was found that for both interfaces the system initially supercools by about 150K (in agreement with some experimental estimates), followed by ordering and crystallization processes. On the (001) surface the growth processes yielded a perfect crystal at a growth velocity of $\sim$ 18m/sec while growth on the (111) surface, driven by the same rate of heat removal, resulted in an imperfect crystal containing an assortment of defects (stacking faults and disordered region), at a growth velocity of $\sim$14m/sec. A slower growth rate of the Si(111) system results in a grown crystal possessing a markedly higher degree of crystalline perfection (stacking faults and a very narrow region of disorder limited to the crystal-vacuum interface). The crystallization rates and their orientational dependences and the crystal face dependence of the maximum growth velocities for the formation of defect-free crystals found in the simulations are in agreement with those deduced from experimental measurements [21]. Furthermore the growth mode found in the simulations emphasizes the dynamical nature of the two interacting phases (solid and melt), the kinetics of ordering processes in the melt in the vicinity of the moving interface [21,22] and the interface morphology. While the (111) microfaceting remains during growth on Si(100), which proceeds in an almost continuous, monotonous manner, growth on Si(111) proceeds in a layer-wise manner, with periods of crystallization interrupted by stages of no apparent movement of the interface. Close inspection reveals that during these stages dynamical self-annealing of defects occurs. As the defective region achieves a high level of structural perfection crystallization resumes [21].

Amorphous Film Growth

Thin film fabrication via direct condensation of vapor atoms onto a substrate at temperatures below the melting point has been used for over 100 years [23,24]. More recently low-energy beam deposition (generally below keV) methods have emerged as important techniques for growth of electronic materials of specified properties [25]. Full atomistic simulations of these processes [26-28] are difficult due to system size and time-span considerations.

Illustrations of results obtained via MD simulations [28] of the growth of amorphous thin films of silicon, using the SW potentials, are shown in Figs. 1 and 2. In previous studies the structural and dynamical and thermodynamical properties of amorphous silicone prepared via cooling of bulk molten silicon have been investigated [29] yielding results in agreement with experimental data. In the current study we focus on growth of amorphous silicon via low energy beam deposition. In Fig. 1 results for deposition of silicon atoms onto a cold (room temperature) substrate are shown. In these simulations the (111) crystalline substrate contained several layers with 49 dynamic atoms per layer (with periodic boundary conditions in the XY directions) and was cooled at the bottom. The directions in the plane of the calculational cell are $[1\bar{1}0]$ and $[10\bar{1}]$ and the outward normal is in the [111] direction (distance is in units of σ = 2.0951 Å). The atomic beam is normal to the surface and atoms are released at random positions from a planar source above the substrate at a rate of 500Δt (Δt = 1.15x10^{-3} psec) with a translational kinetic energy corresponding to $3k_BT_M/2$, where T_M is the melting point of silicon. Real space trajectories (viewed along the $[1\bar{1}0]$ direction) at a late stage of the deposition process are shown in Fig. 1a. In Fig. 1b the temperature gradient in the system at this stage (plotted versus distance along the (111) direction) is shown. The density and 4-fold coordination, C_4 and 3-body potential energy, V_3, profiles shown in Figs. 1c-1e,

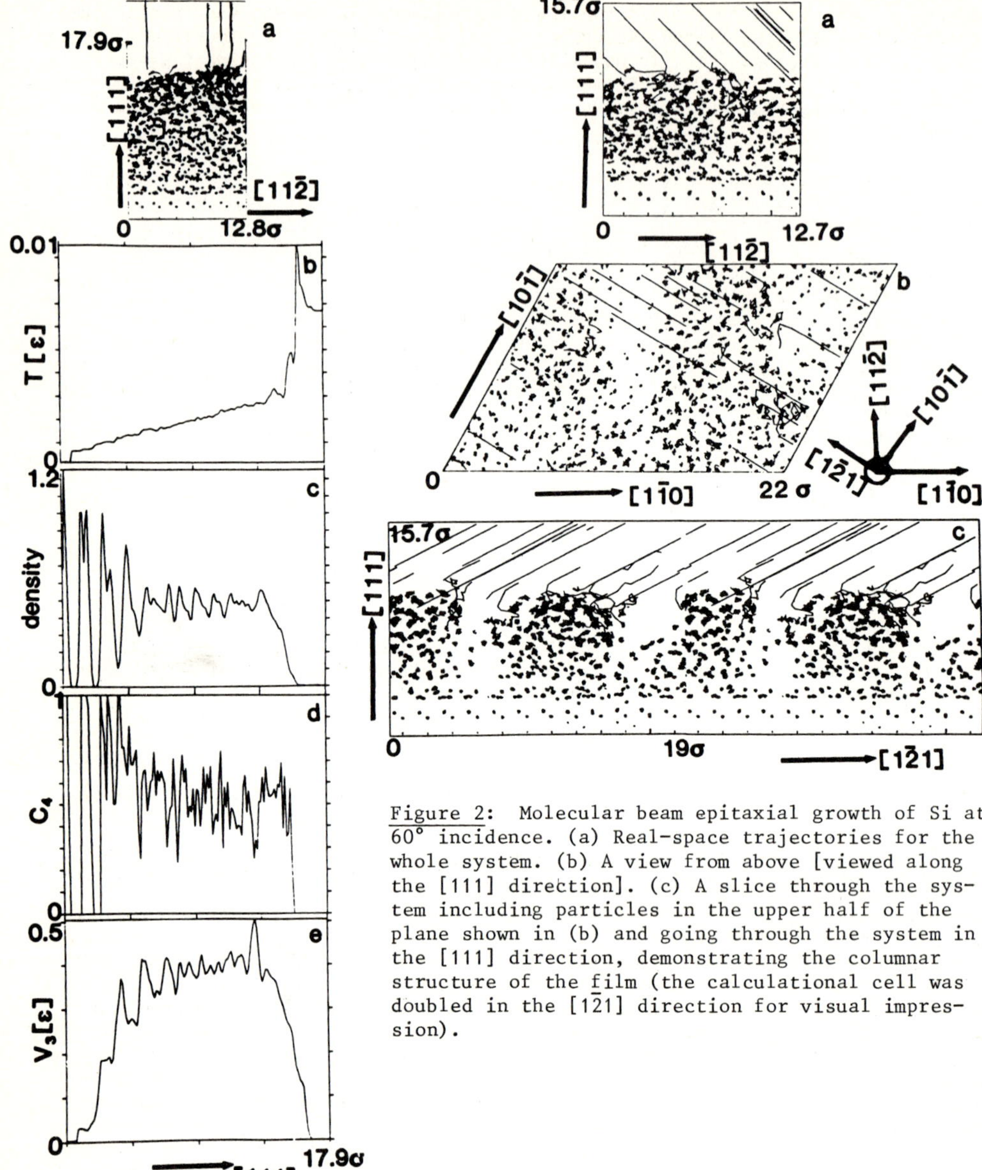

Figure 2: Molecular beam epitaxial growth of Si at 60° incidence. (a) Real-space trajectories for the whole system. (b) A view from above [viewed along the [111] direction]. (c) A slice through the system including particles in the upper half of the plane shown in (b) and going through the system in the [111] direction, demonstrating the columnar structure of the film (the calculational cell was doubled in the [1$\bar{2}$1] direction for visual impression).

Figure 1: Molecular beam epitaxial growth of Si at normal incidence. (a) Real-space particle trajectories (the bottommost 3 layers belong to the crystalline (111) Si substrate). (b) Temperature profile in the system. (c-e) Density, 4-fold coordination (C_4) and 3-body potential energy, V_3, profiles demonstrating characteristics of amorphous Si, and a layered region at the substrate-amorphous-film interface.

respectively, illustrate that in the interface between the substrate crystal (the first 3 layers from the left) and the grown film, the amorphous material is layered and possesses properties intermediate between the crystal and amorphous phase. The effect of the angle of deposition on the characteristics of the grown film are shown in Fig. 2 where results are displayed for deposition at the same rate as that shown in Fig. 1 but with an angle of incidence of 60° from the normal (with the plane of incidence defined by the [111] and [$1\bar{2}1$] directions). In these simulations 96 dynamic atoms per substrate layer were used. A view of the system from the top (along the [111] direction) shown in Fig. 2b reveals regions of smaller density. The columnar microstructure of the grown film is evident from Fig. 2c where a slice through the system containing the upper part of the plane (Fig. 2b) is shown. (Note that the calculational cell was doubled for visual impression in the [$1\bar{2}1$] direction. The periodicity of the columns is simply a consequence of the periodic boundary conditions). Analysis shows that the orientation of the columns and that of the incident beam obey approximately the tangent rule [24] (i.e., $2 \tan\theta_{column} = \tan\theta_{beam}$). The relevance and importance of these studies relates to the remarkable consequences of the microstructure for the film chemical and physical characteristics such as oxygen uptake, reactivity, optical, electrical and mechanical properties [24].

Finite Aggregates

Finite systems, beyond the very small end of the size spectrum, present an immense theoretical challenge since the number of particles in these systems renders the use of molecular science techniques rather cumbersome (or impractical) while their finiteness prohibits the employment of condensed matter methodology based on translational symmetry and complicates, due to size defects, the adaptation of the analytical framework and techniques of statistical mechanics which are formulated on the premise of the thermodynamic limit. Computer simulations alleviate these difficulties opening new avenues for studies of finite aggregates [4].

Employing ionic interaction potentials (see Section 2) we have studied the structure and dynamics of finite alkali-halide clusters [$(NaCl)_n$ for n = 4, 16 and 108] as a function of temperature [30]. Our simulations show that the very nature of the phase transformation and the underlying physical processes involved depend on the size of the system. At the lower end of the size spectrum, the phase space of the system is characterized by a small number of stable configurations (solid isomers) between which the system transforms in a diffusionless manner, with temperature-dependent rates and branching rations. As the system size is increased the number of accessible conformers increases, leading to a hierarchial kinetics of isomerization events which exhibits itself as a broadening of the transition region. For these small clusters coexistence is between solid isomers rather than inter-phase (solid-liquid) coexistence. The latter develops for clusters of sufficient size, characterized by a dense spectrum of accessible states, separated by thermally surmountable barriers. Under these circumstances conventional melting is observed as a sharp transition, the separation of time scales for inter-well and intra-well dynamics ceases, and true solid-liquid coexistence is found.

In this section we illustrate the application of MD simulations to studies of the nature of solid-melt coexistence in a finite system. Results for a silicon cluster containing 1024 atoms and described by the SW potential are shown in Fig. 3.

These results were obtained from long time averages of the system at the melting point of the cluster and show coexistence between a crystalline nucleus, exposed (111) faces, and the melt. As seen from the real-space particle trajectories, the solid nucleus is surrounded by a mantel of melt and exposed (111) crystalline planes (see Fig. 3b). We may ask under what circumstances we should expect such a structure, as opposed to a situation where the solid and melt phases would coexist in

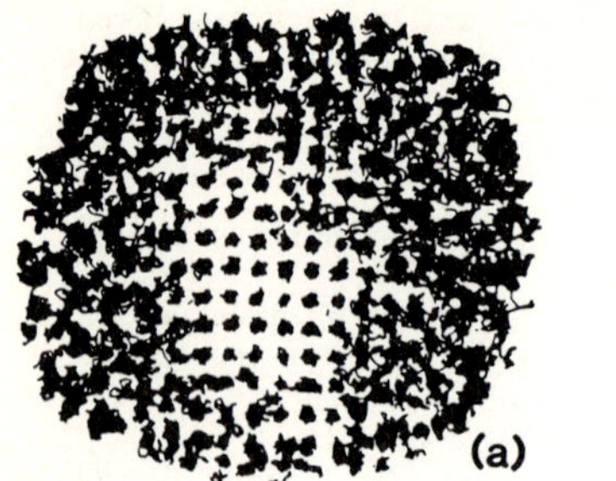

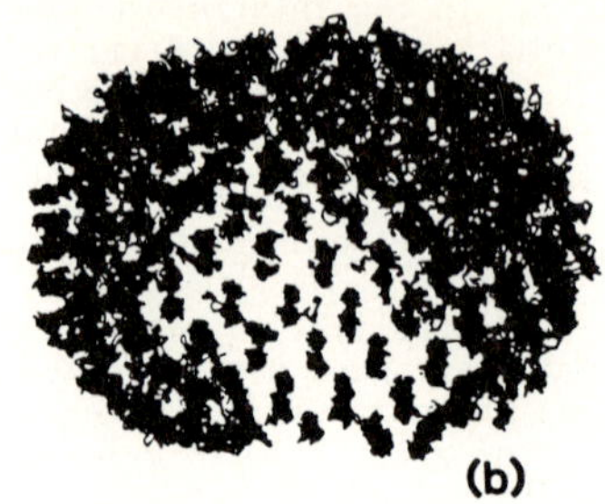

Figure 3: Real space particle trajectories of slices taken at different orientations of a silicon cluster containing 1024 atoms, at the melting point, exhibiting a solid nucleus exposing (111) planes (b).

two adjacent regions connected by a planar interface. Consider the two spherical configurations S and S' shown in Fig. 4, assuming for simplicity that the number of particles in the solid and liquid phase are the same for both and that the solid and liquid densities do not differ appreciably (which implies that the volume fractions of the two phases in S and S' are the same). The difference in free energies for the two systems, $\Delta F_{SS'} \equiv F_{S'} - F_S$, can be written as [31]

$$\Delta F_{SS'} = \pi R^2 \left(1 - \frac{X}{R}\right) [\Delta\sigma \, f\left(\frac{X}{R}\right) + (\sigma_s - \sigma_\ell)(f\left(\frac{X}{R}\right) - 2)] \tag{5}$$

where $\Delta\sigma = \sigma_{s\ell} - (\sigma_s - \sigma_\ell)$ and

$$f\left(\frac{X}{R}\right) = 1 + \frac{X}{R} - \left[2\left(2 + \frac{X}{R}\right)\left(1 - \frac{X}{R}\right)^{1/2}\right]^{2/3} \quad . \tag{6}$$

Note also that the ratio X/R is related to the volume fraction of the solid phase, V_s to the total volume V as $V_s/V = (1 - X/R)^2 \, (2 + X/R)/4$. When X = R the system is all liquid and when X = -R it is all solid.

The criterion for which of the situations S or S' occurs depends on the sign of the expression in square brackets in Eq. (5). This expression is positive (i.e.,

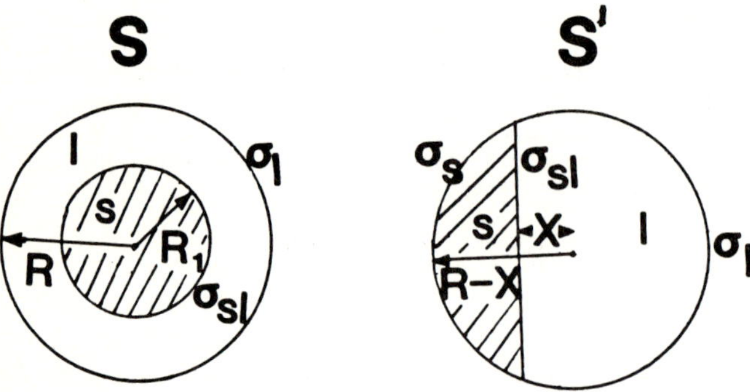

Figure 4: Two alternative configurations S and S' for a cluster at solid-melt coexistence. The solid region (s) is dashed. σ_ℓ, σ_s and $\sigma_{s\ell}$ are the surface tensions for the liquid, solid and liquid-solid interface, respectively.

$F_S < F_{S'}$) when

$$\Delta\sigma > \left[\frac{2-f}{f}\right](\sigma_s - \sigma_\ell) \quad . \tag{7}$$

Since the factor (2-f)/f < 0 as long as the volume fraction of the liquid $(V-V_s)/V <$ 0.95 the condition for relative stability of the S versus the S' configuration, for most situations, depends on the wetting parameter $\Delta\sigma$ (which is negative if the material is self-wetting) and on $\sigma_s-\sigma_\ell$. For Si the values of the interfacial tensions are such that indeed the S configuration of a solid surrounded by the liquid (see Fig. 3) is the stable one. For other materials, such as sodium chloride for example, the reverse is found [31].

Tip-Substrate Interaction in Atomic Force Microscopy

The developments of scanning tunneling microscopy [32] (STM) and of the related atomic force microscopy (AFM) [33] revolutionized our perspectives and abilities to probe the morphological and electronic structure and the nature of interatomic forces in materials, as well as opened new avenues [34] for microscopic investigations and manipulations of technological systems and phenomena such as tribology, litho- graphy and in biochemical applications. In both techniques a sharp tip is brought close to the surface and either the tunneling current (STM) or deflection of the cantilever holding the tip (AFM) are monitored as the surface is scanned.

Of particular interest for the development of these techniques are questions related to the dynamical response of the substrate (and tip) which may result in temporary or permanent modifications of the local properties.

In our simulations [35] the system consists of 4 layers of dynamic Si particles (interacting via SW potentials) with 49 atoms per layer, exposing the (111) surface, and interacting with 2 layers of a static Si substrate of the same structure (calculations with systems of 6 dynamic layers and 81 particles/layer yield very similar results). In accordance with the conventional picture [32-34] the tip is simulated by 4 Si atoms in an initial tetrahedral configuration interacting with 2 layers of a top Si static holder (other tip configurations have also been studied [35]). The 2D calculational cell, defined by the $(1\bar{1}0)$ and $(10\bar{1})$ vectors, is periodically repeated parallel to the (111) plane. The kinetic temperature is controlled via scaling of particle velocities in the bottom layer of the dynamic substrate. Following equilibration at room temperature, with the tip outside the range of interaction, the tip is lowered slowly to a prescribed height. Studies at 3 initial tip heights, h_i (i = 1,3), corresponding to distances of 2.91 Å, 2.345 Å and 1.227 Å between the lowest tip atom and the uppermost layer of the substrate, were performed. h_1, h_2 and h_3 correspond to the attractive, equilibrium, and repulsive regions of the interparticle potential, respectively. To faithfully simulate the laboratory process for structurally (and thus energetically) relaxed tip-substrate configurations care was taken to properly equilibrate the system throughout the scan sequences. Throughout we use σ = 2.0951 Å as the unit of length, and 1.65728 x 10^{-9} N as the unit of force. The reduced units of length X^*, Y^* and Z^*, along the $(1\bar{1}0)$, $(10\bar{1})$ and (111) directions are 12.82σ, 12.82σ, and 5.8333σ, respectively.

In Figure 5 we show real-space particle trajectories viewed from the side along the $(10\bar{1})$ direction (Fig. 5a-c) and from the top (Fig. 5d) for a scan at the low height h_3, with the lowermost tip atom scanning on top of an atomic row of the substrate. The trajectories shown in Fig. 5a and 5d are for a complete scan and

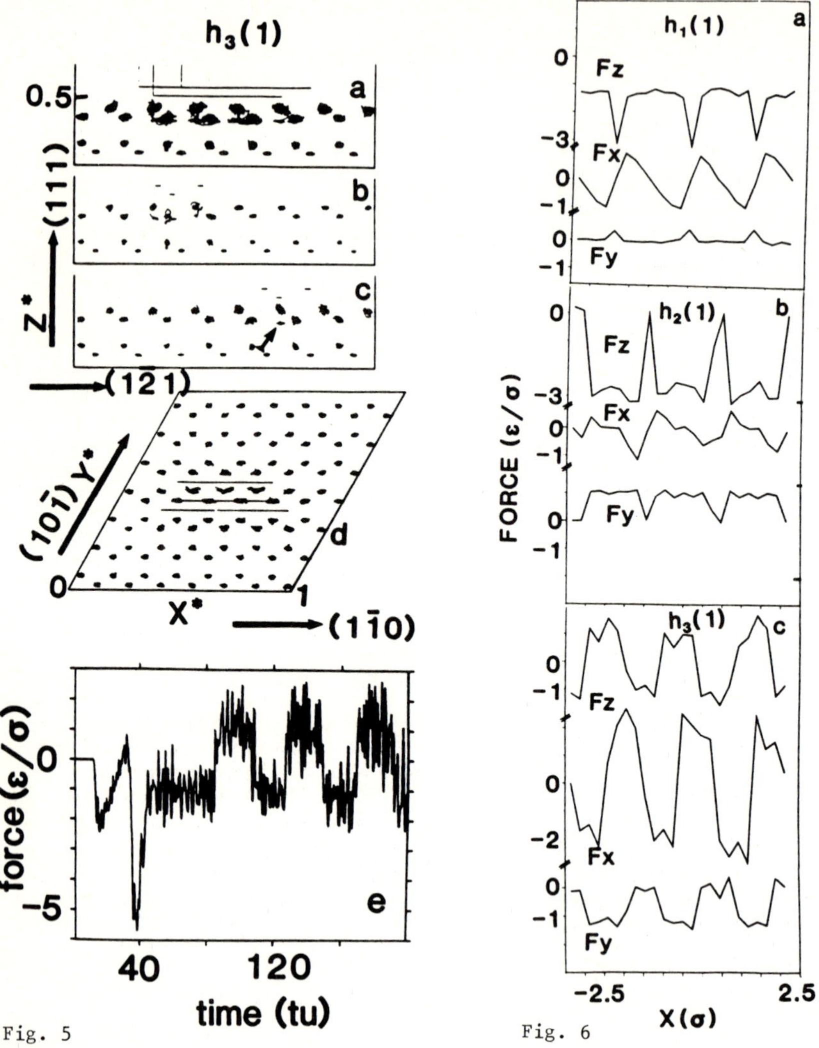

Figure 5: Real space particle trajectories for tip scanning of a Si(111) surface viewed from the side along the [10$\bar{1}$] direction (a-c) and from the top (d) for a $h_3(1)$ scan. (a-d) Complete scan trajectories over 3 surface unit cells. The tip-particles' trajectories are the horizontal lines above the substrate. (c,d) Trajectories at the beginning and end stages of the scan simulations demonstrating transient localized surface defect formation. (e) Total F_z force on the tip atoms versus time, demonstrating the relaxation of the surface ($t \lesssim 40$ tu) followed by periodic oscillations which portray the surface atomic structure. tu = 7.66×10^{-14}s and the integration time step $\Delta t = 0.015$ tu.

Figure 6: Recorded forces on the tip, scanning in the x direction of a Si(111) surface, for 3 tip heights above the surface $h_i(1)$, i = 1,2,3 corresponding to 2.91 Å, 2.345 Å and 1.227 Å, respectively.

those in Fig. 5b and 5c at the beginning and end of the scan (only 3 tip particles out of the 4 are visible from this viewing direction). In this calculation a hard but reactive tip is treated (i.e., static tip atoms) which is the situation encountered in most experiments. As seen in Fig. 5b the interaction with the tip triggers local displacements of substrate particles. In particular, the atom in the top layer right below the tip drops to an interstitial position in the second layer. This is seen clearly in Fig. 5c, where the interstitial atom is marked by an arrow. Note that once the tip advances, the localized defect induced by the tip anneals (compare the left part of the scan in Figs. 5b and 5c). In comparison, locating the tip at $h_1(1)$ (i.e., 2.91 Å), in the region of attractive interaction, results in a localized outward displacement of atoms in the substrate top layer.

A record of the total substrate force on the tip versus time in the direction normal to the surface F_z (with positive values corresponding to outward direction) is shown in Fig. 5e, starting with the tip outside the range of interaction at time t = 0. As the tip is lowered it is initially attracted to the surface, followed by a decrease in the force as the tip enters the repulsive region, and subsequently experiencing a repulsive interaction ($F_z > 0$ at ~ 30 tu). This is then followed by a sharp attraction (~ 40 tu), culminating in a resultant attractive force ($F_z \sim -1$), although the tip at this point is in intimate contact with the surface, which would have been expected to yield a strong repulsion. The dynamical mechanism underlying the observed attraction is the generation of surface defects (see Fig. 5a-d). Thus, the interaction with the tip can induce significant local rearrangement in the substrate and consequently can alter (even reverse the sign) of the resultant recorded forces. Further lateral scanning over substrate atoms is signified by periodic oscillations in the force (see Fig. 5e).

The dependence of the character of the recorded forces vs. position on tip height is illustrated in Figs. 6a-c for scan 1. These records were obtained after the initial equilibration, with the tip lowered to the desired height. Significant variations in the magnitudes and character of the forces are observed depending upon the tip separation from the surface. For example, the F_z component at $h_1(1)$ is overall attractive with sharp negative spikes when the tip is scanning between surface atoms. For the low-tip configuration, Fig. 6c, passage between surface atoms is signified by broad repulsive peaks. It should be emphasized that the recorded forces reflect the dynamical relaxations which result from the tip-substrate interactions. Since these processes result in modifications of the substrate surface structure neglecting to include them in the analysis would have resulted in erroneous structural interpretation of the data. Furthermore, for other scan geometries, such as for a scan between atomic rows, we observed that the tip may cause permanent damage (particularly at the low tip height, h_3) via attachment of substrate particles to the tip. Simulations for various tip geometries, tip sizes and substrate conditions and including tip-atoms dynamics reveal a gallery of tip-substrate induced dynamical processes whose consequences are reflected in the recorded data.

Metallic Systems

As discussed in the previous section simulations of metallic systems require special consideration of the nature of interactions and cohesion in these systems. These considerations led to the development of a new method of simulations [13] which allows for volume and shape variations and incorporates explicitly the dependence on the density of the "volume energy" and the effective pair potentials. Application of this method to studies of a liquid metal (Mg) yielded good agreement with experiments for several properties, including internal energy, density, structure factors, the adiabatic bulk modulus and trends in the diffusion, electrical resistivity and thermopower at several temperatures and pressures [13]. In addition, simulations of the formation and properties of a metallic glass ($Ca_{67}Mg_{33}$) [14], i.e., the solid phase formed when segregation and crystallization are avoided by ultra rapid cooling of the liquid alloy, yielded results [14] in good agreement with neutron scattering data for the static and dynamic structure factors, thus providing structural information and revealing the nature of elementary excitations in disordered media. Furthermore, it was found that the accessible configurational energy of the glass possesses several nearly degenerate potential minima and that transitions between these accessible local potential minima involve a large motion of single atoms accompanied by local structural relaxation of the environment [14,1].

In this section we discuss the dynamics and energetics of disordering and melting of aluminum surfaces [17] using MD simulations in conjunction with the EMT [15].

Bulk melting, which is a first-order transition, involves the growing of the liquid phase at the expense of the solid one with which it coexists. Among the possible nucleation mechanisms and sites for the melting process, which include crystalline defects (such as vacancies, interstitials, dislocations and grain boundaries), impurities and surfaces, the latter appear to be the most likely, due to the weakening of the binding of surface atoms caused by missing neighbors. (In this context we caution against the often occurring confusion between "surface melting" and "surface roughening".) Experimentally it has been found that disordering of the outermost layers of (110) surfaces of Pb [36] and Al [37] begins at temperatures as low as 150K below the bulk melting temperature, T_M, with a gradual thickening of the melted region as T_M is approached. In contrast, melting of the close packed (111) surfaces shows only weak disordering and that at a temperature close to T_M.

In Fig. 7, we exhibit for Al (110) the calculated layer by layer absolute square of the structure factor, $S_\ell(\vec{k},T)$, where ℓ denotes the layer between $z_{\ell-1}$ and z_ℓ averaged over particle positions generated in a constant temperature (T) simulation. The wave vector $\vec{k}$ was chosen as the first allowed Bragg peak $\vec{k} = (2,0,0)2\pi/d$, where d is the nearest neighbor Aℓ-Aℓ distance. Following a slow, monotonic decrease with temperature due to incoherence caused by lattice vibrations, sharp drops, indicating complete loss of order, occur. The incipient loss of order at $T \sim T_M - 150K$ (where the bulk melting temperature T_M is determined to be at $T_M \sim 800K$) is associated with the formation of a liquid (or quasi-liquid) region of a thickness of ~3 atomic layers, whose thickness grows gradually as the temperature is increased. Furthermore, the calculated total scattering intensity of low energy electrons (LEED) as a function of temperature is found to be in good agreement with measured LEED intensities [37]. We have also found that while the (100) surface shows pronounced disordering, the periodic crystalline layer structure on the (111) face is preserved up to temperatures close to the bulk melting point. Coupled with detailed analysis of the atomic trajectories we find that the surface premelting process is triggered by the formation of vacancies at the surface. This provides a clue for understanding the mechanism of the initiation of melting and the observed crystal face anisotropy. Indeed calculations using the EMT yield that the energy required for the formation of a Frenkel pair on the (110) surface is only 0.3 eV while on the (111) surface the energy cost is 1.3 eV and in the bulk it is even larger (4.2 eV).

Stress, Strain and Flow

The structural, mechanical and dynamical responses of material systems to external stresses, on the atomic level, are key issues in developing a fundamental understanding of a number of systems and phenomena of coupled basic and technological interest such as tribology, lubrication, wear, material fatigue and yield, crack propagation, stress-induced phase and structural transformations, and hydrodynamical phenomena, to name just a few.

Traditionally, MD simulations have been employed to systems of fixed shape and size of the periodically replicated calculational cell (i.e., constant volume simulations). More recently methods for simulating systems in which the volume and

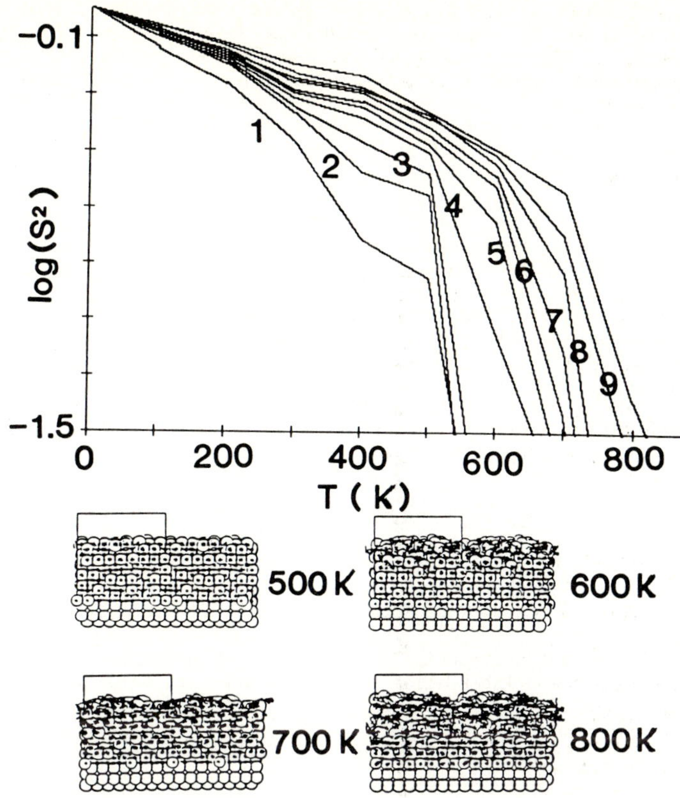

Figure 7: (a) Absolute square of the structure factor for layers 1-10 (1 being the topmost surface layer) of Aℓ (110) versus temperature. The sudden decrease between 500 and 800K is an indication of premelting.

shape of the calculational cell may vary dynamically have been developed [38-41], which open the way to investigations of a large number of materials phenomena in which the dynamical freedom of the system to change volume and/or structure (or phase) is essential [42]. In addition a number of methods have been developed for simulations of flow and hydrodynamical systems [43-47], which allow detailed investigations of these non-equilibrium phenomena.

Recently, we have developed several simulation techniques for studies of systems under non-isotropic stress and applied them to investigations of material systems in interfacial contact exhibiting tribological interactions [44,45]. These simulations provide atomic-scale dynamical information about the stress accumulation and relief mechanisms in tribological systems and demonstrate the crystallographic and materials' parameters dependencies (such as cohesive energies, atomic sizes and interface crystallography) of these processes. In this section we illustrate the application of recently developed methods [46,47] to simulations of shear flow [46].

Studies of non-equilibrium systems, flow in particular, present immense theoretical and technical challenges. Coupled with the basic conceptual problems presented by non-equilibrium phenomena is the technological motivation related to the molecular design of fluids of desired rheological [48] and hydrodynamical characteristics for lubrication purposes. Lubricating fluids are made usually of polymeric molecules (sometimes in colloidal suspension) which are characterized by structural and shear stress relaxation times ranging from milliseconds to minutes, and the relevant flow rates are of the order of the reciprocal of these times. Thus the Weissenburg number [48] for these materials (the product of shear rate and relaxation time) is of the order of unity. While simulations of phenomena on the above time scales are computationally prohibitive, it is possible to study the nature of the rheological processes occurring in these systems via simulations of simpler fluids sheared at the same Weissenburg number [49]. Simple Lennard-Jones systems exhibit such "corresponding state" behavior.

Our studies are aimed at understanding the behavior of fluid films sheared between confining solid slabs, such as the circumstances encountered in many tribological systems. In constructing these simulations we have developed a new MD technique [45-47], based on the generalized constant pressure simulation ideas [40].

To demonstrate the method we display results for steady-state flow of a simple fluid sheared in a planar Couette geometry. The system is composed of two solid slabs confining a liquid. The particles in the solid (s) and liquid (ℓ) materials are characterized by 6-12 Lennard-Jones (LJ) potentials $v(r) = 4\epsilon[(\sigma/r)^{12} - (\sigma/r)^{6}]$, with the well-depth for the inter-liquid particle interaction $\epsilon_{\ell\ell} = \epsilon_{ss}/10$ (where ϵ_{ss} is the well-depth for interaction of particles in the solid region) and the interaction between the solid and liquid particles characterized by $\epsilon_{s\ell} = \sqrt{\epsilon_{ss}\epsilon_{\ell\ell}}$.

In Fig. 8a we display the density profile versus distance along the [001] direction of the system at equilibrium at T = 0.125 (uniform across the system). The layered structure of the solid is evident. Note however, that the liquid adjacent to the solid exhibits pronounced layering [1,2,22]. The density profile of the steady-state system under shear in the x direction shown in Fig. 8b exhibits a similar liquid-layering behavior in the interfacial region. However in this state a nonuniform temperature distribution in the system has developed (see Fig. 8c). The strain rate profile (time averaged) across the system (along the [001] (z) direction) is shown in Fig. 8d (the z axis is normal to the solid surface planes which lie in the xy plane. The system is sheared in the x direction, i.e., the shear rate is $\dot{\gamma}_{xz}$ and is calculated from the particle momenta which include a streaming component [46]). Using the atomic trajectories we calculate the instantaneous atomic stress tensors [50] $\sigma_{\alpha\beta}(i)$, (where $i = 1,\ldots,N$ and $\alpha,\beta = x,y,z$) which together with the local strain rates can be used to obtain the viscosity profile $\zeta(z) = -\sigma_{xz}(z)/\dot{\gamma}_{xz}$ shown in Fig. 8e. We remark that the value of the viscosity in the middle region of the fluid is in good agreement with that calculated for a bulk LJ liquid under similar conditions (see Fig. 2 in ref. [49], note that in our liquid $\epsilon_{\ell\ell} = \epsilon_{ss}/10$). This indicates that even for thin films such as the one used in this simulation, bulk behavior is found. However, we note that unlike the case discussed in [49] our system is allowed to develop a temperature gradient in a natural, dynamic manner (since the temperature is controlled only inside the solid confinements).

The above results demonstrate the system dynamical response at relatively small shear rates. At higher shear rates the interfacial liquid-layering is much less pronounced and in addition non-Newtonian flow occurs. These results and studies of the dependencies of the flow characteristics on the strength of the fluid-solid interaction and the composition of the fluid [46] demonstrate the potential utility of MD simulations in tribological molecular-design problems.

*Acknowledgements: The studies described in this review are the results of collaborative work with several colleagues who should be regarded as coauthors of this review and to whom I am grateful: R. N. Barnett, C. L. Cleveland, W. D. Luedtke, W. L. Ribarsky, S. Sutton, J. K. Norskov and P. Stoltze. These studies were supported by the U.S. DOE grant number FG05-86ER-45234 and the work on AFM and stress simulations in part by the Hughes Aircraft Company, DARPA-Hughes Tribo- logical

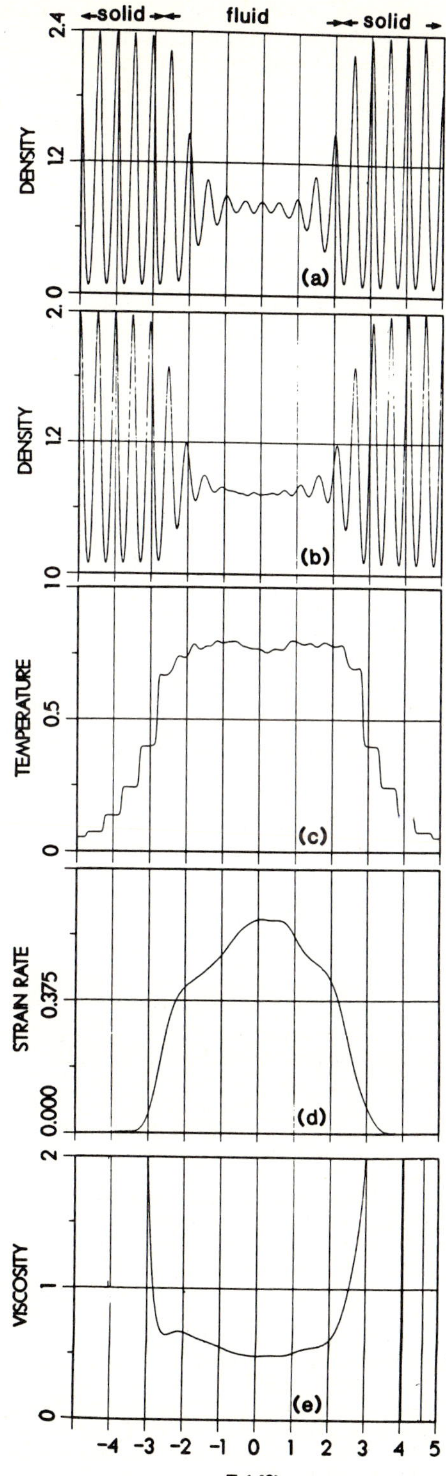

Figure 8: (a) Density profile for a 1260 LJ system at solid-liquid equilibrium ($\varepsilon_{\ell\ell} = \varepsilon_{ss}/10$, see text). (b-e) Density, temperature, strain-rate and viscosity profiles of the system at steady-state under a constant (but spacially varying, see (d)) strain rate. Distance along the z direction (normal to the (001) surface planes) in units of 18.32σ. All quantities are in reduced LJ units (see text). Note layering of the liquid at the interface region, temperature gradient, strain rate and viscosity variations (the values of the viscosity in the solid have been truncated in (e)). The mass m and the parameter σ of the liquid and solid particles are taken to be the same. Results are in reduced units: pressure ($\varepsilon_{ss}\sigma^{-3}$), number density $\rho(\sigma^{-3})$, time ($\sigma m^{1/2}\varepsilon_{ss}^{-1/2}$), distance ($\sigma$), temperature ($\varepsilon_{ss}/k_B$) and viscosity ($m^{1/2}\varepsilon_{ss}^{1/2}\sigma^{-2}$). The number of particles in each of the confining solid slabs is 630, arranged in a face-centered-cubic lattice (9 layers) exposing the (001) face at the interface, and the number of liquid particles is 630. The computational cell is repeated periodically in three-dimensions. The temperature of the system is controlled to T = 0.125 via scaling of particle velocities in the middle regions of the solid slabs.

Fundamentals Program Contract No. F33615-C-5087, and the NSF Tribology Program. Most of the simulations were performed on the CRAY-XMP at NMFECC, Livermore, California, the CDC CYBER-990 at the Georgia Institute of Technology and some at the NSF supercomputer center at Pittsburgh. The assistance of A. Ralston and V. Mallette in the preparation of the manuscript is gratefully acknowledged.

References

[1] U. Landman et al., Mat. Res. Soc. Symp. Proc. Vol. 63 (MRS, Boston, 1985), p. 273; see also other articles in this volume.

[2] F. F. Abraham, Adv. Phys. 35, 1 (1986); J. Vac. Sci. Technol. B2, 534 (1984).

[3] MRS Bull. Volume XIII (2), February 1988, p. 14-39.

[4] U. Landman, R. N. Barnett, C. L. Cleveland, J. Luo, D. Scharf and J. Jortner, in "Few-Body Systems and Multiparticle Dynamics", Ed. D. A. Micha, AIP Conf. Proc. 162 (AIP, New York, 1987), p. 200.

[5] D. W. Heerman, "Computer Simulation Methods" (Springer, Berlin, 1986).

[6] "Computer Simulations of Solids", Eds. C. R. A. Catlow and W. C. Machord (Springer, Berlin, 1982).

[7] U. Landman, "Path-Integral and Real-Time Dynamic Simulations of Quantum Systems", this volume.

[8] G. C. Maitland, M. Rigby, E. B. Smith and W. A. Wakeham, "Intermolecular Forces", (Clarendon, Oxford, 1981).

[9] M. J. Sangster and M. Dixson, Adv. Phys. 25, 247 (1976).

[10] P. N. Keating, Phys. Rev. 145, 637 (1966).

[11] F. H. Stillinger and T. A. Weber, Phys. Rev. B31, 5262 (1985).

[12] See, e.g., W. A. Harrison, "Pseudopotentials in the Theory of Metals" (Benjamin, Reading, Mass., 1966).

[13] R. N. Barnett, C. L. Cleveland and U. Landman, Phys. Rev. Lett. 54, 1679 (1985).

[14] R. N. Barnett, C. L. Cleveland and U. Landman, Phys. Rev. Lett. 55, 2035 (1985).

[15] K. W. Jacobsen, J. K. Norskov and M. J. Puska, Phys. Rev. B 35, 7423 (1987).

[16] See M. Baskas, M. Daw, B. Dodson and S. Foils in reference 3, p. 28.

[17] P. Stoltze, J. K. Norskov and U. Landman, "Disordering and Melting of Aluminum Surfaces", Phys. Rev. Lett. 61, 440 (1988).

[18] See citations in reference 19.

[19] U. Landman, W. D. Luedtke, R. N. Barnett, C. L. Cleveland, M. W. Ribarsky, E. Arnold, S. Ramesh, H. Baumgart, A. Martinez and B. Khan, Phys. Rev. Lett. 56, 155 (1986).

[20] F. F. Abraham and J. Q. Broughton, Phys. Rev. Lett. 56, 734 (1986).

[21] U. Landman, W. D. Luedtke, M. W. Ribarsky, R. N. Barnett, and C. L. Cleveland, "Molecular-Dynamics Simulations of Epitaxial Growth from the Melt. I. Si(100) and II. Si(111)", Phys. Rev. B37, 4637, 4647 (1988).

[22] See also earlier MD studies of liquid-phase epitaxy: U. Landman, R. N. Barnett, C. L. Cleveland and R. H. Rast, J. Vac. Sci. Technol. A3, 1574 (1985); U. Landman, C. L. Cleveland and C. S. Brown, Phys. Rev. Lett. 45, 2032 (1980) and in "Nonlinear Phenomena of Phase Transitions and Instabilities", Ed. T. Riste (Plenum, New York, 1982), p. 379.

[23] M. Faraday, Phil. Trans. 147, 145 (1957).

[24] For a recent review see H. J. Leamy, G. H. Gilmer, and A. G. Dirks, in "Current Topics in Materials Science", Ed. E. Kaldis (North-Holland, Amsterdam, 1980), Vol. 6, Chap. 4.

[25] B. R. Appelton, R. A. Zuhr, T. S. Noggle, N. Herbots and S. J. Pennycook in "Beam-Solid Interactions and Transient Processes", Eds. M. O. Thompson, S. T. Picraux and J. S. Williams (MRS Symp. Proc. 74, Pittsburgh, PA, 1987), p. 45.

[26] M. Schneider, J. K. Schuller and A. Rahman, Phys. Rev. B36, 1340 (1987).

[27] P. A. Taylor and B. W. Dodson, Phys. Rev. B36, 1355 (1987); B. W. Dodson, Phys. Rev. B36, 1068 (1987).

[28] W. D. Luedtke and U. Landman, Phys. Rev. B (to be published, 1988).

[29] W. D. Luedtke and U. Landman, "Preparation and Melting of Amorphous Silicon by Molecular Dynamics Simulations", Phys. Rev. B37, 4656 (1988).

[30] J. Luo, U. Landman and J. Jortner, in "Physics and Chemistry of Small Clusters", Ed. P. Jena, B. K. Rao and S. N. Khanna (Plenum, New York, 1987), p. 201; see also article by R. S. Berry et al., in the above reference p. 185, for MD studies of the melting and freezing of rare-gas clusters.

[31] U. Landman, W. D. Luedtke, C. L. Cleveland and R. N. Barnett (to be published).
[32] G. Binnig and H. Rohrer, IBM J. Res. Develop. 30, 355 (1986).
[33] G. Binnig, C. F. Quate and Ch. Gerber, Phys. Rev. Lett. 56, 930 (1986).
[34] P. K. Hansma and J. Tersoff, App. Phys. Rev. (1986).
[35] U. Landman, W. D. Luedtke and A. Nitzan, Phys. Rev. Lett. (to be published, 1988).
[36] J. W. M. Frenken and J. F. van der Veen, Phys. Rev. Lett. 54, 134 (1985); B. Pluris, A. W. van der Gon, J. W. M. Frenken and J. F. van der Veen, Phys. Rev. Lett. (1987).
[37] P. von Blackenhagen, W. Schommers and V. Voegele, J. Vac. Sci. Technol. A5, 649 (1987).
[38] H. C. Andersen, J. Chem. Phys. 72, 2384 (1980).
[39] M. Parrinello and A. Rahman, J. Appl. Phys. 52, 7182 (1981).
[40] J. R. Ray and A. Rahman, J. Chem. Phys. 80, 4423 (1984).
[41] For an extension to metallic systems see ref. 13.
[42] See reviews by S. Yip and M. Parrinello in "Molecular-Dynamics Simulations of Statistical-Mechanical Systems", Fermi School, XCVII Corso, (Soc. Ital. di Fisica, Bologna, 1986).
[43] See recent reviews by D. J. Evans and G. P. Morriss, Comput. Phys. Rep. 1, 297 (1984), and D. J. Evans and W. G. Hoover, Ann. Rev. Fluid Mech. 18, 243 (1986).
[44] M. W. Ribarsky and U. Landman in "Approaches to Modeling of Friction and Wear", Eds. F. F. Ling and C. H. T. Pan, (Springer-Verlag, New York, 1988), p. 159.
[45] M. W. Ribarsky and U. Landman, "MD Simulations of Stress and Strain Processes", J. Appl. Phys. (1988).
[46] S. Sutton, M. W. Ribarsky and U. Landman, "MD Simulations of Shear Flow", J. Chem. Phys. (1988).
[47] C. L. Cleveland, "New Equations of Motion for Molecular Dynamics Systems that Change Shape", J. Chem. Phys. (1988).
[48] H. A. Barnes, "Dispersion Rheology", (Roy. Soc. Chem., London, 1988).
[49] D. M. Heyes, Mol. Phys. 57, 1265 (1986).
[50] D. Srolovitz, T. Egami and V. Vitek, Phys. Rev. B24, 693 (1981).

Simulations of Oxygen Monolayer and Bilayer Systems

K.M. Flurchick [1] *and R.D. Etters* [2]

[1] ETA Systems, Inc., Computer Services Annex, University of Georgia, Athens, GA 30602, USA

[2] Physics Department, Colorado State University, Fort Collins, CO 80523, USA

Introduction

We present the results of a simulation using a physical model developed to study the equilibrium configurations of adsorbed oxygen on graphite. In the model, an atom-atom potential between the oxygen and graphite was used which depended on the lateral positions of the oxygen molecules as well as the distance above the surface. A cluster model is proposed to study the low density region, for which very little information is available. The nucleation of adsorbed rare gas films on a graphite basal plane has been studied experimentally and interpreted using a two-dimensional cluster model [1,2]. The cluster model used in this study is an extension of the work done by Pan et al. [3] by including lateral variations of the potential in the gas-surface interaction. Zero temperature minimum energy structures of two-dimensional oxygen clusters with the number of O_2 molecules in the cluster, N, ranging from 1 to 16 molecules adsorbed on the graphite basal plane surface are determined. The energies and structures of adsorbed infinite monolayer films are also studied with the energy minimization method which involved a search for possible low temperature superlattices. The acceptability of the superlattices is tested by comparing the superlattice energy to the incommensurate overlayer energy. This calculation, which includes the lateral variations of the gas-surface potential, allows the determination of the orientational epitaxy of the O_2 overlayer. Both the δ [4] and δ' [4] orientational epitaxy were found in these calculations. The ζ_1 [4] and ζ_2 [4] phases were also found in these calculations. Initial studies of adsorbed bilayer systems are also presented.

Experimental and theoretical studies of adsorbed systems are briefly reviewed with particular emphasis on the orientational epitaxy of oxygen adsorbed on graphite. Oxygen on graphite has been experimentally studied by several techniques, including X-ray [5], neutron diffraction [6], LEED measurements[4,7], and specific heat measurements [8,9].

At low coverages $\rho/\rho_0 < 1.6$, Nielsen and McTague [6], using neutron scattering, observed a single diffraction peak and labelled this region the δ phase. In the δ+gas region below $\rho/\rho_0 = 1.00$, the lattice vectors have constant values a=3.252 Å , b=7.985 Å . Between $\rho/\rho_0 = 1.00$ and $\rho/\rho_0 = 1.70$ they decrease monotonically to values a=3.18 Å , b=7.86 Å , and above $\rho/\rho_0 = 1.7$ they are roughly constant. From the observed X-ray diffraction peaks [5], the δ phase clearly cannot be triangular. There is also evidence that the $\vec{a}$ and $\vec{b}$ lattice vectors are not quite orthogonal. The most physically plausible structure is a centered monoclinic lattice, slightly distorted from the centered rectangular structure, with the molecular axes parallel to the graphite surface and to each other.

The LEED results measured by M. Toney et al.[4] and by Fain et al. [7], for the low density δ phase complement the X-ray diffraction results. The exact orientation of the molecule in the plane has not been determined by the LEED study of Toney, but the parallel orientations shown were proposed in the X-ray studies [5,10] and the zero temperature calculations of Etters and Pan [3] and Pan et al. [3], and are physically expected based on steric considerations. The structure is close packed. In addition, Toney has observed orientational epitaxy for O_2 on graphite which changes with coverage and temperature. This is the first time such changes in orientational epitaxy have been observed for a nontriangular lattice and for diatomic molecules. They also report that the O_2 δ phase unit cell aligns at about 3° to the (110) direction of the graphite basal plane at $\rho/\rho_0 \approx 1.2$ and T = 24K.

Springer Proceedings in Physics Vol. 33: **Computer Simulation Studies in Condensed Matter Physics**
Editors: D. P. Landau · K. K. Mon · H-B. Schüttler

M. Toney et al. also report a uniquely different face centered unit cell that they have called δ' . The orientational epitaxy of the δ' phase is described by $\phi_{\delta'}$, and is very different from that of the δ phase. However, because the lattice constants a, b and γ of the δ and δ' phases are so close the X-ray diffraction study would not have been able to distinguish between them. The δ' phase aligns at about 26° to the (110) direction of the graphite basal plane at coverages of about $\rho/\rho_0 \approx 1.7$ and T = 16K.

As the coverage is increased from $\rho/\rho_0 = 1.6$ to $\rho/\rho_0 = 1.9$ at T=15K, there is a linear tradeoff between the δ phase and the ζ phase. The lattice vectors approach close to that of a single triangular plane of bulk β phase. The $\delta - \zeta$ transition corresponds to a structural change in which the O_2 molecules reorient themselves from parallel to perpendicular to the graphite substrate and is first order. This transition is driven by the competition between the oxygen-carbon and oxygen-oxygen van der Waals interactions, the latter becoming dominant at higher coverages. The ζ phase is approximately triangular and the molecules are perpendicular to the surface. From the LEED measurements of Toney and Fain [17], two different ζ phases have been found with a phase transition between them at T=18K. The ζ_1 phase is stable below 18K and is an oblique triangular mesh, with lattice constants a = 3.21 Å and b = 5.81 Å , at T < 16K. The ζ_2 phase is stable above 18K and has a triangular mesh, with lattice constants a = 3.30 Å and b = 5.76 Å , at T = 23K. In addition, a third phase labelled ζ', which is stable over the entire temperature range 12K $\leq$ T. $\leq$ 26K was also reported. This phase has the same lattice constants as the ζ_2 phase but very different orientational epitaxy.

The substrate-adsorbate distance for O_2 is predicted to be about 3.2 Å [3] which is larger than the 1.42 Å intralayer carbon-carbon length. Thus, the molecules are far enough above the surface that they only weakly sense the variations in the basal-plane electron density. This lateral variation in the adsorption on graphite can cause interesting commensurate-incommensurate phase transitions [11 and references therein].

Mathematical Model

To calculate the properties of adsorbed molecular systems the total interaction energy is expressed as a sum over the pair interactions between the molecules in the crystal and between the molecules and the substrate atoms. In this calculation the many-body interactions are neglected, and it is assumed that the graphite structure is not altered by the presence of the O_2 molecules.

For a system of O_2 molecules adsorbed on a graphite substrate, the total potential energy E consists of three parts,

$$E = V_1 + V_2 + V_3. \tag{.1}$$

V_1 is the internal energy among the adsorbed oxygen molecules, assuming that the bond length in an O_2 molecule is not perturbed from the initial value of 1.208Å, V_2 is the adsorption energy, i.e., the interaction energy between the adsorbed molecules and the graphite, and V_3 is the internal energy of graphite solid. It is assumed that the graphite structure is not perturbed by the presence of oxygen and that the atoms in graphite are stationary. Therefore V_3 is a constant for all the calculations performed and will be ignored. Assuming only two body interactions, V_1 can be written as

$$V_1 = \sum_{i<j} V_{O_2-O_2}(\vec{r}_i, \vec{\omega}_i, \vec{r}_j, \vec{\omega}_j), \tag{.2}$$

where $V_{O_2-O_2}$ is the potential energy between molecules i and j, located at $\vec{r}_i$ and $\vec{r}_j$, with orientations $\vec{\omega}_i$ and $\vec{\omega}_j$, respectively. An atom-atom Lennard-Jones potential[12], with a quadrupolar interaction is used for $V_{O_2-O_2}$. For two molecules labeled 1 and 2, $V_{O_2-O_2}$ is written as

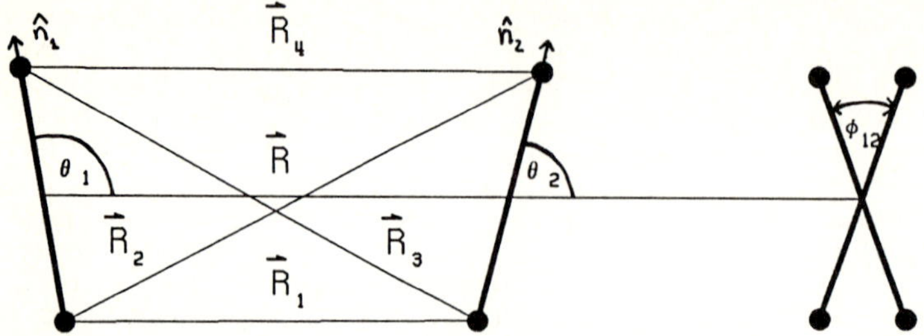

Figure .1: Illustration of the variables used to determine the atom-atom and electric quadrupole energy of an O_2 pair. The figure on the right is the projection of the two molecules onto a plane perpendicular to $\vec{R}$.

$$\begin{aligned} V_{O_2-O_2}\,(\vec{r}_1, \vec{\omega}_1, \vec{r}_2, \vec{\omega}_2) &= \sum_{i=1}^{4} 4\epsilon \left[\left(\frac{\sigma}{R_i}\right)^{12} - \left(\frac{\sigma}{R_i}\right)^{6}\right] \\ &+ \frac{Q^2}{R^5}\Big(1 - 17\cos^2\theta_1\cos^2\theta_2 - 5\cos^2\theta_1 \\ &- 5\cos^2\theta_2 + 1 - 16\sin\theta_1\sin\theta_2\cos\theta_1\cos\theta_2\cos\phi_{12} \\ &+ 2\sin^2\theta_1\sin^2\theta_2\cos^2\phi_{12}\Big)\,, \end{aligned} \tag{.3}$$

where $\sigma = 3.05\text{Å}$, $\epsilon = 54.34\text{K}$ [13], and $Q = -0.39\text{x}10^{-26}$ esu [14]. The variables in [3] are illustrated in Fig. 1. The separation of the two atoms in each molecule is taken to be a constant, 1.208Å [11]. The atom-atom interaction in [3] is the dominant part of the O_2 - O_2 potential. This potential [3] was first suggested by English and Venables [13] in a general study of diatomic molecular crystals, from which they were able to predict crystal structures with some success.

The potential energy of an atom adsorbed on graphite has been put into an analytical form by Steele [15] based on the atom-atom Lennard-Jones potential. This can be expressed as

$$\begin{aligned} \frac{U_s(\vec{r})}{\epsilon_{gs}} &= E_0 + E_1\left(\cos\left(\vec{G}_1\cdot\vec{\tau}\right) + \cos\left(\vec{G}_2\cdot\vec{\tau}\right)\right. \\ &+ \left.\cos\left(\left(\vec{G}_1 + \vec{G}_2\right)\cdot\vec{\tau}\right)\right), \end{aligned} \tag{.4}$$

where

$$E_0 = \frac{4\pi}{a_s}\left(\frac{2}{5}\frac{1}{z^{10}} - \frac{1}{z^4}\right), \tag{.5}$$

$$E_1 = -\frac{4\pi}{a_s}\left[\frac{1}{30}\left(\frac{G}{2z}\right)^5 K_5(Gz) - 2\left(\frac{G}{2z}\right)^2 K_2(Gz)\right] \tag{.6}$$

and $\vec{\tau}$ is relative to the center of a graphite surface hexagon. Here, the lengths are in units of σ_gs, energies are in units of $4\epsilon_{gs}$ and

$$G = \left|\vec{G}_1\right| = \left|\vec{G}_2\right| = 2\pi/(a_1\ \cos 30^\circ)\,. \tag{.7}$$

The quantities a_1, $\vec{G}_1$, and $\vec{G}_2$ are illustrated in Fig. 2. The Lennard-Jones parameters for the gas-surface interaction are determined by the usual averaging, $\epsilon_{gs} = \sqrt{\epsilon \times \epsilon_{C-C}}$ and $\sigma_{gs} = 0.5\,(\sigma + \sigma_{C-C})$ with $\epsilon_{C-C} = 28\text{K}$, and $\sigma_{C-C} = 3.40\text{Å}$.

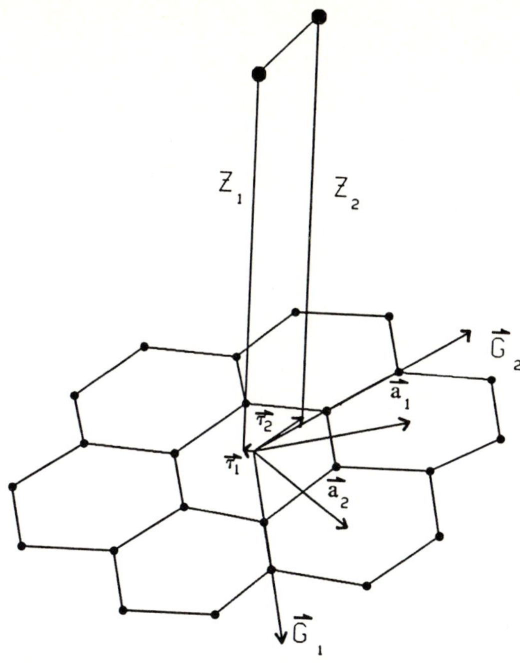

Figure .2: Schematic illustration for an adsorbed oxygen molecule on the graphite basal plane. The $\vec{\tau}_1$ and $\vec{\tau}_2$ are the two-dimensional vectors that locate the projection of each oxygen atom onto the graphite surface. Also, $|\vec{a}_1| = |\vec{a}_2| = 2.46\text{Å}$ and $|\vec{G}_1| = |\vec{G}_2| = 2\pi\,(|\vec{a}_1|\cos 30^{\circ})$

Candidate superlattices are determined from the data supplied from a simulation neglecting the lateral variations of the gas-substrate interaction. The condition of superlattice registry is that an integral number of unit cells start and end at equivalent positions on the graphite. To determine superlattice registry on the graphite basal plane, the O_2 monolayer unit cell is mapped onto the graphite plane using a coordinate system centered on a graphite surface hexagon . As a first approximation, the O_2 monolayer unit cell, with lattice parameters X_0 and Y_0, in the x and y direction, respectively, is determined using only the E_0 term from Steele's expansion of the gas-surface interaction [15]. A large O_2 cell, with lattice vectors of length $N_X\ X_{E0}$ and $N_Y\ Y_{E0}$, is made of integer multiples of the E_0 unit cell. This large O_2 cell is mapped onto the graphite substrate and the E_0 unit cell lattice parameters are distorted so the larger cell made of multiples of the E_0 unit cell registers on the graphite substrate. The distorted O_2 cell lattice vectors that characterize the dimension of the large O_2 cell are related to the E_0 unit cell by $N_x X = N_x(X_{E0} + \epsilon_x)$ and $N_y Y = N_y(Y_{E0} + \epsilon_y)$, where ϵ_x and ϵ_y are the distortions of the E_0 unit cell lattice parameters in the x or y direction. The range of values for ϵ_i is less than 0.1Å per unit cell. This acceptance value is determined from the maximum distortion that can arise from the lateral variations in the substrate potential acting on a single O_2 molecule. The number of unit cells in the two directions are in general different. The final determination for superlattice acceptance is that the energy per particle calculated over the superlattice was lower than the energy per particle of the low density phase determined using the E_0 interaction alone.

The equilibrium lattice configuration of infinite monolayer oxygen films adsorbed on graphite is determined by minimizing the total energy of the system using "Pattern Search" [16]. It is an easily programmed method of minimizing a function of several variables by changing one parameter at a time. To determine whether or not a particular minimum energy configuration is the absolute or a local minimum, different initial configurations and step sizes were used. In our case, for different starting values, the method reproduced the same energy minimum to within 0.005 percent.

The superlattice mapping was performed with maximum values of N_x and N_y limited to 20 in order to limit the number of molecules in a particular superlattice due to computational restrictions. This limitation gave 780 atoms for the largest superlattice allowed. Treating the orientational and translational degrees of freedom the same, this resulted in a maximum of 4680 independent variables to minimize. In this simulation, the time consuming step is the determination of the energy at each step of the minimization. The minimization itself involved very little overhead, even with the large number of degrees of freedom present in this calculation.

Adsorbed bilayer films were also studied by considering the special case where the second oxygen layer was considered to be commensurate with the first layer, and each O_2 layer was considered to be incommensurate with the graphite substrate. Hence, only the E_0 term from the expansion in (4) is included in the gas-surface interaction potential. In addition, the condition of commensurability between adlayers fixes the center of mass positions, in the x-y plane, of the O_2 molecules of the second layer with respect to the first layer. However, the O_2 molecules of the second layer are not orientationally constrained, and the distance between the two layers is also a free parameter. The O_2 - O_2 interaction is calculated using (2) assuming a two-sublattice model for each O_2 layer. (A two-sublattice model contains two inequivalent sites per unit cell.)

Results and Conclusions

The minimum energy structure of two-dimensional adsorbed O_2 clusters were calculated using the $V_{O_2-O_2}$ interaction in [3], plus an adsorption energy calculated using the E_0 and E_1 terms in [4]. Of particular interest is the single O_2 adsorbed on graphite. English and Venables [13], in their work with rare gas atoms, found that when the lateral variations of the substrate potential are included, rare gas atoms adsorbed at the center of a graphite surface hexagon. It was assumed that this was the case for diatomic molecules as well. In this work, the effect of the lateral variations in the substrate potential on a single O_2 molecule was investigated by translating the molecule across the graphite hexagon. The minimum energy position of the center of mass of a single O_2 molecule, measured from an origin at the center of a graphite surface hexagon is (x,y) = (0.55 Å , 0.00 Å), with the molecule parallel to the x direction. One oxygen atom is located near the single atom adsorption site (the center of the graphite hexagon) and the other atom is near a saddle point halfway between two carbon atoms along a hexagon edge at a distance of 0.55 Å from the center of the graphite hexagon. This is a favorable position because the distance between the oxygen atoms of the O_2 molecule (1.208 Å) is nearly equal to the distance from the center of a hexagon to the center of an edge $\left(1.423\,(\cos 30^\circ)\,\text{Å} \;=\; 1.232\text{Å}\right)$. Because of the symmetry of the graphite substrate, the O_2 molecule has six different positions with the same minimum energy.

The minimum energy structures for O_2 molecular clusters adsorbed on graphite, for cluster sizes N = 2 to 16 shows that the center of masses of the molecules in the clusters do not lie in a plane parallel to the substrate, but have an oscillatory variation in the z coordinate of the center of mass. The difference between the z coordinate of the O_2 molecule farthest from the substrate and the O_2 molecule closest to the substrate in a particular cluster is denoted by Δz and is equal to 0.037 Å. Δz increases monotonically from N=2 to N=12 and then begins to oscilliate but it appears that the mean value still increases. The O_2 molecular axes lie in an x-y plane above the substrate. It is of interest to note that a significant number of the molecules are nearly at single O_2 minimum energy positions relative to the substrate. As the clusters increase in size, the O_2 molecules move farther from the single O_2 minimum energy positions.

There are no experimental results to compare directly with the calculations of the O_2 adsorbed clusters, however, there is the previous calculation of Pan et al. [3]. The calculations of Pan included only the E_0 term in the substrate potential. When the lateral variations of the substrate potential are included, the configurations distort slightly.

The effects of the lateral variations in the gas-surface interaction potential for an infinite adsorbed film were calculated by evaluating the E_0 and E_1 terms from Steele's expansion, based on monolayer films in superlattice registry with the substrate. A superlattice, defined in the mathematical section above, has fixed corners, i.e., the corner O_2 molecules have a fixed distance from each other, namely N_xX or N_yY, but the O_2 molecule has complete rotational freedom. In this way only the center of mass coordinates of one corner molecule are required in the superlattice minimization calculation. The corner molecule also has rotational freedom. In addition, the entire superlattice can translate relative to the substrate, thus comprising an infinite adsorbed film. The molecules along the edges of the superlattice were constrained in that opposite sides of the superlattice are the same, so a minimization calculation includes only two of the four edges. All of the molecules in the interior of the superlattice had complete translational and rotational freedom.

The lattice constants of the low density monolayer unit cell calculated using just the E_0 term from equation 4 are $X_{E0} = 8.07$Å and $Y_{E0} = 3.32$Å. The minimum energy per particle for this structure, neglecting the lateral variations of the substrate potential, is E = -8.0506 x 4 ϵ , at $\rho/\rho_0 = 1.17$. Superlattices for low density O_2 monolayer films were determined by mapping the O_2 monolayer onto the graphite substrate as described in the mathematical model section. The superlattice vectors are given by N_xX and N_yY, respectively, and because of the large mismatch between the O_2 and graphite lattice vectors, the superlattices are large. In this zero temperature calculation, there are a number of superlattices of different sizes with Θ_1 near 7.63° (see Table 1). The superlattice that is closest to the experimental results is the minimum energy structure, $N_x = 13$ by $N_y = 4$ with $\Theta_1 = 7.59°$ and $\Theta_2 = 9.83°$. The lattice parameter b is 0.4% larger than the experimental measurement. The skewness of the superlattice is 92.4° which is within the temperature-dependent values reported by Toney et al.[4]. The difference between the superlattice monolayer energy and the energy of the monolayer calculated neglecting the lateral variations of the substrate potential is about 5.87K per molecule.

The δ' phase has mesh constants that are about the same as those in the densest δ phase, but with a very different orientational epitaxy. The lowest energy structure found is the $N_x = 17$ by $N_y = 18$ superlattice with $\theta_1 = 33.89°$ and $\theta_2 = 34.31°$. The lattice parameter a differs by about 4.7%, and b differs by about 2.8%, from the measured values of Toney. This superlattice is slightly skewed ($\gamma = 90.4°$), but this is within the experimental uncertainty reported.

As the packing of O_2 molecules on the substrate increases, the structure of the monolayer undergoes an orientational change from lying parallel to the substrate to being perpendicular to the substrate [3]. Table 2 lists the high density non-rectangular superlattices ($\gamma \neq 90°$) obtained. The lowest energy structure is the $N_x = 14$ by $N_y = 17$ superlattice with $\theta_1 = 12.22°$ and $\theta_2 = 10.07°$. The lattice parameter a differs by about

TABLE 1 **The O_2 superlattice parameters for non-rectangular low density superlattices. lattice parameters X and Y are in Å . The energy is in units of 4 ϵ = 217.34K.**

N_X	N_Y	Θ_1 (degrees)	Θ_2 (degrees)	X (Å)	Y (Å)	ENERGY (4 ϵ)
4	13	7.59	9.83	8.0811	3.3327	-8.0739
13	8	16.47	16.10	8.1061	3.3325	-8.0780
14	10	26.88	26.33	8.0926	3.3342	-8.0773
15	8	16.43	16.10	8.0514	3.3325	-8.0804
13	10	26.63	26.33	8.0594	3.3342	-8.0782
20	10	26.92	26.33	8.0160	3.3342	-8.0770
18	17	33.89	34.31	8.0823	3.3441	-8.0806
18	17	40.26	40.89	8.0741	3.3220	-8.0805
19	15	40.28	39.83	7.9924	3.3352	-8.0972

TABLE 2 The O_2 minimum energy superlattice parameters for high density non-rectangular superlattices. Note the energy is in units of $4\epsilon = 217.34K$. The lattice parameters X and Y are in Å .

N_X	N_Y	Θ_1 (degrees)	Θ_2 (degrees)	X (Å)	Y (Å)	ENERGY (4ϵ)
7	13	12.22	9.83	5.7641	3.3327	-7.6363
17	13	10.07	9.83	5.7429	3.3327	-7.6471
12	8	14.39	16.10	5.7253	3.3325	-7.6462
13	8	16.63	16.10	5.7381	3.3325	-7.6496
18	10	26.63	26.33	5.8207	3.3342	-7.6386
16	13	27.69	27.16	5.7411	3.3219	-7.6456
19	16	41.74	43.90	5.7372	3.3325	-7.6368
9	19	48.07	47.65	5.7379	3.3352	-7.6432

3.1%, and b differs by about 0.9%, from the measured values of Toney. $\gamma = 87.85°$ as compared to $\gamma = 89.12 \pm 0.4°$ from Toney et al.[4]. The difference in energy between the incommensurate O_2 monolayer calculated using just the E_0 term from Steele's expansion and the energy found including the lateral variations in the substrate potential for the $N_x = 14$, $N_y = 17$ superlattice is 10.063K. This is near the lower limit of the experimental temperature range.

The energy per particle versus surface coverage for the monolayer [3] and the bilayer is shown in Fig. 3, for the case when the two O_2 layers are considered to be commensurate with one another. The circles represent the energy per particle of the monolayer and the triangles denote the energy per particle of the bilayer. As seen in Fig. 3, bilayer formation is energetically favored for surface coverages $\rho \geq 2.1\rho_0$. At this first minimum, the unit cell of the first and second layer is a=3.16 Å , b=5.15 Å . The height above the substrate of the first layer is z=3.2 Å , and the height above the substrate of the second layer is z=5.96 Å . The O_2 molecules of the first layer are about 10° out of the x-y plane and the molecules of the second layer are about 40° out of the x-y plane. For densities between $1.9\rho_0 \leq \rho \leq 3.4\rho_0$, the molecules of the two layers are oriented nearly perpendicular to each other, with the molecules of the first layer about 8° – 10° out of the x-y plane.

For densities below $\rho = 3.4\rho_0$, the second layer is moving away from the substrate as the surface coverage increases, but the first layer remains a constant height above the substrate. The second layer is moving away from the first layer as the packing of each layer is increasing. The distance between the two layers varies from $2.37Å \leq z_{12} \leq 3.10Å$, as the density varies from $1.9 \leq \rho/\rho_0 \leq 3.4$. At $\rho = 3.4\rho_0$ the bilayer undergoes a structural transistion. The molecular axes of the molecules of the second layer begin to orient themselves perpendicular to the substrate and also increase their distance from the substrate, while the molecular axes of the molecules of the first layer remain slightly out of the plane at about 10°. The orientation transistion of the O_2 bilayer is completed by $\rho = 3.8\rho_0$. There is very little experimental work concerning O_2 bilayer films adsorbed on graphite and no theoretical calculations available. The bilayers described by Mochrie et al. [18] have a first layer unit cell with lattice vectors a=3.3 Å , b=5.72 Å and a second layer unit cell with lattice vectors, a=3.24 Å and b=5.61 Å at a surface coverage of $\rho = 3.4\rho_0$ and a temperature of T=39.7K, near the melting temperature of monolayer oxygen. Using the lattice parameters above, the density of the first layer is $\rho = 1.670\rho_0$ and the density of the second layer is $\rho = 1.746\rho_0$. The results presented in this work indicate that as the O_2 surface density increases, the formation of a bilayer is energetically favored near the density reported by Mochrie. However, it is clear additional work needs to be done using the incommensurate bilayer model presented above.

In summary, the lateral interaction gives rise to a lower energy superlattice structure than that given by the incommensurate monolayer calculated using only E_0. In addition,

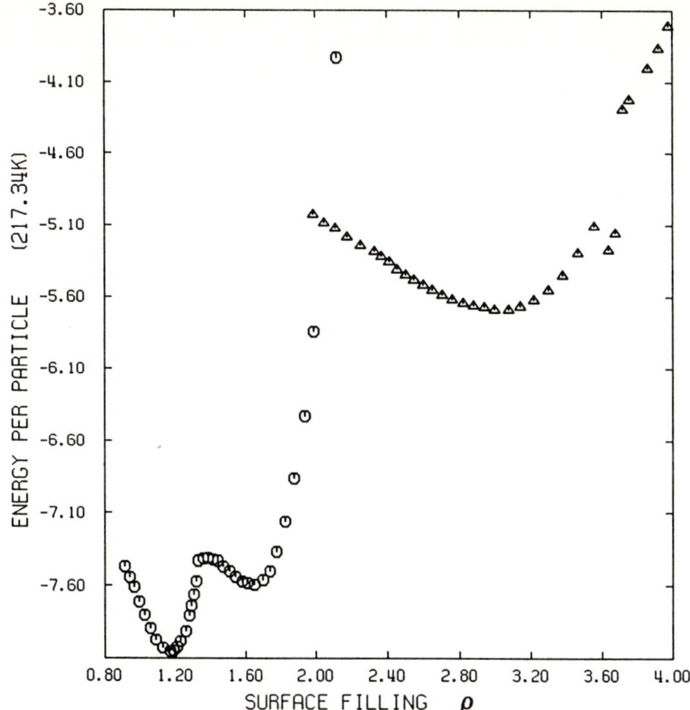

Figure .3: Energy per particle versus surface coverage ρ for the monolayer (circles) and the commensurate bilayer (triangles). The filling is in units of the $\sqrt{3} \times \sqrt{3}$ graphite commensurate structure and the energy is in units of $4\epsilon = 217.34$K.

the centered parallelogram structure and the orientational epitaxy of O_2 monolayers adsorbed on graphite can be explained via the lateral variations in the gas-surface interaction potential. This work shows a minimum in the potential energy surface of the adsorbed monolayer at 0K. There is the question as to whether these superlattices, calculated at zero degrees Kelvin, are stable at the experimental temperatures ($15\text{K} \leq T_{ex} \leq 24\text{K}$). To answer this question satisfactorily, details of the energy well and the height of the saddle points are necessary. This is not within the scope of the present work. The difference between the binding energy of the superlattice and the binding energy of the incommensurate monolayer, and the experimental temperature is insufficient to determine the stability of the superlattice. In the case of the δ and δ' phases, the long range order of the superlattice may be destroyed by the high temperature at which the measurements were made but some remnant of the superlattice epitaxy is possible, as discussed above. In the case of the ζ_1 and ζ_2 phases, the difference between the binding energy of the superlattice and the binding energy of the incommensurate monolayer is near the lower limit of the experimental temperatures and may only be slightly perturbed from the superlattice structures presented here.

References

1. J. A. Venables and D. J. Ball, Proc. Roy. Soc. London, **A322** , 331 (1971).
2. M. Weissmann and N. V. Cohan, J. Chem. Phys., **72** , 4562 (1980).

3. R.D. Etters and Ru-Pin Pan, Phys. Rev. Lett., **45** , 645, (1980); R.P. Pan, R.D. Etters, K. Kobashi and V. Chandresekharan, J. Chem. Phys., **77** , 1035 (1982).

4. M.F. Toney, D. Diehl and S.C. Fain Jr., Phys. Rev.,**B27** , 6413 (1983).

5. P.A. Heiney, P.W. Stephens, S.G.J. Mochrie, J. Akimitsu, R.J. Birgeneau and P.M. Horn, Surf. Sci. **125** , 539 (1983).

6. M. Nielsen and J. P. McTague, Phys. Rev. **B19** , 3096 (1979).

7. S.C. Fain, Jr., M.F. Toney and R.D. Diehl, **Proc. Fifth Int. Conf. on Solid Surfaces**, to be published.

8. R. Marx and R. Braun, Solid State Comm., **33** , 229 (1980); **Proc. Fourth Int. Conf. on Solid State Surfaces and the European Conf. on Surface Science**, Sept. 22 - Sept. 26 (1980). pp.100-103

9. J. Stoltenberg and O. E. Vilches, Phys. Rev., **B22** , 20 (1980).

10. P. W. Stephens, P. A. Heiney, R. J. Birgeneau, P. M. Horn, J. Stoltenberg, and O. E. Vilches, Phys. Rev. Lett., **45** , 1959 (1983).

11. J. Villain and M.B. Gordon, Surf. Sci. **125** , 1 (1983).

12. J.E.Lennard-Jones, Proc. Roy. Soc. (London), **A106** , 463 (1924); P.M. Morse, Phys. Rev., **34** , 57 (1929).

13. C. A. English and J. A. Venables, Proc. Roy. Soc.(London), **A340** , 57 (1974)

14. D.E. Strogyn and A.P. Strogyn, Molec. Phys. **11** , 371 (1966).

15. W. A. Steele, Surface Science **36** , 317 (1973).

16. D. J. Wilde and C. S. Brightler, **Foundations of Optimization**, Prentice-Hall, Englewood Cliffs, N.J. (1967).

17. M. F. Toney and S.C. Fain, Phys Rev. **B15** , in press.

18. S.G.J. Mochrie, M. Sutton, J. Akimitsu, R.J. Birgeneau, P.M. Horn, P.Dimon, and D.E. Moncton, Surf. Sci., in press.

Part II

Quantum Systems

Classical and Quantum Simulations of Quasi Two-Dimensional Condensed Phases: Krypton and Helium on Graphite

F.F. Abraham

IBM Research Division, Almaden Research Center, 650 Harry Road, San Jose, CA 95120, USA

1. Introduction

Employing the computer simulation methods of classical and quantum statistical mechanics, we have studied the thermodynamics structures of rare-gas krypton and helium monolayer films physisorbed on graphite. While the adsorbate-adsorbate and adsorbate-substrate interactions are short-range and simple, the incommensurability of the unrelaxed rare-gas film and the graphite substrate gives rise to a relaxed modulated film structure exhibiting the symmetry of the substrate face with characteristic length scales spanning tens to thousands of angstroms. The distinct advantage of computer simulation is that these microstructures may be directly "seen" at the atomistic level. In Sect. 2, we discuss the simulation of classical krypton on graphite where the incommensurate solid phase is investigated using over one hundred thousand atoms and where the spatial patterns span dimensions ranging from tens to thousands of angstroms. In Sect. 3, we describe the simulation of quantum helium on graphite using the Feynman path integral Monte Carlo method. In this case, we can study systems of about one hundred atoms, but we have simulated a rich variety of different phases seen experimentally.

2. The Incommensurate Honeycomb Phase of Krypton on Graphite

A high-density surface layer of krypton physisorbed on graphite is a good model system to study the properties of the incommensurate solid phase and the commensurate-incommensurate (C-IC) transition. The krypton-carbon interaction favors regularly spaced adsorption sites at the graphite surface, and it is well known from experiment [1,2] that monolayer krypton on graphite forms a solid in registry with the underlying substrate. As a consequence of the size of the krypton atom, only one-third of the adsorption sites are occupied, and there exist three energetically degenerate commensurate sublattices. With increasing coverage, it is no longer possible for all of the krypton atoms to occupy adsorption sites, and the system becomes more and more incommensurate, approaching a lattice constant representative of bulk krypton. However, in the transition region, the krypton solid is significantly modulated due to the krypton-graphite interaction. The nature of the incommensurate phase of krypton on graphite and its transition to the $\sqrt{3} \times \sqrt{3}$ R30° commensurate phase are being extensively investigated by laboratory experiment [1-9] by theory [10-20] and by computer simulation [21-27].

Theories have been developed for the incommensurate phase at zero temperature which are based on the model of incommensurate domain walls separating large commensurate regions such that the distance between domain walls is large compared to the width of the walls. For krypton on graphite, the domain walls may be in three different directions because of the hexagonal substrate structure. If wall intersections are

Springer Proceedings in Physics Vol. 33: **Computer Simulation Studies in Condensed Matter Physics**
Editors: D. P. Landau · K. K. Mon · H-B. Schüttler

energetically unfavorable, a *striped* phase might be expected where walls are only in one direction. However, VILLAIN [12,13] has noted that a *honeycomb* array of walls has a degeneracy in which the hexagons of the array can expand or contract without changing the total wall length or the number of nodes, i.e., wall crossings. Hence, Villain argues that this additional contribution to the entropy stabilizes the honeycomb phase relative to the striped phase.

Using molecular dynamics, we have simulated the structure of the incommensurate krypton film adsorbed on graphite as a function of temperature and coverage. We have used systems as large as 103,041 and 161,604 krypton atoms [23]. With this many atoms, graphite substrate dimensions up to 1700Å are realized and are comparable to present-day laboratory capabilities. We observe that the incommensurate phase consists of commensurate islands separated by a interconnecting network of incommensurate domain walls, the structure of this network being a sensitive function of temperature and coverage. At low temperatures, the honeycomb network of domain walls is observed for all coverages. An incommensurate striped phase is not seen. With increasing temperature, distortions from the perfect honeycomb structure become more prevalent. At high temperatures, the individual domain walls fluctuate significantly from the symmetry directions while possessing boundary roughness and a greater wall thickness.

In our studies, we have adopted the Lennard-Jones 12:6 pair potential to represent the interaction between the various atoms of the krypton-graphite system:

$$u(r) = 4\varepsilon\left\{\left(\frac{\sigma}{r}\right)^{12} - \left(\frac{\sigma}{r}\right)^{6}\right\}, \tag{2.1}$$

where r is the interatomic separation. The range of interaction is normally cut off at some distance to lessen the computational burden of summing over all atoms, the cutoff typically being at 2.5σ. When using the Lennard-Jones potential, we normally express quantities in terms of *reduced units*. Lengths are scaled by the parameter σ, the value of the interatomic separation for which the LJ potential is zero, and energies are scaled by the parameter ε, the depth of the minimum of the LJ potential. Reduced temperature is therefore kT/ε. Simple pairwise additivity of the atomic interactions is assumed, and the carbon atoms defining the semi-infinite solid are fixed at their lattice sites. In this case, the total potential energy U for a given configuration of N krypton atoms $\vec{r}(i)$, i = 1, ... N, above a *fixed* configuration of carbon atoms defining the graphite semi-infinite solid $\vec{R}(j)$, j = 1, ..., ∞, has the form

$$U = \sum_{i>j}^{N} u_{Kr-Kr}(|\vec{r}(i) - \vec{r}(j)|) + \sum_{\substack{i=1\\ \{Kr\}}}^{N} \sum_{\substack{j=1\\ \{C\}}}^{\infty} u_{Kr-C}(|\vec{r}(i) - \vec{R}(j)|). \tag{2.2}$$

The krypton-graphite interaction may be evaluated analytically by expanding this potential as a Fourier sum in the surface reciprocal lattice vectors in order to lessen the computational burden.

Figure 1 shows the principal results of our simulations for the 103,041 atom system. The incommensurate and commensurate regions are shown as solid black and solid white areas, respectively; in actual fact, we plotted the incommensurate atoms as points but the lack of graphical resolution merged the points to make a solid region. We first consider the case where the temperature is fixed at the low value of 0.05 and the coverage is varied. We note that for all coverages a honeycomb network of domain walls is established, the network consisting of straight walls with smooth boundaries which are aligned with the three symmetry directions of the graphite substrate. The commensurate regions form an array of honeycomb domains, the individual hexagons not being identical in size and shape. This honeycomb domain structure with *breathing* freedom is direct confirmation

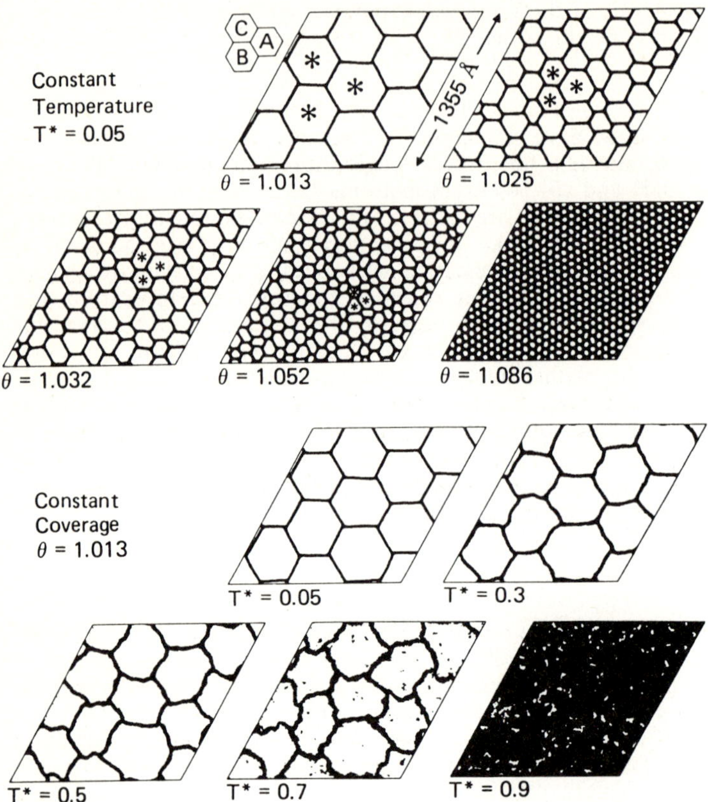

Fig. 1. Pictures of the domain-wall network for an equilibrium configuration of the incommensurate phase as a function of coverage θ at fixed temperature T = 0.05 and as a function of temperature at a fixed coverage θ = 1.013.

of Villain's picture of the incommensurate phase, and this is the first direct observation of this structure. At fixed temperature, the percentage of krypton atoms that are commensurate (%C) decreases linearly with increasing coverage (90%, 80%, 75%, 60% and 37%, respectively), while the domain-wall thickness remains essentially constant at 18Å. This %C decrease is associated with an increase of the total length of domain walls, and this gives rise to smaller and more numerous commensurate domains. We simulated as large a system as we felt was practical within the constraint of our computer resources – a 161,604 krypton atom system. The temperature and coverage are 0.05 and 1.005, respectively. In Fig. 2, we again see the incommensurate honeycomb structure at this low temperature.

One can visualize two extremes for the atomic microstructure of the incommensurate state: (1) the krypton atoms are near registry except for thin domain walls [28], or (2) the krypton monolayer is a lattice that is weakly modulated by the substrate field [29]. In Fig. 3, we present the honeycomb picture for T* = 0.05, θ = 1.025 and for the three commensurate-radius cutoff criteria δ = 0.1, 0.2 and 0.3. We conclude that at this low temperature, the krypton lattice is modulated according to the McTague-Novaco picture. With increasing coverage, this modulation will decrease, i.e., when the wall separation approaches the thickness of an individual wall.

Returning to Fig. 1, we now consider the series of simulations where the coverage was held fixed at 1.013. With increasing temperature, the overall appearance of the incommensurate phase remains that of the domain-wall network which becomes increasingly distorted, the walls becoming broader and the wall boundaries roughening considerably. Also, we note the marked increase in the wall thickness with increasing temperature, the wall thickness at $T^* = 0.7$ being approximately twice the wall thickness at $T^* = 0.05$. This is consistent with the gradual decrease of percentage of commensurate atoms (90%, 88%, 85%, 81%, respectively). At the highest temperature, the system is mainly incommensurate since the krypton film has melted into a liquid.

To classify the domain walls for krypton on graphite, one can distinguish between two configurationally distinct types of walls, which we will call *heavy (h)* and *light (l)* walls [17]. In some other references, e.g., KARDAR and BERKER [18], these walls are labelled "super heavy" and "heavy" walls, respectively. The classification is determined by the orientation of the particular wall and by the two different sublattices separated by this wall. Analyzing the pictures in Fig. 1, we find only heavy walls at the lower temperatures. At temperatures above $kT/\varepsilon = 0.5$, the wall orientation can be characterized as a distribution with peaks around the three symmetry directions. Those walls which can be classified according to the above discussed scheme are almost exclusively ($\gtrsim 95\%$) heavy walls. Only sometimes some short wall segments fulfill the condition required for light walls. In Fig. 4, we note the beautiful temporal *breathing* of the honeycomb domain structure, and this is the first direct observation of this temporal behavior.

To study the energetics of the weakly incommensurate phase, we have simulated a quasi two-dimensional system of 20,736 krypton atoms on a graphite substrate, and the results are discussed in reference [30].

Computer simulations of classical [molecular] physisorbed films are now being undertaken and are leading to unique insights into their associated structural phase

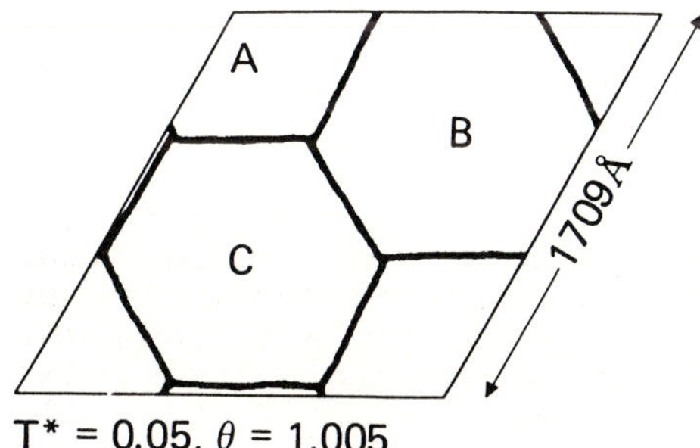

Fig. 2. A picture of the domain-wall network for an equilibrium configuration of the incommensurate phase simulated using 161,604 krypton atoms on graphite at a coverage $\theta = 1.005$ and temperature $T^* = 0.05$. The percentage of commensurate atoms %C is 96.

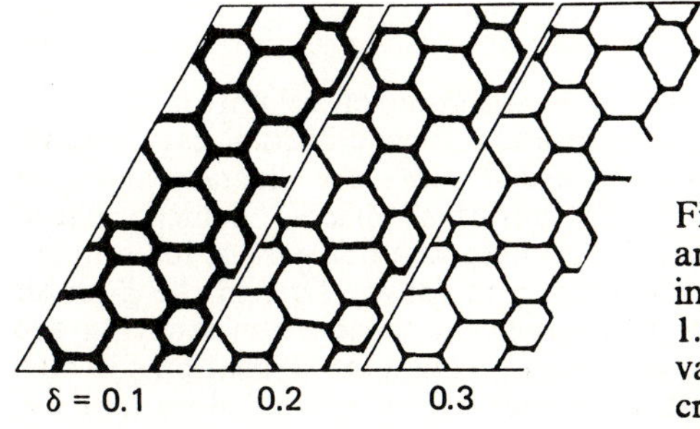

Fig. 3. Pictures of the domain-wall network for an equilibrium configuration of the incommensurate phase at a coverage $\theta = 1.025$ and temperature $T^* = 0.05$, for various values of the commensurate-radius cutoff criterion δ.

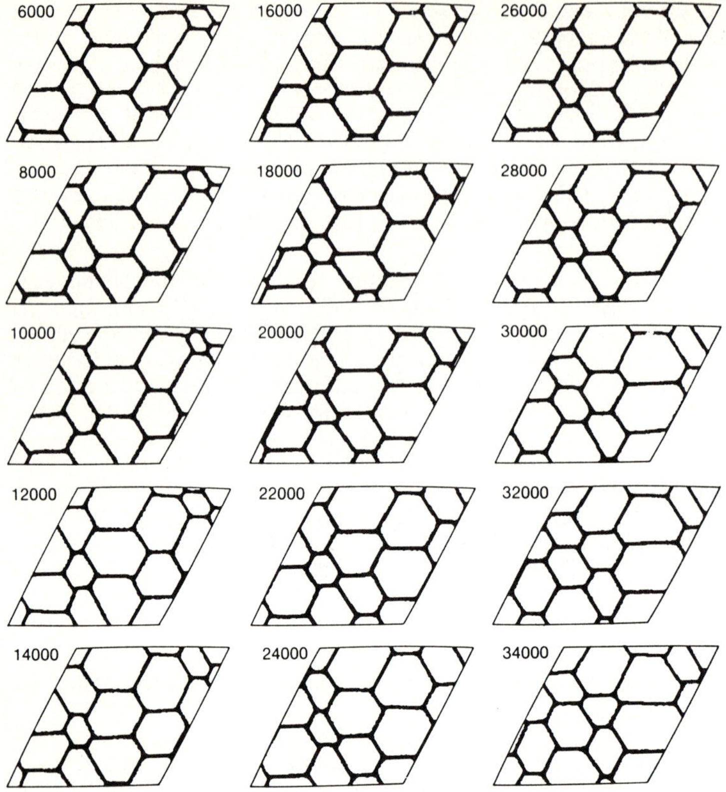

Fig. 4. Temporal "breathing" of the equilibrium honeycomb domain-wall structure. The numbers denote time in units of 100 time-steps.

transitions (e.g., MIGONE et al. [31]; TALBOT et al. [32]; SOKOLOWKSI and STEELE [33]; Piper et al. [34]; PETERS and KLEIN [35,36]). For example, the effect of compressing monolayers of N_2 physisorbed on graphite has been studied by PETERS and KLEIN [36]. They find that axial compression of a commensurate ordered herringbone monolayer initially generates striped domain walls about 35Å wide rather than a uniform incommensurate structure. Also, orientational order in the domain walls persists to higher temperatures than in the commensurate regions. The molecular structure provides a greater richness of ordering phenomena.

3. The Incommensurate Striped Phase of Helium on Graphite [37]

This study was largely inspired by the beautiful phase diagram of helium on graphite which was presented by SCHICK [38] as his interpretation of experimental data, and by our desire to see if quantum Monte Carlo [39,40] could verify the diagram and give important information on the microstructure of the adsorbed phases. Schick's phase diagram is presented in Fig. 5, but we note that the β-phase was not part of his original picture. In fact, our expectation was to find a typical high-density fluid in the β-phase region. Up to the present, there have been a few significant successes of studying the quantum many-body problem at finite temperature using computer simulation [39,40]. However, we would highlight the very impressive He^4 simulation of Bose-Einstein

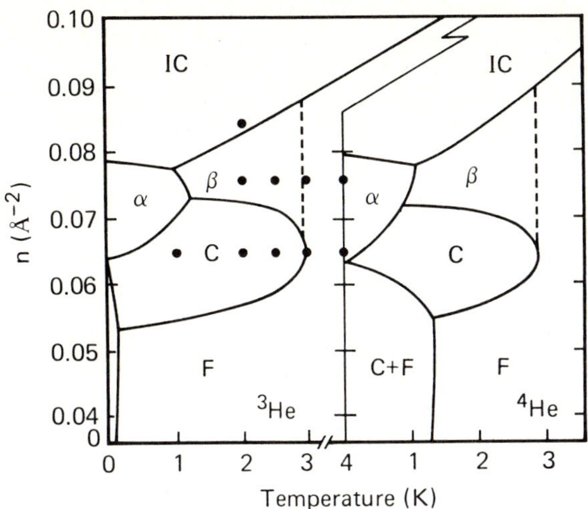

Fig. 5. Phase diagram for He^3 and He^4 on graphite [38]. The beta-phases The β-phases were not a part of the original published diagrams and were taken from [42]. The solid circles denote temperature-density locations were the Monte Carlo simulations were performed.

condensation by CEPERLEY and POLLOCK [41]. We selected He^3 for our study so as to attack the Fermi statistics problem which can plague attempts to obtain reasonable averages basic to the Monte Carlo sampling procedure. Only after we had implemented a scheme for accounting for Fermi statistics did we learn from computer experimentation that exchange was not important at the temperatures and densities of our study. This can be appreciated by the observation that the phase diagrams of He^3 and He^4 on graphite (Fig. 5) are very similar.

We will demonstrate that the phases of He^3 adsorbed on graphite can be accurately simulated by the Feynman path-integral Monte Carlo method, realistic potential functions for the substrate-adsorbate and adsorbate-adsorbate interactions, and a three-dimensional geometry. The fluid, commensurate solid, incommensurate solid, and reentrant fluid (β) phases are found and are in agreement with the experimental phase diagram. The microscopic structure of the re-entrant fluid is observed to be a striped domain-wall liquid, in agreement with the experimental interpretation of MOTTELER [42] and the striped helical Potts model calculation of HALPIN-HEALY and KARDAR [43]. It was only after our simulation of the *fluid* in the β-phase region that we were made aware of the fact that it had been suggested earlier [44].

In the Feynman path-integral representation, a single quantum particle is isomorphic to a classical cyclic polymer chain of M beads in which each bead j interacts with its neighboring beads $j-1$ and $j+1$ through a harmonic force constant $mM/\hbar^2\beta^2$ and experiences a reduced external potential $V(\vec{r}(j))/M$ (the particle mass is m, and β is inverse temperature.) The harmonic coupling arises from the free particle propagator f describing the quantum mechanical contribution of kinetic energy to the density matrix:

$$f\left(\vec{r}(j+1),\vec{r}(j);\frac{\beta}{M},L\right) \equiv \left(\frac{Mm}{2\pi\beta\hbar^2}\right)^{3/2} \times \sum_n \exp\left(-\frac{Mm}{2\beta\hbar^2}(\vec{r}(j+1)-\vec{r}(j)-n\cdot L)^2\right). \tag{3.1}$$

This representation is exact only in the limit of M going to infinity. For a given temperature and density, one has to determine empirically that M beyond which the thermodynamic properties do not effectively change. The lower the temperature, the larger M must be. We have adopted a higher order correction to this "high temperature approximation." It takes the form of a simple modification to the potential energy V [45]:

$$V'(\vec{r}(j)) = V(\vec{r}(j)) + \frac{1}{24}\frac{\hbar^2}{m}\left(\frac{\beta}{M}\right)^2\left(\frac{\partial V}{\partial \vec{r}(j)}\right)^2. \tag{3.2}$$

We refer the reader to the papers of TAKAHASHI and IMADA [45-47] for a detailed description of the path-integral Monte Carlo method that we implemented. Unless the He^3 atoms were treated as distinguishable, direct calculation of the determinant of free particle propagators was performed in evaluating the Monte Carlo weight function |W| for the spinless fermion system of N atoms [48]:

$$W = (N!)^{-M}\prod_{j=1}^{M} \det A(j+1, j) \exp\left(-\frac{\beta}{M}\sum_{j=1}^{M} V'(j)\right), \tag{3.3}$$

and

$$\{A(j+1)\}_{k,l} = f(\vec{r}_k(j+1), \vec{r}_l(j)),\ 1 \le k, l \le N, \tag{3.4}$$

where $V'(j) = V'(\vec{r}_1(j), ..., \vec{r}_N(j))$ is defined by (3.2), and $A(j+1,j)$ is an $N \times N$ matrix of the one particle propagator matrix elements $f(\vec{r}_k(j+1), \vec{r}_l(j))$. The absolute value of W is taken since the weight function in importance sampling should be positive. Two kinds of displacements of coordinates are adopted for importance sampling; a "microscopic" displacement of an individual bead and a "macroscopic" displacement of all of the beads in a cyclic polymer chain according to the recipe of TAKAHASHI and IMADA [45]. One Monte Carlo move is defined as N attempted macroscopic displacements, each one made after M trials of the microscopic displacements. Primitive displacement parameters were adjusted so that the acceptance ratios for microscopic and macroscopic displacements are approximately one-half.

In our Monte Carlo simulations, the number of atoms varied from 36 to 42, depending on the coverage of interest, chosen dimensions of the graphite substrate, and compatibility with periodic conditions for the initialized triangular lattice of a helium solid and the graphite lattice. Periodic boundary conditions were imposed at the four faces of the computational cell which pass through the sides of the basal plane at normal incidence to the surface. A reflecting wall was placed at the top of the computational box beyond the second layer height, but no atom was promoted to the second layer in any of the simulations. We adopted the Lennard-Jones 12-6 pair potential to represent the interaction between helium atoms and helium-carbon atoms, and the potential parameters are taken from COLE and KLEIN [49]. Similar parameters have been shown to describe the phase diagram of classical rare-gas atoms on graphite very well [50]. The graphite was modeled as a semi-infinite rigid solid.

By experimentation, we found that 96 beads are required to obtain energy convergence at 2.0K, and we used bead numbers of 84, 72 and 48 for temperatures of 2.5K, 3.0K and 4.0K, respectively. We also learned that fermion exchange is unimportant for temperatures equal to and above 2.0K for the coverages of our study. In Fig. 5, the solid circles denote temperature-density locations where the Monte Carlo simulations were performed. In Fig. 6, the probability density contour plots for the helium beads are presented for a coverage of unity and for temperatures of 2.0K, 2.5K, 3.0K and 4.0K, respectively. The distribution was obtained by averaging over 10,000 Monte Carlo moves after at least 10,000 previous moves were made from the initialized triangular lattice. The

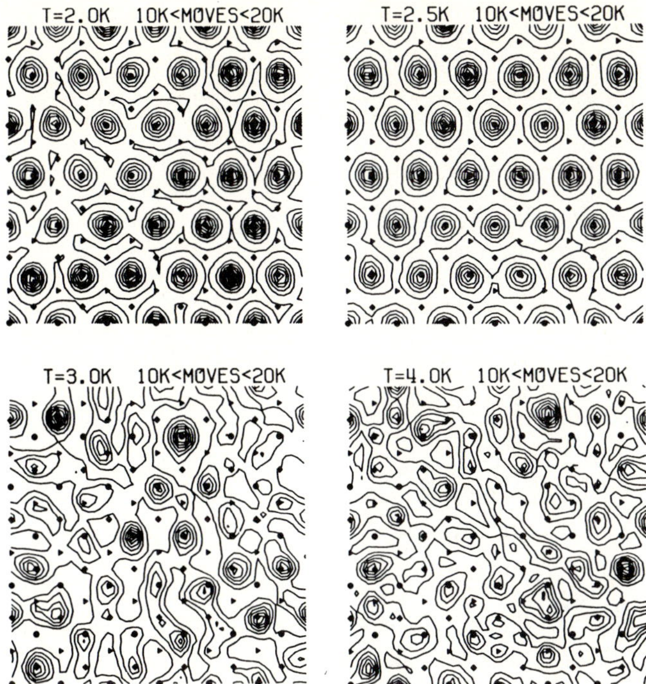

Fig. 6. The probability density contours for the helium beads averaged over 10,000 Monte Carlo moves. The coverage is unity, and the temperatures are 2.0K, 2.5K, 3.0K and 4.0K, respectively. The solid circles, triangles and diamonds denote the three-fold energetically degenerate adsorption sites on graphite for a commensurate solid.

averaging interval is given above each picture. As a consequence of the size of the helium atom, only one-third of the adsorption sites can be occupied by a commensurate solid, and there exist three energetically degenerate commensurate sublattices. The solid circles, triangles and diamonds denote these commensurate sublattices. We see that the commensurate solid state exists at 2.0K and at 2.5K and that the fluid state exists at 3.0K and 4.0K, in agreement with experiment (Fig. 5). In Fig. 7, the probability density contour plots for the helium beads are presented for the temperature of 2.0K and the coverages of 0.83, 1.00, 1.17 and 1.33, respectively. We see the low density fluid, commensurate solid, high-density reentrant fluid, and high-density incommensurate solid as we pass from low to high coverage, and these phases appear to be in agreement with the experimental phase diagram (Fig. 5). In particular, the microscopic structure of the reentrant fluid is observed to be a striped domain-wall liquid, in agreement with the experimental interpretation of MOTTELER [42] and the striped helical Potts model calculation of HALPIN-HEALY and KARDAR [43]. The shaded stripes pictorially denote the striped domain walls. The wall thickness is approximately one to two helium atomic diameters and is consistent with the prediction of theory [43].

It is essential to treat much larger size systems in order to achieve the needed detail required for a quantitative analysis. In particular, the microscopic structure of the domain-wall network is needed. Also, lower temperature simulations will allow a thorough examination of the phase diagram, e.g., the α-phase. We are presently implementing algorithms requiring fewer beads and amenable to parallel computation in hopes that we can approach systems of a thousand atoms using a special purpose array computer [50].

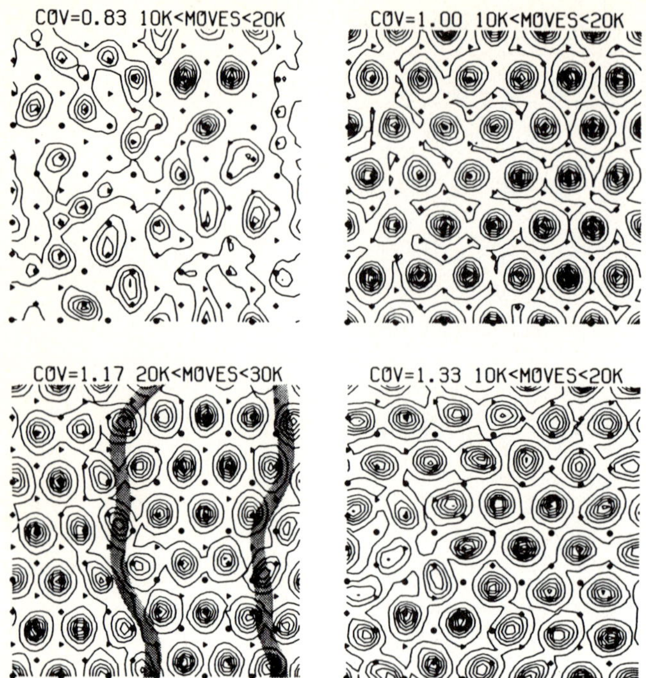

Fig. 7. The probability density contours for the helium beads averaged over 10,000 Monte Carlo moves. The temperature is 2.0K, and the coverages are 0.83, 1.00, 1.17 and 1.33, respectively. The shaded stripes pictorially denote the striped domain walls. See Fig. 6 for additional details.

References

1. O. E. Vilches: Ann. Rev. Phys. Chem. 31, 463 (1980)
2. A. Thomy, X. Duval, J. Regnier: Surf. Sci. Rep. 1, 1 (1981)
3. M. Nielsen, J. Als-Nielsen, J. Bohr, J. P. McTague: Phys. Rev. Lett. 47, 582 (1981)
4. M. D. Chinn, S. C. Fain: Phys. Rev. Lett. 39, 146 (1977)
5. S. C. Fain, M. D. Chinn, R. D. Diehl: Phys. Rev. B 21, 4170 (1980)
6. P. W. Stephens, P. Heiney, R. J. Birgeneau, P. M. Horn: Phys. Rev. Lett. 43, 47 (1979)
7. P. W. Stephens, P. A. Heiney, R. J. Birgeneau, P. M. Horn: D. E. Moncton, G. S. Brown, Phys. Rev. B 29, 3512 (1984)
8. R. J. Birgeneau, E. M. Hammonds, P. Heiney, P. W. Stephens, P. M. Horn: In Ordering in Two Dimensions, ed. by S. K. Sinha (Plenum, New York 1980)
9. D. E. Moncton, P. W. Stephens, R. J. Birgeneau, P. M. Horn, G. S. Brown: Phys. Rev. Lett. 46, 1533 (1981)
10. P. Bak, D. Mukamel, J. Villain, K. Wentkowska: Phys. Rev. B 19, 1610 (1979)
11. H. Shiba: J. Phys. Soc. Jpn. 46, 1852 (1979)
12. J. Villain: In Ordering in Strongly Fluctuating Condensed Matter Systems, ed. by T. Riste (Plenum, New York 1980), p.221
13. J. Villain: In Ordering in Two Dimensions, ed. by S. K. Sinha (North-Holland New York 1980), p.123
14. P. Bak: Rep. Prog. Theor. Phys. 45, 587 (1982)
15. S. N. Coppersmith, D. S. Fisher, B. I. Halperin, P. A. Lee, W. F. Brinkman: Phys. Rev. Lett. 46, 549 (1981)

16. S. N. Coppersmith, D. S. Fisher, B. I. Halperin, P. A. Lee, W. F. Brinkman: Phys. Rev. B 25, 349 (1982)
17. D. A. Huse, M. E. Fisher: Phys. Rev. Lett. 49, 793 (1982)
18. M. Kardar, A. N. Berker: Phys. Rev. Lett. 48, 1552 (1982)
19. J. Villain, M. B. Gordon: Surf. Sci. 124, 1 (1983). This gives an up-to-date review and comparison with experiment.
20. M. Schoebinger, S. W. Koch: Z. Phys. B 53, 233 (1983)
21. F.Hanson, J. P. McTague: J. Chem. Phys. 72, 6363 (1980)
22. F. F. Abraham, S. W. Koch, W. E. Rudge: Phys. Rev. Lett. 49, 1830 (1982)
23. F. F. Abraham, W. E. Rudge, D. J. Auerbach, S. W. Koch: Phys. Rev. Lett. 52, 445 (1984)
24. S. W. Koch, F. F. Abraham: Helv. Phys. Acta 56, 755 (1983)
25. S. W. Koch, W. E. Rudge, F. F. Abraham: Surf. Sci. 145, 329 (1984)
26. R. J. Gooding, B. Joos, B. Bergerson: Phys. Rev. B 27, 7669 (1983)
27. J. S. Whitehouse, D. Nicholson, N. G. Parsonage: Mol. Phys. 49, 829 (1983)
28. F. C. Frank, J. H. van der Merwe: Proc. Roy. Soc. London 198, 205 (1949)
29. J. P. McTague, A. D. Novaco: Phys. Rev. B 19, 5299 (1979)
30. M. Schoebinger, F. F. Abraham: Phys. Rev. B 31, 4590 (1985)
31. A. D. Migone, H. K. Kim, M. H. W. Chan, J. Talbot, D. J. Tildesley, W. A. Steele: Phys. Rev. Lett. 51, 192 (1983)
32. J. Talbot, D. J. Tildesley, W. A. Steele: Mol. Phys. 51, 1331 (1983)
33. S. Sokolowski, W. A. Steele: Mol. Phys. 54, 1453 (1985)
34. J. Piper, J. A. Morrison, C. Peters: Mol. Phys. 53, 1463 (1984)
35. C. Peters, M. L. Klein: Mol. Phys. 54, 895 (1985)
36. C. Peters, M. L. Klein: Phys. Rev. B 32, 6077 (1985)
37. F. F. Abraham, J. Q. Broughton: Phys. Rev. Lett. 59, 64 (1987)
38. M. Schick: In Phase Transitions in Surface Films, ed. by J. G. Dash and J. Ruvalds (Plenum, New York 1981), p.68
39. B. J. Alder, D. M. Ceperley, E. L. Pollock: Acc. Chem. Res. 18, 268 (1985)
40. D. M. Ceperley, B. J. Alder: Science 231, 555 (1986)
41. D. M. Ceperley, E. L. Pollock: Phys. Rev. Lett. 56, 351 (1986)
42. F. C. Motteler: "A heat capacity study of p-H_2 monolayers on graphite," Ph.D. Dissertation (University of Washington, Seattle, Washington 1986).
43. T. Halpin-Healy, M. Kardar: Phys. Rev. B34, 318 (1986).
44. S. C. Fain: private communication. The author is indebted to Professor Fain for bringing references 42 and 43 to his attention.
45. M. Takahashi, M. Imada: J. Phys. Soc. Jpn. 53, 963 (1984)
46. M. Takahashi, M. Imada: J. Phys. Soc. Jpn. 53, 3765 (1984)
47. M. Imada, M. Takahashi: J. Phys. Soc. Jpn. 53, 3770 (1984)
48. In reference 45, it is stated that the determinate evaluation scales as the cube power of the number of particles. However, in the Monte Carlo procedure, only one row is changed for an attempted displacement, and an algorithm for the determinate evaluation has been devised that scales as the square of the number of particles (Nimrod Megiddo, IBM ARC, San Jose, California, private communication).
49. M. W. Cole, J. R. Klein: Surf. Science 124, 547 (1983)
50. F. F. Abraham: Adv. Phys. 35, Sect. 4.2, 1 (1986)

Path-Integral and Real-Time Dynamics Simulations of Quantum Systems

U. Landman *

School of Physics, Georgia Institute of Technology,
Atlanta, GA 30332, USA

1. Introduction

While classical molecular dynamics (MD) and Monte Carlo (MC) simulations have become standard theoretical tools in studies of condensed matter and material science [1], treatments of quantum systems via computer simulations are more scarce. Early proposals for studies of quantum systems have been hampered by prohibitive computational demands. However, progress in the area of computers coupled with theoretical advances brought about renewed endeavors in quantum simulations [1-8] resulting in the formulation, implementation and application of several methods to studies of quantum field theory, many-fermion systems, quantum fluids and crystals, models of nuclear matter, electronic structure of molecules, electron localization in fluids and small clusters and quantum adsorption systems. Among the methods which were developed we note [1-8] the quantum Monte Carlo (QMC) methods, Green's function Monte Carlo (GFMC), Quantum Path-Integral Molecular Dynamics (QUPID), Path-Integral Monte Carlo (PIMC), and time-dependent self-consistent-field (TDSCF) methods in conjunction with the fast Fourier transform (FFT) method for the solution of the time-dependent Schrodinger equation and with classical molecular dynamics. In this chapter we focus on the QUPID and MD-TDSCF techniques. Following brief reviews of the methods in the next section, we turn to several illustrations of their use in recent studies. The examples which we choose are drawn from investigations performed in our laboratory. For entrance into the growing body of literature on these subjects the reader is referred to recent reviews [7,8] and to the cited references.

2. Methods

a. Quantum Path-Integral Molecular Dynamics

Recent developments [7,9-14] of simulation methods of quantum systems rest upon the path-integral formulation [15] of quantum statistical mechanics. Given a Hamiltonian for the system (considered for simplicity to consist of one quantum particle, an electron, interacting with a set of classical ones) the method is based on an isomorphism between calculation of the quantum partition function, Z, of the system given by $Z = \mathrm{Tr}[\exp(-\beta H)]$, and the classical calculation of a partition function, for an isomorphic system, given by

$$Z = \lim_{P\to\infty} \left[\frac{mP}{2\pi\hbar^2\beta}\right]^{3P/2} \int \exp(-\beta V^e_{eff})\, d\tau \quad , \tag{1}$$

where the effective potential is given by

$$V^e_{eff} = \sum_{i=1}^{P} \left[\frac{Pm}{2\hbar^2\beta^2} (\vec{r}_i - \vec{r}_{i+1})^2 + \frac{V_e(\vec{r}_i)}{P} \right] \quad , \tag{2}$$

where it is understood here that $\vec{r}_{(P+1)} \equiv \vec{r}_1$ and the integration in Eq. (1) is taken

Springer Proceedings in Physics Vol. 33: **Computer Simulation Studies in Condensed Matter Physics**
Editors: D. P. Landau • K. K. Mon • H-B. Schüttler

over the volume $d\tau = d^3r_1 \ldots d^3r_P$. $V_e(\vec{r})$ is the interaction potential between the electron and the classical particles.

In the isomorphic classical problem, the electron is mapped onto a closed chain (necklace) of P pseudo-classical particles (beads) with nearest-neighbor harmonic interactions whose strength is determined by the mass (m), the inverse temperature (β) and the number of beads (P). The average electronic energy of the system can then be evaluated from the relation

$$\langle E\rangle = -\frac{\partial}{\partial\beta}\ell nZ_P = K_e + \frac{1}{P}\langle\sum_{i=1}^{P} V_e(\vec{r}_i)\rangle \tag{3}$$

where the pointed brackets indicate averages over the Boltzmann distribution as defined in Eq. (1). The electron kinetic energy estimator is given by [11,13]

$$K_e = \frac{3}{2\beta} + \frac{1}{2P}\sum_{i=1}^{P}\langle\frac{\partial V_e(\vec{r}_i)}{\partial\vec{r}_i}\bullet(\vec{r}_i-\vec{r}_P)\rangle . \tag{4}$$

The first term on the r.h.s. of Eq. (15) is the free particle contribution, K_{free}, and the second term K_{int} is the contribution from the interaction with the classical particles.

In the quantum path-integral molecular dynamics (QUPID) method, statistical sampling over the Boltzmann distribution is replaced by the equivalent [11] averaging over the phase-space trajectories generated via classical MD by the Hamiltonian

$$H = \sum_{i=1}^{P}\frac{m^*\dot{\vec{r}}_i^2}{2} + \sum_{i=1}^{N}\frac{M_I^*\dot{\vec{R}}_I^2}{2} + \sum_{i=1}^{P}\left[\frac{Pm}{2\hbar^2\beta}(\vec{r}_i-\vec{r}_{i+1})^2 + \frac{1}{P}V_e\right] + V_I . \tag{5}$$

The masses m^* and M_I^* can be chosen to be arbitrary, since in classical systems configurational averages do not depend on the masses appearing in the kinetic energy term. V_I is the interatomic interaction potential.

Extensions of the path-integral Monte Carlo method for treatments of many-boson and many-fermion systems have been made and applied [5,6,17,18].

b. Time-Dependent SCF

While the quantum simulation methods mentioned above open new avenues for investigations of the ground state and thermal equilibrium properties of many systems of interest, they are not ideally suited for studies of real-time dynamical processes. Nevertheless recent progress toward this goal, which may prove to be of practical value, has been made [7,19-21]. An alternative approach is based on direct investigation of the time-dependent Schrödinger equation in conjunction with the development of the fast Fourier transform (FFT) algorithm for integrating the time dependent Schrodinger equation [22,23]. Since, in this method, the computational effort increases exponentially with the number of degrees of freedom, approximate methods have to be used. Among those, the time-dependent self-consistent field (TDSCF) [8,24,25] emerges as an important tool.

In the TDSCF Hartree approximation [25,26] the wavefunction for a system of N coupled quantum particles is written as a product of the single particle wavefunctions $\phi_i(\vec{r},t)$ which satisfy the equations of motion

$$i\hbar \dot{\phi}_j = H_j\phi_j + \langle V(\{\vec{r}\})\rangle_j \, \phi_j, \qquad (j = 1,\ldots,N) \tag{6}$$

where a dot represents differentiation with respect to time, $H_j(\vec{r}_j)$ is the single particle Hamiltonian for particle j and where $V(\{\vec{r}\})$ is the inter-particle interaction. $\langle V(\{r\})\rangle_j$ denotes the quantum-mechanical expectation value of V over all coordinates except $\vec{r}_j$. Using this approximation leads to significant saving of computing effort. While there is no systematic way to estimate the accuracy of this approximation, there is a growing base of evidence, including comparisons to exact results, supporting the adequacy of the approximation in many situations. Failure of TDSCF procedures may often be traced to using a wrong set of coordinates (to be made separable by the approximation) or to situations where multiple configurations have to be explicitly taken into account [8,28-31].

A further simplification which leads to additional saving in computing effort is to limit the quantum dynamical description to a small subsystem, while describing the dynamics of the rest of the system by classical (or sometimes semiclassical) mechanics [8,26,27]. The equations of motion for such mixed quantum-classical systems are given by

$$i\hbar\dot{\psi}(\vec{r}_Q;\vec{r}_C,t) = H_Q(\vec{r}_C)\ \psi(\vec{r}_Q;\vec{r}_C,t), \tag{7a}$$

$$H_Q(\vec{r}_C) = K_Q + V(\vec{r}_Q,\vec{r}_C)\ , \tag{7b}$$

$$m_C\ \ddot{\vec{r}}_C = -\left\langle \frac{\partial V(\vec{r}_Q,\vec{r}_C)}{\partial \vec{r}_C} \right\rangle - \frac{\partial U(\vec{r}_C)}{\partial \vec{r}_C}\ , \tag{7c}$$

where $\vec{r}_Q$ and $\vec{r}_C$ denote collectively the sets of quantum and classical coordinates respectively, U is the interaction potential between the classical constituents of the system and $\langle V\rangle$ is the expectation value of the quantum-classical coupling potential taken with the instantaneous quantum wavefunction.

An appealing feature of the mixed quantum - classical TDSCF description is that it makes it possible to include temperature in the simulation. This is done by imposing thermal equilibration on the classical part of the system in one of several possible ways. In our simulations [27] this was done by occasionally randomizing the velocities of randomly chosen classical particles according to an appropriate Maxwellian distribution [32].

3. Sample Results

a. Quantum Path-Integral Molecular Dynamics Simulations

In this section we demonstrate the use of QUPID simulations in studies of electron attachment and localization modes to atomic and molecular clusters.

Clusters constitute a bridge between molecular and condensed matter systems, whose exploration provides ways and means for the elucidation of quantum, dynamic and chemical size effects in large finite systems [33-36]. The following physical and chemical phenomena associated with the attachment of an excess electron to clusters are of considerable interest:

(1) The parentage problem. In many interesting cases the excess electron is attached to a cluster whose individual atomic or molecular constituents do not form a stable negative ion.

(2) Localized and extended states of the excess electron.

(3) Bulk and surface states of the excess electron.

(4) Cluster reorganization accompanying electron attachment.

(5) Cluster isomerization induced by electron attachment.

We have utilized the quantum path integral molecular dynamics method and the MD-TDSCF method [27] to explore the compositional, structural and size dependence of electron localization in ionic [14,37] and hydrogen-bonded [38-40] clusters, which provide novel information regarding the energetics and dynamics [27] of electron localization modes in finite systems.

Ionic Clusters and Electron Induced Isomerization

We have chosen to study electron localization in alkali-halide clusters since the interionic interactions in these clusters [45] and adequate models of the interaction between an electron and the ionic constituents are available [11,14,37,41].

Simulations of electron attachment, at $T = 300^{o}K$, to positively charged clusters, $[Na_nCl_m]^{+(n-m)}$ with m = n-1 or n-2, have revealed [14]: (1) Internal electron localization in moderately large clusters containing an anion vacancy accompanied by a small nuclear configurational rearrangement, thus establishing the dominance of short-range attractive interactions for this process. (2) Surface state localization in moderately large, vacancy free clusters. For $[Na_{14}Cl_{13}]^{+}$, localization around a surface Na^{+} ion resulting in partial neutralization is found. (3) In small clusters novel effects of dissociative electron attachment and cluster isomerization induced by electron localization are predicted.

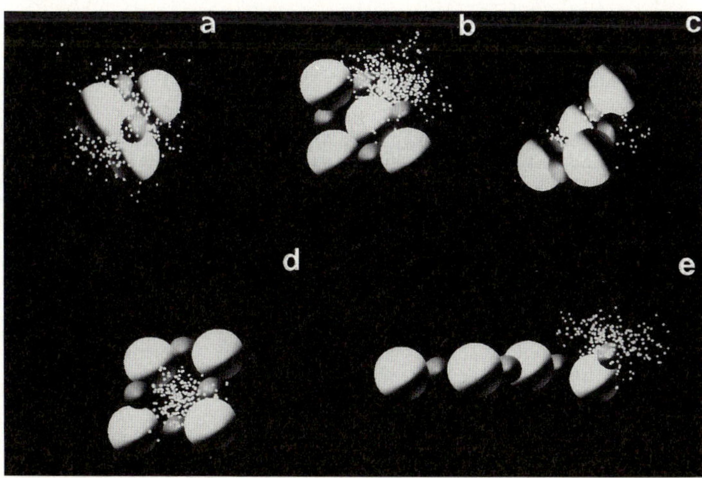

Figure 1. Cluster configurations for $e\text{-}Na_4Cl_4$. (a) 50K, (b) 575K, (c) 750K, (d,e) 1000K. Large balls represent Cl^- anions and small dark balls represent Na^+. Small dots represent the electron distribution.

The effect of an excess electron attached to a neutral ionic cluster on the sequence of isomerizations which the cluster undergoes as a function of temperature is illustrated [37] in Fig. 1 for $[Na_4Cl_4]^-$ where the equilibrium configura- tions at various temperatures are shown [37]. We observe a tendency toward spatial localization of the electron upon increase in T. Furthermore, the presence of the excess electron induces two types of configurational modifications which are either quantitatively or qualitatively different from those in the neutral cluster:

(i) The localized excess electron can play the role of a pseudonegative ion, with appreciable kinetic energy, which is overwhelmed by the electron ion, with appreciable kinetic energy, leading to new nuclear configurations which have no counterpart in the neutral cluster [42].

(ii) The partial neutralization of a single positively charged ion by the excess electron results in the appearance of the high-T configuration of the neutral parent cluster at substantially lower temperatures for the negatively charged cluster. The first effect is observed in the intermediate T range, while both effects are exhibited in the high-T domain.

Excess Electrons in Polar Molecular Clusters

Non-reactive electron solvation in ammonia [43,44], water [45] and other polar fluids [46] provided a theoretical challenge in the general area of electronic states of liquids, elucidating the effects of long-range and short-range attractive electron-solvent interactions and the role of local solvent reorganization on the structure, energetics, charge distribution and optical properties of the excess electron.

The advent of supersonic and cluster beams which allow accurate investigations of size effects and energetics of electron attachment to well characterized systems adds a new dimension to the research on electron solvation in polar fluids and to the general issues of size effects on the mode of electron localization and the minimal cluster sizes which sustain bound states of excess electrons. These studies revealed that while the onset of stability of negatively charged $(H_2O)_n^-$ clusters occurs for n $\gtrsim$ 11 [47-51], negatively charged $(NH_3)^-$ clusters were observed only for n $\gtrsim$ 30 [52]. This pronounced difference cannot be explained in terms of the variation in the molecular dipole moment (1.85D and 1.47D for H_2O and NH_3, respectively) since the electron-molecule interaction consists of additional important contributions [38,39,53] and furthermore the mechanism and energetics of solvation involve solvent reorganization governed by intermolecular interactions beyond the dipolar interactions.

A key issue in modelling the system is the choice of interaction potentials. The molecular degrees of freedom were treated classically (no significant isotopic dependence has been observed) employing known intra- and inter-molecular interactions for water and ammonia [38-40,54]. The electron-molecule interaction was described via a pseudopotential, constructed [35,38-40] in the spirit of the density functional method and consisting of Coulomb, polarization, exclusion and exchange contributions. The energetics of the system can be expressed in terms of the electron vertical and adiabatic binding energies, EVBE and EABE, respectively, where the former is the electron ionization energy of the negatively charged cluster with no cluster reorganization (such as measured by photoelectron spectroscopy) and EABE = EVBE + E_c, where E_c is the cluster reorganization energy, i.e., the difference between the equilibrium intra- and inter-molecular energies of the negatively charged cluster and the corresponding neutral. Energetic stability is inferred from the magnitude and sign of EABE (which at the limit of $n \to \infty$ is the heat of electron solvation in the bulk) with EABE < 0 corresponding to an energetically stable bound state.

The results of our studies are summarized in Fig. 2, where the binding energies are plotted versus $n^{-1/3}$ (full and empty symbols correspond to EVBE and EABE, respectively, circles and squares to interior and surface states, respectively). We observe first that for small to intermediate size $(H_2O)_n$ ($12 \leq n < 64$) clusters stable well-bound electron localization occurs via surface states (SS) while localization in interior states (IS) requires cluster sizes $n \gtrsim 64$. The electron charge distributions of these bound states, characterized by the radius of gyration R_g, varies from $R_g \sim 6a_o$ for the SS of the n=12 cluster to $R_g \sim 4.0a_o$ of the IS for n = 128. Furthermore the EVBE's of the SS states for the $12 \leq n < 18$ water clusters are in quantitative agreement with the measured values [49].

We remark that the discovery of the energetically favored surface mode of localization in small and intermediate $(H_2O)_n$ clusters [35,38,39] points to the

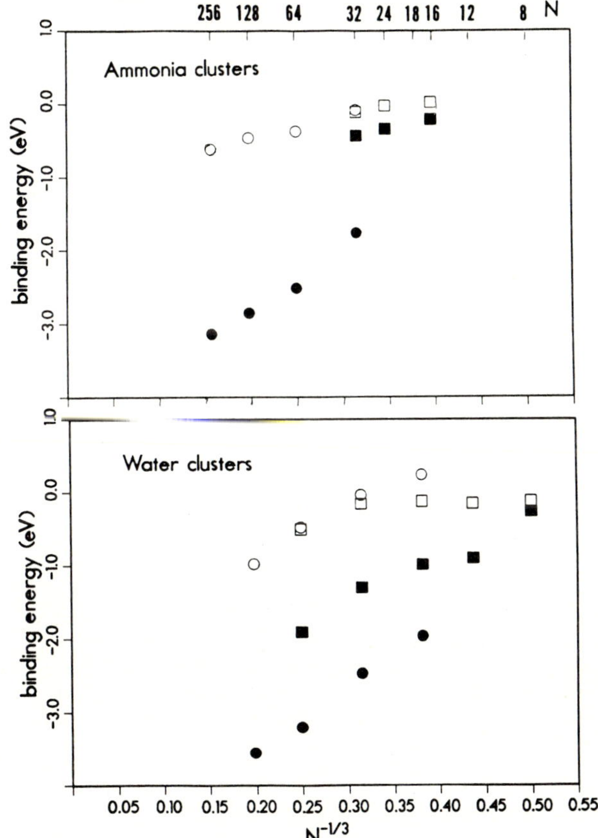

Figure 2. Calculated excess electron binding energies (in electron volts) versus $n^{-1/3}$ for $(NH_3)_n^-$ (in a) and $(H_2O)_n^-$ (in b) clusters. Circles and squares correspond to interior and surface states respectively. Full and empty symbols correspond to the vertical (EVBE) and adiabatic (EABE) binding energies respectively. The ammonia results were obtained at T = 189.5K and those for water at T = 300K for $n \geq 32$ and T = 79K for n < 32. Note the weaker binding in ammonia relative to water, in particular for the surface states, and the linear dependence on $n^{-1/3}$ for the internal states.

inadequacy of previous models [55] where a cavity-like (internal) binding geometry was assumed and which have failed to explain the observed energetic stability of these clusters. The measurement of the transition from surface to internal localization upon increasing cluster size (signified by a marked increase in EVBE) is left as an experimental challenge.

In contrast to the case of the $(H_2O)_n^-$ cluster we observe from Fig. 2 that the onset of stable electron attachment to ammonia clusters (signified by EABE<0, open squares and circles for SS and IS, respectively) occurs via internal localization requiring $n \gtrsim 32$ molecules, and is not preceded by well-bound surface localized states for smaller clusters. In addition, attempts to bind an electron via surface states to n>32 clusters resulted in a transition to the internally localized states.

Thus we conclude that electron localization in $(H_2O)_n^-$ clusters occurs via two modes depending on the cluster size (surface states for 8<n<64) with a transition from surface to internal localization for $n \sim 64$. The interplay between the competing effects of the interaction of the electron with the molecular constituents, the kinetic energy of localization and the intermolecular interactions (underlying cluster reorganization) precludes the formation of stable well-bound surface states in ammonia clusters, resulting in the onset of excess electron binding via internal localization for $n \gtrsim 32$, in agreement with experimental observations.

Finally, as evident, the energies of the internal states of both materials (circles) exhibit linear dependence versus $n^{-1/3}$. An adaptation [39,56] of a dielectric continuum model [57] to finite clusters yields expressions for EVBE and EABE as a function of cluster size which are in agreement with the simulation results and experimental data in the limit of bulk systems [58].

b. FFT and MD-TDSCF Simulations of Spectra and Dynamics

The FFT and TDSCF methods described in Section 2 can be employed in several modes:

(i) Given a configuration of the atomic constituents the FFT method can be used in order to obtain the electric ground and excited state spectra for an excess electron. When used in conjunction with QUPID simulations as a source for an equilibrium ensemble of configurations, such calculations yield spectroscopic information for the equilibrium ensemble [27,59,60]. The evaluation of the eigenvalue spectra can be achieved in one of two ways: a) Start with an arbitrary initial wavefunction and integrate the time-dependent Schrödinger equation using appropriate time increments as required by the energy scale of interest and for a long enough time as implied by the required energy resolution. Using the resulting wave function at time t, $\psi(t)$, form the correlation function $C(t) \equiv \langle\psi(o)\psi(t)\rangle$ and Fourier transform it with respect to time, i.e., $I(E) = \int C(t) \exp(-iEt/h)dt$ [22].

b) Start with an arbitrary initial wavefunction and integrate the time-dependent Schrödinger equation in imaginary time $t = -i\beta'$ from $\beta' = 0$ to a value β. The resulting wave function $\psi(\vec{r},\beta)$ is given by a sum of exponentials

$$\psi(\vec{r},\beta) = \sum_n a_n\phi_n(\vec{r}) \exp(-E_n\beta/\hbar) \qquad (8)$$

where E_n are the eigenvalues and $\phi_n(\vec{r})$ are the eigenfunctions. As $\beta \to \infty$ the sum on the r.h.s. of eq. (8) is dominated by the lowest energy eigenvalue E_o (for bound states $E_n < 0$). In general this procedure will lead to the state of lowest energy for any given initial symmetry. Having obtained the state of lowest energy, the procedure can be used to obtain excited state energies and wavefunctions by projecting out [61] the lower (previously found) states during the (imaginary) time evolution.

(ii) When the system is composed of two subsystems whose dynamics are characterized by widely separated time scales, we encounter a situation which is normally treated within the adiabatic or Born - Oppenheimer approximation. For such situations [27] we introduce a new numerical method which exploits this time-scale separation and is optimally suited for numerical simulations. The method which we term the adiabatic Simulation Method (ASM) is a numerical implementation of the adiabatic approximation where the dynamics of the slow subsystem is followed while the fast subsystem is constrained to stay in one state (which changes adiabatically as the slow subsystem evolves in time). This restriction is implemented by a repeated application of the time evolution operation in imaginary time to the wavefunction of the fast subsystem followed by appropriate (numerical) projection operators.

This method allows determination of spectroscopic data and investigations of dynamical processes for coupled quantum-classical systems at finite temperature, in either equilibrium (where the ensemble averages are obtained via averaging over time) or non-equilibrium situations. To demonstrate the method (applied to the ground state dynamics, i.e., the GSD version of the ASM) we show in Fig. 3 contour plots of the ground state and 3 first excited states (of p-like character) for $(H_2O)_{64}^-$ at 300K. The fluctuations in the electronic energy difference, ΔE induced by the nuclear motions are the origin of the broad absorption spectrum characteristic to the hydrated electron. Such spectroscopic information can be obtained via ensemble averaging or equivalently via a single time trajectory of sufficient time span to assure that the accessible phase-space is sampled adequately. Such histograms showing the distribution of E_g, E_x and ΔE are displayed in Fig. 4 for $(H_2O)_{64}^-$. We have found that the correction to the line-shape given in Fig. 4c due to the transition-dipole matrix element between the ground and excited electronic states is very small.

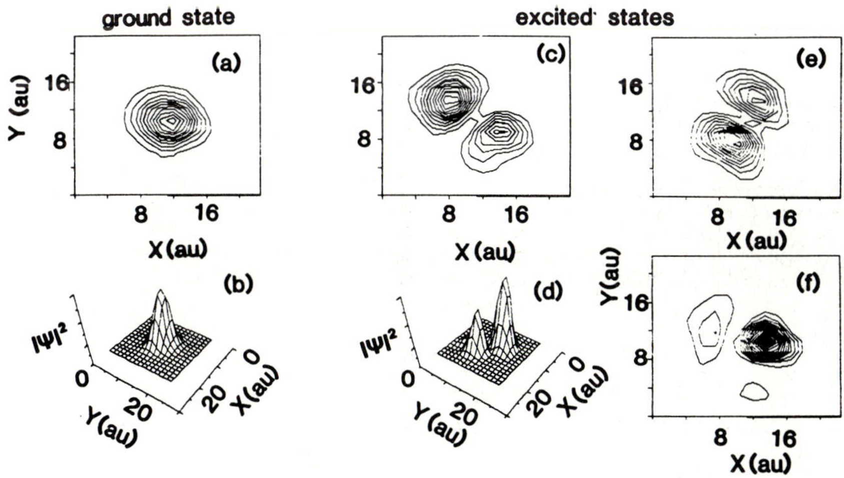

Figure 3: Two- and three-dimensional contours of the excess electron density in the ground (a,b) and three lowest excited states (c-f) in the $(H_2O)_{64}^-$ cluster. The three-dimensional contour in (d) corresponds to the two-dimensional one shown in (c). Note the p-character of the excited states. All contours are calculated in the xy plane at the middle of the z-axis of the calculational cell. Distances are in atomic units.

Several features displayed in Fig. 4 should be noticed. First the broad energy distributions in the ground and excited state energies are an indication of the large sensitivity of the electronic energy to the cluster configurations. This sensitivity is due mostly to the potential energy terms [27].

Secondly, the fact that E_g and E_x fluctuate largely in unison is the origin of the substantially narrower distribution of the gap energy (ΔE). ΔE is seen to peak at $\sim$ 0.075 hartree ($\simeq$ 2.04 eV) for $(H_2O)^-_{64}$ with a small shift to the red for $(H_2O)^-_{128}$. For an electron in bulk water the peak of the absorption band is at 1.73 eV.

Finally, the width of the ΔE distribution (Fig. 4c) is of the order of 0.01 hartree = 0.27 eV. The experimentally observed width in bulk water is 0.8 eV. This larger width reflects the fact that three non-degenerate p-states contribute to the lineshape while in the above GSD simulation we evaluated only the lowest of them for each configuration.

Indeed, having observed that the three lowest p-like excited states (Fig. 3) span an energy range of $\sim$ 0.5-0.7 eV, and assuming that the width (of 0.27 eV) is the same for all of them the theoretical estimate of the absorption band width comes close to the experimental one. (The small differences observed in our calculations of the band width results for the $(H_2O)^-_{64}$ and $(H_2O)^-_{128}$ clusters suggest that finite size effects on these widths are small in this cluster size range justifying the comparison to the bulk water data.)

(iii) In the third mode of operation the full real-time dynamics of the quantum and classical components of the system are followed by a combination of classical MD with the TDSCF formalism (Eqs. 7) [27,62].

As an example we use the water dimer $(H_2O)^-_2$ in which the excess electron binding energy was estimated via QUPID simulations [35] to be in the range 3-25 meV, depending on the cluster configuration. Since an adiabatic separation of the electronic and nuclear motions cannot be assumed here, the ASM computational scheme cannot be used. Instead we have simulated the full dynamical evolution of the $(H_2O)^-_2$ cluster in the TDSCF approximation, under the added assumption that the water molecular motions can be treated, as before, classically.

With regard to the first approximation we should note that the structure of the negatively charged water dimer, with a very diffuse electron charge density weakly bound to the water dimer via primarily dipolar interactions [35,63,64] supports the assertion that the TDSCF approximation may be used in this situation. Separation of the electronic and nuclear motions may be achieved here not because they evolve on different timescales but because they occur on different length scales.

Fig. 5 shows, for $(H_2O)^-_2$ at T = 20K the time evolution of several quantities: the classical potential energy U, the total classical energy $E_{C\ell}$, the kinetic (K_q) potential (V) and total (E_q) quantum energies of the electron (V is of course the energy of interaction between the electron and the classical system), the spatial spread of the electron charge distribution ($R_g = (\langle r^2\rangle - \langle r\rangle^2)^{1/2}$) and the total dipole of the water molecule dimer. In this calculation we used a 16^3 grid with grid unit equal to 9 a_o, and have monitored the position and width of the electron wavepacket to make sure that it remains well within the calculational grid throughout the evolution. We remark that such a large grid cell is appropriate for the description of the very diffuse, weakly bound, electronic wavefunction. The time length of the trajectory shown is 8. x 10^4 au $\cong$ 2 psec.

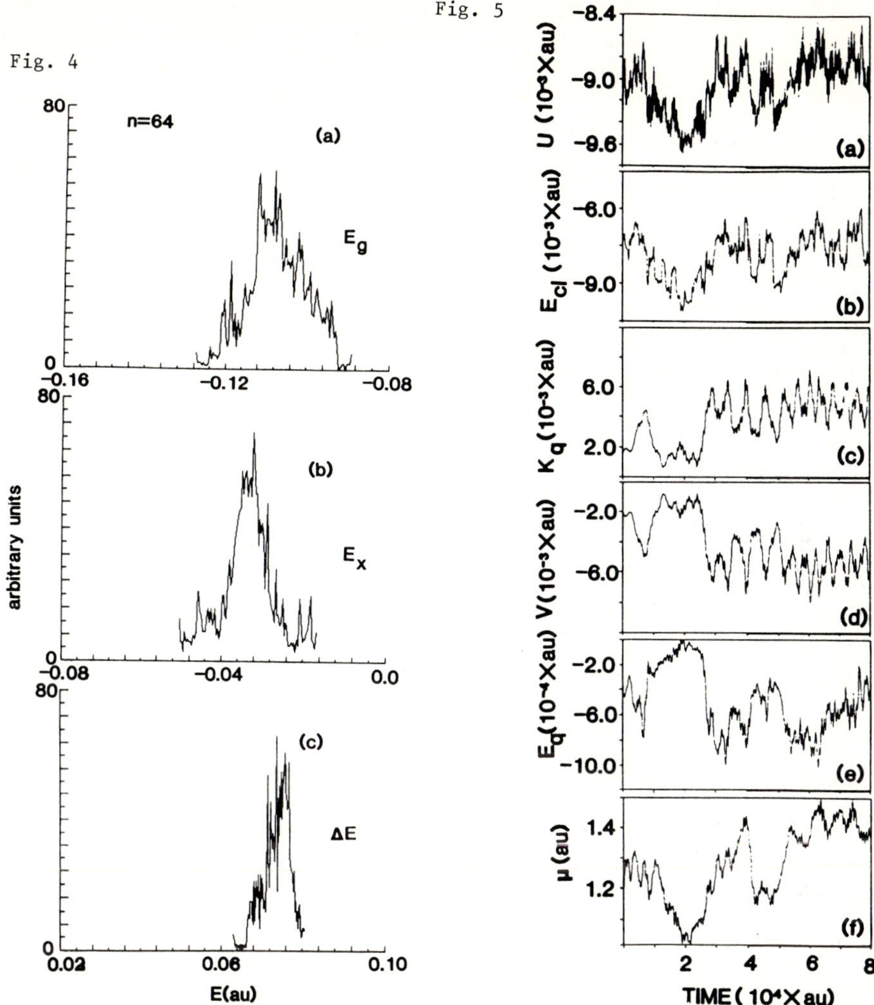

Figure 4: Histograms of the frequency of occurrence of different energy values, obtained from GSD simulations of $(H_2O)^-_{64}$ at 300K for the ground (E_g) and excited (E_x) state energies and the difference between them (ΛE). The breadth of the energy distributions correspond to the widths of the energy levels and that of ΛE corresponds to the width of the absorption spectrum for the ground to first excited state transition. The broadening of the levels is due to time fluctuations (at finite temperature) in the cluster molecular configuration.

Figure 5: Results obtained from TDSCF dynamical simulations of $(H_2O)^-_2$. The molecular potential energy (U), total classical energy ($E_{C\ell}$), kinetic energy of the excess electron (K_q), interaction energy between the excess electron and the molecular cluster (V), total energy of the excess electron (E_q) and total dipole moment of the water molecules (μ) are shown in (a-f). Time in au = 2.4189×10^{-17} s, and all quantities in atomic units. Note the correlation between the increases in the magnitude of the excess electron binding energy (E_q) and the cluster dipole, as the nuclear configurations fluctuate.

The results in Fig. 5 clearly exhibit the correlation between the nuclear motion and the electronic energies. In particular the strong effect of the total cluster dipole on the electron binding energy is evident. The electronic energy fluctuates between values very close to zero and values below -8×10^{-4} Hartree ($\sim$ -22 meV) as the structure of the cluster alternates between high and low dipole configurations. The existence of such configurations of the negatively charged water dimer at 20K was already demonstrated by the QUPID simulations [35]. Inspection of Fig. 5 shows that the transition time between the high and low dipole configurations is roughly in the range 0.1-1 psec.

From the time records of the total interaction potential energy (U), Fig. 5a, between the atomic constituents of the dimer and of the total (classical) energy ($E_{C\ell}$, Fig. 5b) we observe that the structural transitions from the low-dipole to high-dipole configurations are accompanied by a decrease in the magnitude of the interatomic interactions (compare Figs. 5a,b and Fig. 5f). We remark that the low-dipole configuration is the one closer to that of the neutral water dimer.

*Acknowledgements: The studies described in this review are the results of collaborative work with several colleagues who should be regarded as coauthors of this review and to whom I am grateful: R. N. Barnett, C. L. Cleveland, J. Jortner, N. R. Kestner, A. Nitzan and D. Scharf. These studies were supported by the U. S. DOE grant number FG05-86ER-45234 (to UL) and in part by the US-Israel Binational Science Foundation, Jerusalem (to JJ and UL, and to AN). The assistance of A. Ralston and V. Mallette in the preparation of the manuscript is gratefully acknowledged.

References

[1] For reviews see the article by U. Landman "Molecular Dynamics Simulations in Material Science and Condensed Matter Physics", the article by F. F. Abraham, and other contributions in this volume.
[2] D. M. Ceperley and M. H. Kalos in "Monte Carlo Methods in Statistical Physics", Ed. K. Binder (Springer-Verlag, Berlin, 1979), p. 145.
[3] K. E. Schmidt and M. H. Kalos in "Applications of the Monte Carlo Method in Statistical Physics", Ed. K. Binder (Springer-Verlag, Berlin, 1984), p. 4.
[4] Articles in "Monte Carlo Methods in Quantum Problems", Ed. M. H. Kalos (Reidel, Dordrecht, 1984).
[5] B. J. Alder, D. M. Ceperley and E. L. Pollock, Acc. Chem. Res. 18, 268 (1985).
[6] D. M. Ceperley and B. J. Alder, Science 231, 555 (1986).
[7] B. J. Berne and D. Thirumalai, Ann. Rev. Phys. Chem. 37, 401 (1986).
[8] R. Kosloff, "Time Dependent Quantum Mechanical Methods for Molecular Dynamics", to appear in J. Phys. Chem. (1988).
[9] D. Chandler and P. G. Wolynes, J. Chem. Phys. 79, 4078 (1981).
[10] D. Chandler, J. Phys. Chem. 88, 3400 (1984).
[11] M. Parrinello and A. Rahman, J. Chem. Phys. 80, 860 (1984).
[12] D. De Raedt, H. Sprik and H. L. Klein, J. Chem. Phys. 80, 5719 (1984).
[13] M. F. Herman, E. J. Bruskin and B. J. Berne, J. Chem. Phys. 76, 5150 (1982).
[14] U. Landman, D. Scharf, and J. Jortner, Phys. Rev. Lett. 54, 1860 (1985).
[15] R. P. Feynman and A. R. Hibbs, Quantum Mechanics and Path Integrals (McGraw-Hill, New York, 1965).
[16] L. S. Schulman, Techniques and Applications of Path Integrals (Wiley, New York, 1981).
[17] M. Takahashi and M. Imada, J. Phys. Soc. Jpn. 53, 3765 (1984); ibid. 53, 963 (1984); ibid. 53, 3770 (1984).
[18] F. F. Abraham and J. Q. Broughton, Phys. Rev. Lett. 59, 64 (1987).
[19] J. Chang and W. H. Miller, J. Chem. Phys. 87, 1648 (1987).
[20] J. D. Doll, R. D. Colson and D. L. Freeman, J. Chem. Phys. 87, 1641 (1987).
[21] A. L. Nichols III and D. Chandler, Excess Electrons in Simple Fluids IV. Real Time Behavior, Preprint.
[22] M. D. Feit, J. A. Fleck, Jr., and A. Steiger, J. Comput. Phys. 47, 412 (1982); M. D. Feit and J. A. Fleck, Jr., J. Chem. Phys. 78, 301 (1983); ibid. 80, 2578 (1984).

[23] D. Kosloff and R. Kosloff, J. Comput. Phys. 52, 35 (1983).
[24] P. A. M. Dirac, Proc. Comb. Phil. Soc. 26, 376 (1930).
[25] A. D. Mclachlan, Mol. Phys. 7, 139 (1964).
[26] D. Kumamoto and R. Silbey, J. Chem. Phys. 75, 5164 (1981).
[27] R. N. Barnett, U. Landman and A. Nitzan, Phys. Rev. A (1988) and "Dynamics and Spectra of a Solvated Electron in Water Clusters", J. Chem. Phys. (August 15, 1988).
[28] N. Makri and W. H. Miller, J. Chem. Phys. 87, 5781 (1987).
[29] J. Kucar, H. D. Meyer and L. S. Cederbaum, Chem. Phys. Lett. 140, 525 (1987).
[30] R. H. Bisseling, R. Kosloff, R. B. Gerber, M. A. Ratner, L. Gibson and C. Cerjan, J. Chem. Phys. 87, 2760 (1987).
[31] Z. Kotler, R. Kosloff and A. Nitzan (to be published).
[32] J. R. Fox and H. C. Anderson, J. Phys. Chem. 88, 4019 (1984). In our MD-TDSCF simulations we use the stochastic thermalization method for constant temperature simulations described by these authors, and the classical equations of motion are integrated using the velocity form of the Verlet algorithm.
[33] J. Jortner, Ber. Bunsenges, Physik. Chem. 88, 188 (1984).
[34] J. Jortner, D. Scharf and U. Landman in "Proceeding of the Enrico Fermi Summer School on Excited State Spectroscopy in Solids", XCVI Corso, p. 438 (1987).
[35] U. Landman, R. N. Barnett, C. L. Cleveland, D. Scharf and J. Jortner, J. Phys. Chem. 91, 4890 (1987); Int. J. Quant. Chem. 21, 573 (1987).
[36] J. Jortner, D. Scharf and U. Landman, "Molecular Clusters", in "Elemental and Molecular Clusters", Eds. G. Benedek, T. P. Martin, and G. Pacchioni (Springer-Verlag, Berlin, 1988).
[37] D. Scharf, U. Landman and J. Jortner, J. Chem. Phys. 87, 2716 (1987).
[38] R. N. Barnett, U. Landman, C. L. Cleveland and J. Jortner, Phys. Rev. Lett. 59, 811 (1987).
[39] R. N. Barnett, U. Landman, C. L. Cleveland and J. Jortner, J. Chem. Phys. 88, 4421, 4429 (1987).
[40] R. N. Barnett, U. Landman, C. L. Cleveland, N. R. Kestner and J. Jortner, J. Chem. Phys. 88, 6670 (1988).
[41] D. Scharf, J. Jortner and U. Landman, Chem. Phys. Lett. 130, 5504 (1986).
[42] J. Luo, U. Landman and J. Jortner in "Physics and Chemistry of Small Clusters", Eds. P. Jena, B. K. Rao and S. N. Khanna (Plenum, New York, 1987), p. 201.
[43] P. P. Edwards, Advances in Inorg. and Radiochem. 25, 135 (1982).
[44] W. Weyl, Ann. Phys. 197, 601 (1863).
[45] E. J. Hart and J. W. Boag, J. Am. Chem. Soc. 84, 4090 (1962).
[46] "Solutions Metal-Ammonia", Eds. G. Lepoutre and M. Sienko (Benjamin, New York, 1964); "Metal-Ammonia Solutions", Eds. J. J. Lagowski and M. Sienko (Butterworth, London, 1970). "Electrons in Fluids", Eds. J. Jortner and N. R. Kestner (Springer-Verlag, Berlin, 1973); Canad. J. Chem. 55, 1795-2277 (1977).
[47] H. Haberland, H. Langosch, H. G. Schindler and D. R. Worsnop, Surface Sci. 156, 157 (1985).
[48] H. Haberland, H. G. Schindler and D. R. Worsnop, J. Chem. Phys. 81, 3742 (1984).
[49] J. V. Coe, D. R. Worsnop and K. H. Bowen, J. Chem. Phys. (to be published).
[50] M. Knapp, O. Echt, D. Kreisle and E. Recknagel, J. Chem. Phys. 85, 636 (1986).
[51] M. Kanpp, O. Echt, D. Kreisle and E. Racknagel, J. Phys. Chem. 91, 2601 (1987).
[52] H. Haberland, H. G. Schindler and D. R. Worsnop, Ber. Bunsenges. Phys. Chem. 88, 270 (1984).
[53] J. Schnitker and P. J. Rossky, J. Chem. Phys. 86, 3462 (1987).
[54] R. N. Barnett, U. Landman, C. L. Cleveland, N. R. Kestner and J. Jortner, "Excess Electrons in Ammonia Clusters", Chem. Phys. Lett. (to be published, 1988).
[55] M. Newton, J. Chem. Phys. 58, 5833 (1973); A. N. Rao and N. R. Kestner, J. Chem. Phys. 80, 1587 (1984).
[56] R. N. Barnett, U. Landman, C. L. Cleveland and J. Jortner, "Size Dependence of the Energetics of Electron Attachment to Large Water Clusters", Chem. Phys. Lett. 145, 382 (1988).
[57] J. Jortner, J. Chem. Phys. 30, 839 (1959).
[58] M. Newton, J. Phys. Chem. 79, 2795 (1975).
[59] J. Schnitker, K. Motakabbir, P. J. Rossky and R. Friesher, Phys. Rev. Lett. 60, 456 (1988).

[60] D. F. Coker and B. J. Berne, "Excess Electronic States in Fluid Helium", J. Chem. Phys. (1988).
[61] R. Kosloff and H. Talezer, Chem. Phys. Lett. 127, 223 (1986).
[62] A. Selloni, P. Carenvali, R. Car and M. Parrinello, Phys. Rev. Lett. 59, 823 (1987).
[63] A. Wallqvist, D. Thirumalai and B. J. Berne, J. Chem. Phys. 85, 1583 (1986).
[64] D. J. Chipman, J. Phys. Chem. 82, 1980 (1978).
[65] J. R. Reimers and R. D. Watts, Chem. Phys. 85, 83 (1984).

Superfluidity of a Two-Dimensional Bose–Coulomb Gas

D. Peters and B. Alder

Lawrence Livermore National Laboratory, Livermore, CA 94550, USA

We have observed that superfluidity occurs in Monte Carlo path integral calculations of a two-dimensional Bose-Coulomb gas when the thermal de Broglie wavelength, λ, is comparable to the interparticle distance "a." By this criterion, utilizing only the experimentally determined charge carrier density and effective mass for the new high-T_c superconductors, superconducting transition temperatures in the range of 100-200 K are obtained, the range reflecting the experimental uncertainties in these two observables.

1 Introduction

This work was inspired by the high temperature superconductors. Early on, it appeared from optical work on polycrystalline material that the paired charges existed above the temperature of the superconducting transition, T_c. This meant that these pairs might undergo a superfluid transition at T_c. A back of the envelope calculation led to a temperature comparable to the observed one assuming: (1) that electrons (or holes) form pairs above T_c which act as bosons, (2) that the pairs are separated by distances that are large compared their own dimensions, so that they can be treated as double point charges, (3) that the pairs are confined to move in the copper oxygen planes in a relatively uniform background, because the electron affinities of copper and oxygen might be close, (4) that the superconductivity criterion for these pairs should be similar to that for superfluidity in He^4, namely that the de Broglie wavelength is comparable to the interparticle spacing, even though the interaction potential is very different and (5) that the charge density and the effective mass can be reliably inferred from experiment. To verify assumption 4 above, we used a path integral quantum Monte Carlo code to quantitatively study the superfluid/superconducting phase transition of the two-dimensional (2D) Bose-Coulomb gas. Present theoretical belief is that the superfluid transition is qualitatively different in two and three dimensions, namely that in two dimensions it is associated with the unbinding of vortex pairs. These questions are discussed in the following sections.

2 Path Integral Monte Carlo

As is well known, thermodynamic behavior is determined through the trace of the density matrix

$$\rho(\vec{R},\vec{R}';\beta) = \langle \vec{R} | e^{-\beta H} | \vec{R}' \rangle \quad ,$$

Springer Proceedings in Physics Vol. 33: **Computer Simulation Studies in Condensed Matter Physics**
Editors: D. P. Landau • K. K. Mon • H-B. Schüttler

which can be written in the equivalent form

$$= \int \ldots \int \rho(\vec{R},\vec{R}_1;\tau)\rho(\vec{R}_1,\vec{R}_2;\tau)\ldots\rho(\vec{R}_{s-1},\vec{R}';\tau)\, d\vec{R}_1 d\vec{R}_2 \ldots d\vec{R}_{s-1} \tag{1}$$

in terms of the density matrices, $\rho(R_i,R_j;\ \tau)$, where the coordinates R represent the 2N (in 2D) particle coordinates and each density matrix in the integrand is at a higher effective temperature, $\tau = \beta\ /s\ =\ 1/kTs$, where s is an arbitrary integer [1]. The advantage is that for high enough temperatures (large s), excellent approximations to the exact density matrix are known. Quantum statistics can be introduced by the appropriate permutations, P, of the particles' positions in the density matrix, so that, for bosons

$$\rho_{bosons}(\vec{R},\vec{R}';\beta) = (N!)^{-1}\sum_{P} \rho(\vec{R},P\vec{R}';\beta) \quad .$$

The density matrix cast in the form (1) is a multi-dimensional integral, best evaluated numerically by Monte Carlo techniques. The high temperature density matrix can be written in pair-product form:

$$\rho(\vec{R},\vec{R}';\tau) = \prod_{i=1}^{N} \rho_0(\vec{r}_i,\vec{r}_i';\tau) \prod_{i\neq j}^{N} \rho_{int}(\vec{r}_{ij},\vec{r}_{ij}';\tau) \quad , \tag{2}$$

where ρ_0 is the free-particle term

$$\rho_0(\vec{r}_i,\vec{r}_i';\tau) = \frac{2\pi m}{h^2\tau} \exp\left(\frac{-(\vec{r}_i - \vec{r}_i')^2\, 2\pi^2 m}{h^2\tau}\right)$$

and ρ_{int} represents the effects of the particle-particle interactions. The coordinates r_i and r_{ij} are the positions of particle i and the separations between particles i and j, respectively.

The high temperature two-body density matrix must satisfy the Bloch equation:

$$\frac{\partial\rho(\vec{r}_{ij},\vec{r}_{ij}';\tau)}{\partial\beta} = \left(\vec{\nabla}'^2 + V(\vec{r}_{ij}')\right) \rho(\vec{r}_{ij},\vec{r}_{ij}';\tau) \quad , \tag{3}$$

where $V(r_{ij}')$ is the two-body potential. For the Coulomb (1/r) potential with periodic boundary conditions, the standard technique is to use Ewald sums to deal with the periodic images. To numerically handle such a sum efficiently, the potential $\phi(r)$ between two particles and all of their images is broken up into short range and long range parts [2]:

$$\Phi(\vec{r}) = \sum_{\vec{L}} f(\vec{r} - \vec{L}) + \sum_{\vec{k}} g(\vec{k}) \exp(i\vec{k} \cdot \vec{r}) \quad , \tag{4}$$

where the sums are over all lattice vectors, $\vec{L}$, and all reciprocal lattice vectors, $\vec{k}$, and where

$$f(r) = \frac{\mathrm{erfc}(\sqrt{\alpha}\, r)}{r} \quad ,$$

$$g(k) = \begin{cases} \dfrac{2\pi}{k\Omega}\,\mathrm{erfc}\left(\dfrac{k}{2\sqrt{\alpha}}\right) , & k \neq 0 \\ -\dfrac{2}{\Omega}\sqrt{\dfrac{\pi}{\alpha}} , & k = 0 \end{cases}$$

Ω is the volume of the unit cell and α an adjustable parameter. The complementary error function in f(r) effectively cuts off the Coulomb potential beyond some distance, chosen to be half the unit cell size by adjusting α, thus reducing the sum over all $\vec{L}$ in (4) to only the $\vec{L}$=0 term. The function f(r) now represents a short range potential which is a strong function of particle positions, while the k-space sum contains the long range interactions, which depend only weakly on particle positions. The density matrix obtained by solving (3) is different from its semiclassical form only if V(r) varies rapidly over the thermal de Broglie wavelength. Hence, the important quantum mechanical effects on the density matrix can be treated by including only the rapidly changing part of the potential, f(r), in place of V(r). After dividing by the appropriate free-particle term, this solution is called $\rho_{int}^{sr}(r_{ij},r_{ij}'; \tau)$, where "sr" denotes that it has been obtained by using only the short range part of the Coulomb interaction. At the temperatures of interest here, the slowly varying part of the Coulomb interaction,

$$U_{lr}(\vec{r}_{ij}) = \sum_{\vec{k}} g(\vec{k}) \exp(i\vec{k}\cdot\vec{r}) \quad ,$$

is well represented in the semiclassical endpoint approximation:

$$\rho_{int}^{lr}(\vec{r}_{ij},\vec{r}_{ij}';\tau) = \exp\left[-\frac{\tau}{2}\left(U_{lr}(\vec{r}_{ij}) + U_{lr}(\vec{r}_{ij}')\right)\right] \quad ,$$

where again the free-particle part has been divided out. The full N-particle high temperature density matrix is thus written as

$$\rho(\vec{R},\vec{R}';\tau) = \prod_{i=1}^{N} \rho_0(\vec{r}_i,\vec{r}_i';\tau) \prod_{i\neq j}^{N}\left[\rho_{int}^{sr}(\vec{r}_{ij},\vec{r}_{ij}';\tau)\,\rho_{int}^{lr}(\vec{r}_{ij},\vec{r}_{ij}';\tau)\right] \quad .$$

In practice the Bloch equation, (3), is solved by an iterative matrix squaring technique [3] which generates ρ_{int}^{sr} from its form in the high temperature semiclassical limit.

Once a good approximation for the high temperature density matrix is obtained, the integral in equation (1) is evaluated by a Metropolis Monte Carlo method [4,5]. One can think of the s positions of each particle in the integrand of (1) as forming a polymer whose links consist of interactions determined by the pair density matrix. For the diagonal elements of the density matrix, the polymers close on themselves. The system thus contains N polymer rings, each of length s. Sampling of phase space consists of snipping out a section of one or more polymers and attempting to replace them with another section containing new, trial positions. Exchange processes, which are of basic importance to superfluidity, are included by allowing the polymers to "cross link". The onset of superfluidity is marked by a sudden increase in the number of these exchanges. Increasing the efficiency of sampling this contorted phase space involves a number of other techniques which have been previously described [4,5].

3 Results

In three dimensions the superfluid transition occurs when the thermal de Broglie wavelength,

$$\lambda = \sqrt{\frac{h^2}{8\pi^2 m k_B T}} ,$$

is comparable to the average interparticle spacing, "a", and is relatively insensitive to the interaction potential as judged by similar transition temperatures for the ideal gas and the actual He^4 system. The numerical work on small sized systems extends these observations to two dimensions by showing that the ratio λ/a at T_c is of order unity and is remarkably insensitive to the particle-particle interaction, even for the long range Coulomb potential (see Table 1). The superfluid transition temperature was studied for the 2D Bose-Coulomb system at two different densities, r_s = 30 and r_s = 10 , where $r_s = a/a_o$, and a_o is the effective Bohr radius,

Table 1. Values of the ratio of the de Broglie wavelength to the interparticle radius, λ/a, at T_c for various systems

System	Dimension	No. of particles	a[Å]	λ/a
He^4 (a)	3	64	2.22	0.75
He^4 (b)	2	25	2.71	0.87
Coul. (r_s=10)	2	16	0.22	0.76
Coul. (r_s=30)	2	16	0.66	0.78

(a) refs. [4,5]

(b) E.L. Pollock and D.M. Ceperley: unpublished

$$a_o = \frac{h^2 \varepsilon}{4\pi^2 m q^2} ,$$

where ε is the dielectric constant of the material, and m and q are the effective mass and charge of the bosons.

The superfluid/superconducting phase transition was observed in three different ways. The first was a kink in the total energy vs. temperature curve (Fig. 1). The second was a rapid increase in the number of larger "ring exchange" cyclic permutations at the temperature at which the energy showed a kink (Fig. 2).The transition temperature is shown to be the point at which the probabilities of the largest ring exchanges that were included exceeded those of the smallest one. The third measure of the existence of superfluidity was the determination of the superfluid density (Fig. 3), through the winding number [6]. The winding number is directly related to the cyclic permutations. Figure 3 shows that the behavior of the superfluid density in two and three dimensions for these small systems is very similar and that the superfluid transition temperature differs from the "universal jump" condition predicted by KOSTERLITZ and THOULESS (KT) [6] for two-dimensional systems with short range interactions,

$$\frac{\rho_s(T_c^-)}{T_c} = 8\pi k_B \left(\frac{m}{h}\right)^2 .$$

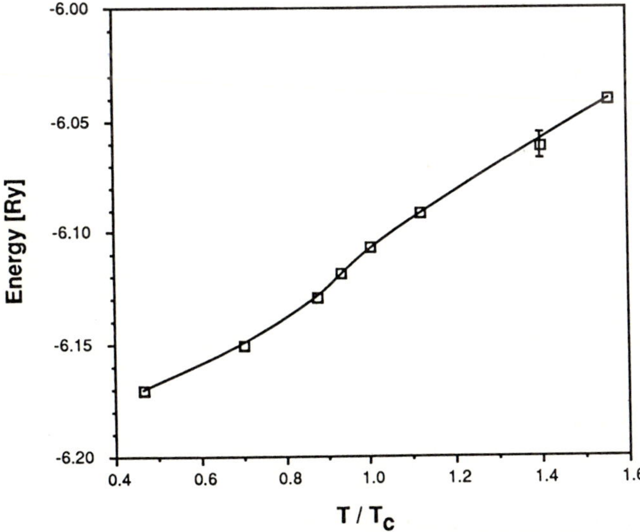

Fig. 1. Total energy vs. temperature for the two-dimensional Coulomb system at a density of r_s = 30, with bosons of mass = $6m_e$ and charge = 2e. The transition temperature, T_c, under these conditions is 29,000 K. A typical error bar is indicated by the vertical line.

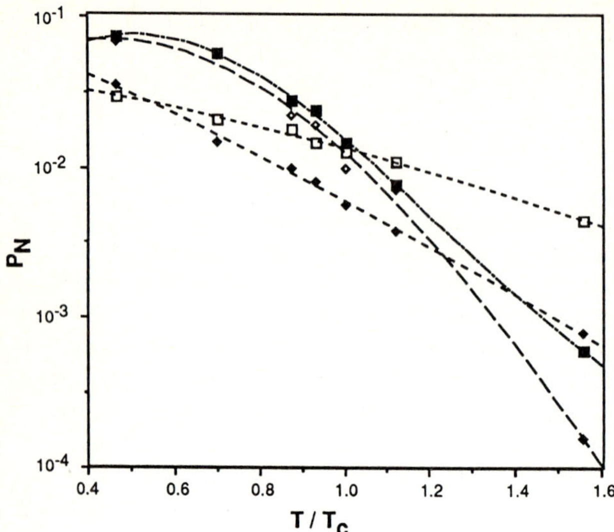

Fig. 2. The probability, P_N, that any given particle is participating in a cyclic ring exchange of length N as a function of temperature. P_2 = (□), P_3 = (♦), P_4 = (■), P_5 = (◇).

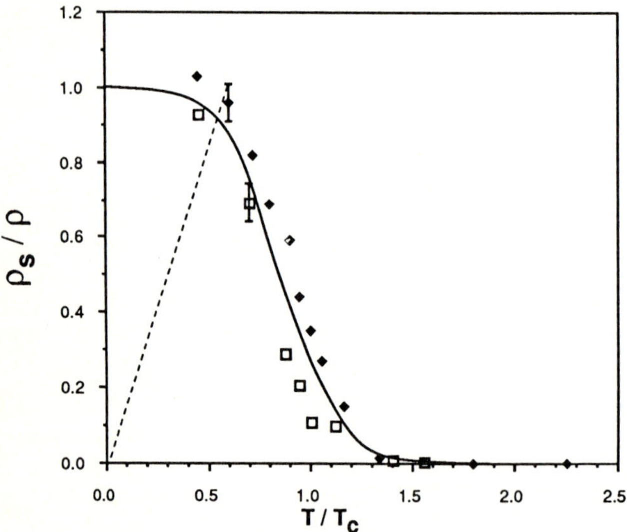

Fig. 3. Superfluid fraction vs. temperature for the two-dimensional Coulomb system at a density of r_s = 30 (□) and the two dimensional He^4 system (♦). The Kosterlitz-Thouless universal transition condition is represented by the point where the dashed line intersects the solid curve (hand drawn through the data). Typical error bars are indicated by the vertical lines.

In the KT theory, paired vortices are predicted to unbind at this point, destroying the superfluidity. Work is underway to determine the scaling to macroscopic results by examining different size systems as well as whether the number scaling is similar to the X-Y model where the KT theory seems to have been verified for larger systems [7].

4 Relation to High T_C Superconductors

Since the above results indicate that the superfluid transition temperature appears to be primarily a function of the effective mass and number density of the charge carriers, and only weakly dependent on the details of the interactions between the bosons, it is possible to predict T_c based on experimental observations of these quantities. If we take $\lambda/a = .76$ as the criterion for the onset of superconductivity, the predicted transition temperature is

$$k_B T_c = \left(\frac{h^2}{45.8}\right)\frac{1}{ma^2} \quad .$$

In the superconductor $YBa_2Cu_3O_{7-y}$ the charge carrier density is measured to be $3x10^{21}$ cm^{-3} [8]. With a unit cell length perpendicular to the conducting planes of $c=11.666x10^{-8}$ cm [9] and two conducting planes/unit cell, a planar charge density of $1.75x10^{14}$ cm^{-2} can be established. Since two charge carriers (holes) bind together to form a boson, the area density of bosons is $8.75x10^{13}$ cm^{-2} in each conducting plane. This leads to a boson disk radius of a = 6.03 Å. If the holes in each pair are separated by a smaller distance, as for example if they reside on nearest neighbor copper and oxygen atoms, it is not unreasonable to consider the bosons as doubly charged point particles. These bosons would likely interact at large distances through a screened Coulomb potential, and have a short range interaction not unlike that of two helium atoms. For this value of "a" and two values of the effective mass that represent the extremes of the measured uncertainty [10], the values of T_c using the criterion $\lambda/a = 0.76$ are given in Table 2. Considering the crudity of the model, it is remarkable that the superconducting transition temperatures are on the order of the calculated

Table 2. Predicted transition temperatures for doubly charged bosons for two different effective masses at a density corresponding to a = 6.03 Å

m^*_{holes}	m^*_{bosons}	T_c [K]	r_s
5	10	215	76
10	20	108	152

values. In some sense these temperatures are an upper limit to the superconducting transition for the copper oxide materials unless the charge densities and effective masses can be drastically changed from what is presently possible. If KT theory proves to be correct, the predicted transition temperatures would be roughly 50% lower.

It remains to investigate whether under these conditions the system is in the fluid phase since the 2D Coulomb system at low densities transforms into a crystal (Wigner crystallization). From previous Green's function Monte Carlo calculations at 0 K for the three-dimensional Bose-Coulomb gas [11], this transition was determined to be at an r_s of 160, which roughly translates (by taking a 2/3 power) to an $r_s \cong 30$ in two dimensions. In the high temperature classical limit, the one-component plasma has been determined to crystallize when $\Gamma \cong 145$, where

$$\Gamma = \frac{q^2}{k_B T a \varepsilon}$$

is the ratio of the mean potential energy to the kinetic energy [12]. This translates to a value of $r_s = 120$. Wigner crystallization for the actual temperature lies somewhere in between these two extremes, estimated to be at a value of $r_s \cong 100$, close to the estimated r_s values given in Table 2. The values of r_s in Table 2 were determined by dividing the 6.03Å radius value by the effective Bohr radius of the doubly charged particles, using the appropriate effective mass in a medium of dielectric constant 6. In future calculations we hope to determine the Wigner crystallization density accurately, but it is clear that at lower densities of charge carriers, superfluidity would be inhibited by Wigner crystallization unless the tendency for these systems to crystallize is prevented because the non-uniform periodic background potential is incommensurate with the Wigner crystal lattice.

Acknowledgements

We wish to thank David Ceperley and Roy Pollock for instruction in the use of their path integral program as well as many helpful discussions.

This work was supported by the U.S. Department of Energy and Lawrence Livermore National Laboratory under contract No. W-7405-Eng-48.

1. R.P. Feynman: In Statistical Mechanics (Benjamin, Reading, Mass. 1972)
2. J.P. Hansen: Phys. Rev. A 8, 3096 (1973)
3. R.G. Storer: J. Math. Phys. 9, 964 (1968)
4. D.M. Ceperley and E.L. Pollock: Phys. Rev. Lett. 56, 351 (1986)
5. E.L. Pollock and D.M. Ceperley: Phys. Rev. B 36, 8343 (1987)

6. J.M. Kosterlitz and D.J. Thouless: J. Phys. C 6, 1181 (1973)
7. J. Tobochnik and G.V. Chester: Phys. Rev. B 20, 3761 (1979)
8. N.P. Ong, Z.Z. Wang and J. Clayhold: In Proc. Int. Workshop on Mechanisms of Superconductivity, 1061 (Plenum, New York, NY, 1987)
9. T. Siegrist et. al.: Phys. Rev. B 35, 7137 (1987)
10. V.Z. Kresin and S.A.Wolf: In Proc. Int. Workshop on Mechanisms of Superconductivity, 287 (Plenum, New York, NY, 1987)
11. D.M. Ceperley and B.J. Alder: Phys. Rev. Lett. 45, 566 (1980)
12. R.C. Gann, S. Chakravarty and G.V. Chester: Phys. Rev. B 20, 32 (1979)

Quantum Monte Carlo Studies of the Holstein Model

R.T. Scalettar[1], *N.E. Bickers*[2], *and D.J. Scalapino*[3]

[1]Department of Physics, University of Illinois, Urbana, IL61801, USA
[2]Institute for Theoretical Physics, University of California, Santa Barbara, CA93106, USA
[3]Department of Physics, University of California, Santa Barbara, CA93106, USA

Abstract

In this paper we study the competition between superconducting and charge density wave (CDW) formation in a system of interacting electrons and phonons by solution of the Eliashberg equations and by quantum Monte Carlo. We show that within the Eliashberg formalism, self energy corrections suppress the pair field susceptibility at high temperatures and that only below a certain cross-over temperature does one see enhancement. By conducting simulations at these low temperatures, we report the first observation of enhanced superconducting susceptibilities in numerical simulations of a model with dynamic electron-electron interactions.

1. Introduction

The molecular crystal, or "Holstein" model [1],

$$H = -t\sum_{i,j,\sigma}(c^{\dagger}_{i,\sigma}c_{j,\sigma} + c^{\dagger}_{j,\sigma}c_{i,\sigma}) + g\sum_{i} n_i(a_i + a^{\dagger}_i) + \omega_0\sum_{i} a^{\dagger}_i a_i - \mu\sum_{i\sigma} n_{i\sigma} \tag{1}$$

is one of the simplest Hamiltonians describing the interactions of electrons and lattice vibrations. Here $c^{\dagger}_{i,\sigma}$ creates an electron of spin σ at site i, $n_i = \sum_{\sigma} c^{\dagger}_{i,\sigma}c_{i,\sigma}$ is the charge density on site i, and $a^{\dagger}_i$ is the creation operator for a local phonon mode of frequency w_0 at site i. While this model is probably too crude to admit any detailed physical realizations, the coupling of the breathing mode of the oxygen octahedron about the Bi ions in $BaPb_{1-x}Bi_xO_3$ would have this sort of coupling to the charge density on the Bi atom.

Typically, one expects in inorganic materials that the electron transfer energy t and electron-phonon coupling constant g are larger than the phonon frequency w_0. These materials are thus in the "adiabatic" limit where the phonons move much less rapidly than the electrons. However in some organic materials which have relatively high frequency phonon modes and electron transfer takes place by tunneling described by a small t, or more generally in models of interactions of electrons with excitons or plasmons, one might be interested in the case where w_0 is comparable or even exceeds t and g.

Despite its simplicity, the properties of the Holstein model are not rigorously known. Qualitatively, there is the possibility of two types of phase transitions. The first is a superconducting one, in which pairs of electrons experience an effective attraction mediated by the retarded electron-phonon coupling. The crude physical picture is that an electron distorts its ionic environment and because the phonons move slowly, this distortion lingers in time even after the electron has hopped to another lattice site. A second electron thus has the opportunity to move into this favorable environment, following the first electron. Within perturbation theory we can estimate the strength of this attraction as the square of the electron-phonon coupling constant multiplied by the phonon propagator:

$$V \approx g^2 \frac{2\omega_0}{\omega^2 - \omega_0^2 + i\delta} \tag{2}$$

Springer Proceedings in Physics Vol. 33: **Computer Simulation Studies in Condensed Matter Physics**
Editors: D. P. Landau • K. K. Mon • H-B. Schüttler

corresponding to the diagram in which two electrons exchange a phonon line. For $\omega < \omega_0$ we have $V < 0$. The usual BCS argument then gives a transition temperature

$$T_c \approx \omega_0 e^{-\frac{1}{|V|N(0)}} , \tag{3}$$

where $N(0) \approx \frac{1}{8t}$ is the density of electronic states and one might crudely approximate V by its zero frequency value.

However, the system is also susceptible to the formation of a CDW-Peierls state, in which pairs of electrons occupy alternate lattice sites. Treating the lattice statically, one can integrate out the phonon degrees of freedom to get an effective on-site attraction between electrons with characteristic strength

$$U \approx \frac{-2g^2}{\omega_0}. \tag{4}$$

(Naturally, this is the same result obtained by setting $\omega = 0$ in Eq. (2).) The tendency of the system to CDW formation is greatest at half-filling where there is an optimal number of electrons to form a CDW free from defects. Further, on a square lattice where the Fermi surface nests at half-filling, this tendency is even more enhanced. Naturally, inclusion of an on-site Coulomb repulsion will tend to destabilize this type of order.

Quantum Monte Carlo techniques have been increasingly employed for the study of such systems of interacting bosons and fermions [2]. These methods are particularly advantageous to employ since they permit studies of systems in the intermediate coupling regimes where analytic treatments are least reliable, and do so in a manner in which all approximations are well controlled. The most serious drawback of these approaches is the difficulty in studying spatially large systems and reaching low temperatures [3]. It appears that for magnetic phase transitions such as that occurring in the Hubbard model, these limitations are not crucial [4]. However, despite recent advances, transition temperatures for superconductivity may be too low to see clearly with quantum Monte Carlo. Thus, a secondary issue of interest in this paper will be to understand the feasibility of studying superconducting phase transitions with simulations.

The plan of the remainder of this paper is as follows: In Sec. 2 we describe briefly the results of studies of the Holstein model by solution of the self-consistent Eliashberg equations. In Sec. 3 we discuss the Monte Carlo technique we have employed. Finally, we summarize this work in Sec. 4.

2. The Eliashberg Equations

As an electron moves through a crystal lattice, it is "dressed" by its interactions with lattice vibrations. Here we are interested in the superconducting pair field susceptibility

$$\begin{aligned} \chi_{sc} &= \int_0^\beta \langle \Delta(\tau)\Delta^\dagger(0)\rangle d\tau \\ \Delta^\dagger(\tau) &= \frac{1}{\sqrt{N}} \sum_i c_{i\uparrow}^\dagger(\tau) c_{i\downarrow}^\dagger(\tau) . \end{aligned} \tag{5}$$

This susceptibility provides a measure of the tendency for s-wave pairing and diverges at T_c. Neglecting vertex corrections, χ_{sc} can be calculated by summing the ladder graphs

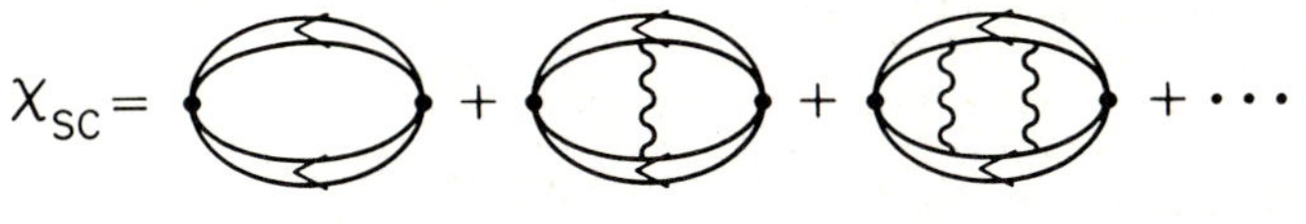

$\chi_{SC} =$ [diagram] + [diagram] + [diagram] + $\cdots$ (6*a*)

with the dressed single particle propagator given by

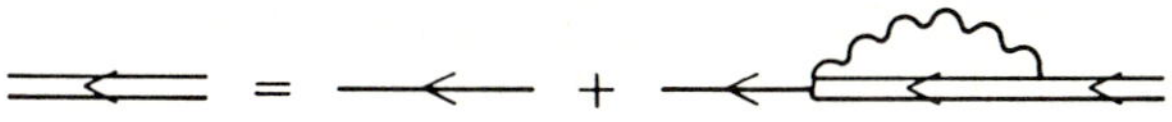

[diagram] = [diagram] + [diagram] (6*b*)

This is the same type of approximation used by Eliashberg below T_c, and and we will refer to these as the Eliashberg equations.

We have solved these equations numerically in the approximation where the dressed phonon line is replaced by the non-interacting phonon propagator. This leaves out processes in which a phonon creates and absorbs electron-hole pairs which have the effect of softening the phonon mode and driving the CDW transition. Indeed, we will see in our simulations that the superconducting phase transition predicted within this treatment can be preempted by CDW formation.

In Figs. 1a,b,c we show the superconducting pair field susceptibility obtained from the Eliashberg equations (6) as a function of temperature for three different values of the frequency ω_0. The coupling $g = 1$ and the chemical potential is chosen so that the average occupation per site, $\langle n \rangle = 0.8$. The full curve is for the noninteracting, $g = 0$, value. We see that at high temperatures χ_{sc} is suppressed, but then at low temperatures is enhanced over the noninteracting value. As ω_0 is decreased, the critical temperature T_c where χ_{sc} diverges increases, but the cross-over temperature actually becomes smaller.

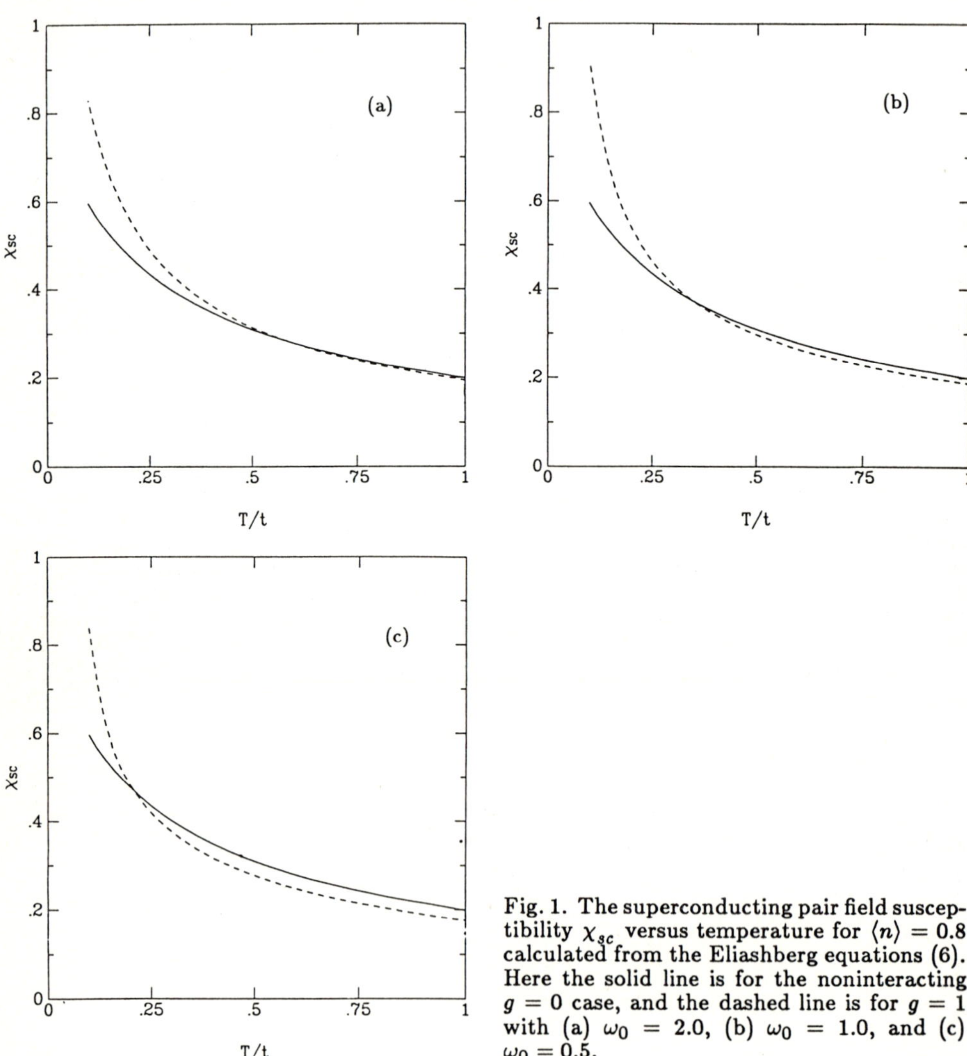

Fig. 1. The superconducting pair field susceptibility χ_{sc} versus temperature for $\langle n \rangle = 0.8$ calculated from the Eliashberg equations (6). Here the solid line is for the noninteracting $g = 0$ case, and the dashed line is for $g = 1$ with (a) $\omega_0 = 2.0$, (b) $\omega_0 = 1.0$, and (c) $\omega_0 = 0.5$.

3. Quantum Monte Carlo Approach

Because of limitations of space, we present here only an outline of the Monte Carlo technique we have employed. Further details are found in Ref. 3, and a more general background in Ref. 6.

In performing numerical simulations, we are interested in calculating various correlation functions at finite temperature

$$\langle A \rangle = \frac{\text{tr Ae}^{-\beta \text{H}}}{\text{tr e}^{-\beta \text{H}}} \quad . \tag{7}$$

We convert the trace into a path integral by discretizing the imaginary time interval, $\beta = N_\tau \Delta\tau$, and separating the exponential of the Hamiltonian into the product of the exponential of the kinetic and potential energies.

$$\begin{aligned}
Z &= \text{tr e}^{-\beta \text{H}} = \text{tr } (\text{e}^{-\Delta\tau \text{H}})^{\text{N}_\tau} \approx \text{tr } (\text{e}^{-\Delta\tau \text{H}_1} \text{e}^{-\Delta\tau \text{H}_2} \text{e}^{-\Delta\tau \text{H}_3})^{\text{N}_\tau} \\
H_1 &= -t \sum_{ij\sigma} (c_{i\sigma}^\dagger c_{j\sigma} + c_{j\sigma}^\dagger c_{i\sigma}) \\
H_2 &= -\mu \sum_i n_i + g\sqrt{2\omega} \sum_{i\tau} n_i x_{i\tau} \\
H_3 &= \frac{\omega^2}{2} \sum_{i\tau} x_{i\tau}^2 + \frac{1}{2(\Delta\tau)^2} \sum_{i\tau} (x_{i\tau+1} - x_{i\tau})^2
\end{aligned} \tag{8}$$

where we have replaced the time derivative of the phonon coordinate by a finite difference [7].

Since the trace is over exponentials of a quadratic form in the fermion operators, it can be performed analytically. (If quartic electron-electron interaction terms were present, they could be eliminated by the introduction of Hubbard-Stratonovich fields [6].) The result of the trace is the product of two determinants (one for each spin σ):

$$Z = \int \prod_{i\tau} dx_{i\tau} e^{-\Delta\tau H_3} \ \text{detM}_\uparrow(\text{x}) \ \text{detM}_\downarrow(\text{x}) \quad . \tag{9}$$

Explicit expressions for M_σ are given for various models in [3,6]. M is a very sparse matrix of dimension NN_τ where N is the spatial volume, however detM can also be expressed as the determinant of a dense matrix of dimension N. Because the phonon field couples in the same way to both electron spin species, the fermion determinants are equal and their product forms a positive definite quantity which, along with the exponential of the "bosonic" part of the action H_3, can be used as a Boltzmann weight for the phonon fields. In general these determinants need not be related, and one encounters the difficult problem of cancellations between contributions of different sign. This is a question which requires further study [8].

In order to avoid evaluating the determinants, we replace them by Gaussian integrals over an auxiliary field ϕ:

$$Z = \int \prod_{i\tau} dx_{i\tau} d\phi_{\uparrow i\tau} d\phi_{\downarrow i\tau} e^{-(\Delta\tau H_3 + \phi_\uparrow O_\uparrow^{-1} \phi_\uparrow + \phi_\downarrow O_\downarrow^{-1} \phi_\downarrow)} \tag{10}$$

where $O = M^T M$.

Equation 10 now represents the partition function of three classical fields x, $\phi_\uparrow$, $\phi_\downarrow$, with a nonlocal interaction resulting from the matrix inverse. Configurations of these fields can be generated with the appropriate weight by an algorithm which combines heat bath updates of the ϕ fields with molecular dynamics evolution of the phonon variables. The heat bath step consists of generating two vectors R_σ of Gaussianly distributed random numbers, $p(R_{i\sigma}) = ce^{-R_{i\sigma}^2}$, and setting

$$\phi_\sigma = M_\sigma^T R_\sigma \quad . \tag{11}$$

Molecular dynamics evolution of the Hubbard-Stratonovich field x proceeds by the introduction of a momentum p conjugate to x

$$Z = \int \prod_{i\tau} dx_{i\tau} d\phi_{\uparrow i\tau} d\phi_{\downarrow i\tau} dp_{i\tau} e^{-\frac{1}{2}\sum_{i\tau} p_{i\tau}^2} e^{-(\Delta\tau H_3 + \phi_\uparrow O_\uparrow^{-1}\phi_\uparrow + \phi_\downarrow O_\downarrow^{-1}\phi_\downarrow)} \quad . \tag{12}$$

This clearly has no effect on correlation functions involving the x and ϕ fields. The fields x and p are then evolved according to a discretized version of Hamilton's equations.

$$\begin{aligned} x'_{i\tau} &= x_{i\tau} + p_{i\tau} dt \\ p'_{i\tau} &= p_{i\tau} - (\Delta\tau \frac{dH_3}{dx_{i\tau}} - \sum_\sigma \phi_\sigma O_\sigma^{-1} \frac{dO_\sigma}{dx_{i\tau}} O_\sigma^{-1} \phi_\sigma) dt \end{aligned} \tag{13}$$

dt is the step size for the discretization. In practice we actually employ a leapfrog method for the numerical integration of the equations of motion. The momenta are refreshed periodically by drawing them from a Gaussian distribution $e^{-\frac{1}{2}p_{i\tau}^2}$.

All the steps in the algorithm scale as the volume of the system, that is, the product of the spatial volume N and the number of imaginary time slices N_τ. The computation time is also proportional to the number of iterations required to invert the matrix O. The eigenvalue spectrum of O is not very sensitive to spatial volume, which is what makes possible the simulation of spatially large systems. It is, however, strongly dependent on temperature and coupling, and care must be taken to condition the matrix in order to reach low temperatures and strong couplings [3].

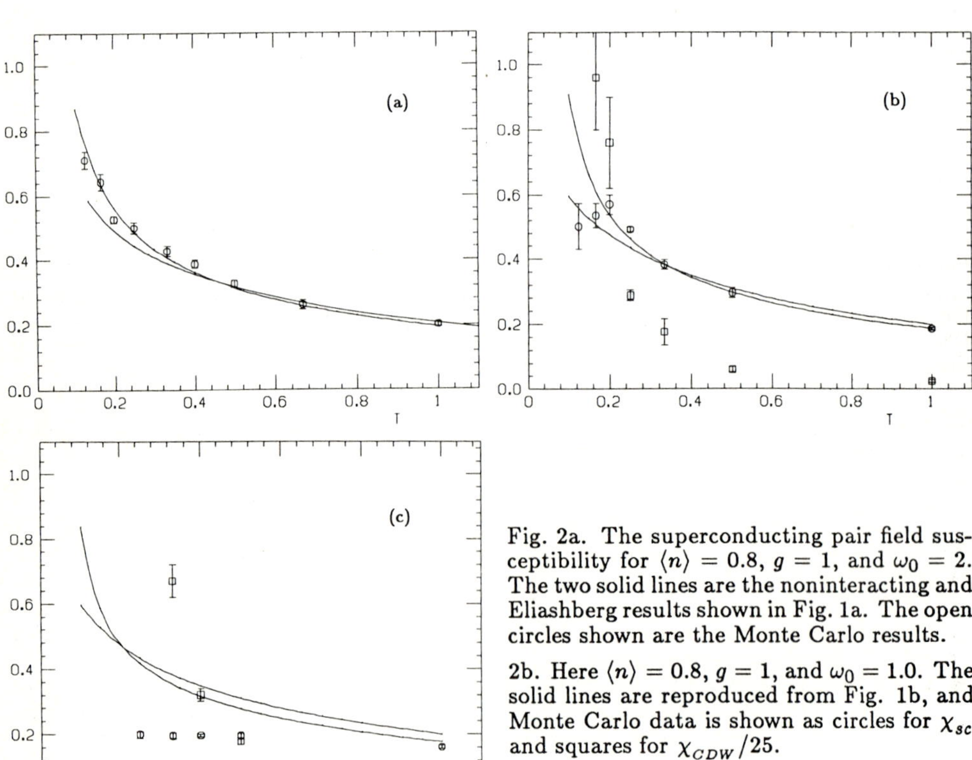

Fig. 2a. The superconducting pair field susceptibility for $\langle n \rangle = 0.8$, $g = 1$, and $\omega_0 = 2$. The two solid lines are the noninteracting and Eliashberg results shown in Fig. 1a. The open circles shown are the Monte Carlo results.

2b. Here $\langle n \rangle = 0.8$, $g = 1$, and $\omega_0 = 1.0$. The solid lines are reproduced from Fig. 1b, and Monte Carlo data is shown as circles for χ_{sc} and squares for $\chi_{CDW}/25$.

2c. Here $\langle n \rangle = 0.8$, $g = 1$, and $\omega_0 = 0.5$. The solid lines are reproduced from Fig. 1c.

Physical observables such as the energy, and the magnetic and superconducting susceptibilities can be readily expressed in terms of quantities which appear in the evolution of the fields x, ϕ. We employ the identity

$$2\langle R_{i\tau}(O^{-1}\phi)_{j\tau'}\rangle = 2\langle R_{i\tau}(M^{-1}R)_{j\tau'}\rangle = M^{-1}_{i\tau j\tau'} \tag{14}$$

which holds since $\langle R_{i\tau}R_{j\tau'}\rangle = \frac{1}{2}\delta_{ij}\delta_{\tau\tau'}$. Also M^{-1} is the fermion Greens function, $M^{-1}_{i\tau j\tau'} = \langle c_i(\tau)c_j^\dagger(\tau')\rangle$ [6], hence we can obtain all quantities by reducing them to expectation values of products of Green's functions and performing the above average.

In Fig. 2a,b,c we show χ_{sc} as a function of temperature for the same sets of parameters used in Fig. 1. The two curves are again the interacting value within the Eliashberg treatment and the noninteracting result. The open circles are values obtained by numerical simulations. The open squares show the CDW susceptibility

$$\begin{aligned} \chi_{CDW} &= \int_0^\beta \langle Q(\tau)Q(0)\rangle d\tau \\ Q(\tau) &= \frac{1}{\sqrt{N}}\sum_i (-1)^i n_{i\tau} \end{aligned} \tag{15}$$

divided by a normalization factor of 25 for display purposes. We see that for $\omega_0 = 2$, the CDW instability does not develop and the Eliashberg prediction for the behavior of χ_{sc} is followed quite closely. However, for $\omega_0 = 1$, the CDW instability preempts the superconductivity. For $\omega_0 = 0.5$ the tendency to lock into alternating doubly occupied and unoccupied sites is so great the χ_{sc} is highly suppressed at all temperatures.

4. Conclusions

We have seen that numerical simulations can be used to study superconducting phase transitions with dynamical interactions. An approximate numerical technique based on the Eliashberg equations is seen to be qualitatively correct in some parameter regimes, but to miss the essential physics in others. The initial suppression of the pair field susceptibilities by the interactions makes it essential to reach low enough temperatures in the simulations to make a statement concerning superconductivity in this model.

Acknowledgments

We would like to thank R.L. Sugar for many useful discussions. This work was supported in part by the United States Department of Energy under Grants DE-FG03-85ER45197 and DE-AT03-81ER40029; by the National Science Foundation under Grants DMR86-12860 and DMR86-15454; and by NSF Grant PHY82-17853 supplemented by funds from NASA. We also thank the San Diego Supercomputer Center.

References

1. T. Holstein, Ann. Phys. 8, 325 (1959).
2. In Proceedings of the Conference on the Frontiers of Quantum Monte Carlo, ed. by J.E. Gubernatis, in J. Stat. Phys. 43, 729 (1986).
3. R.T. Scalettar, D.J. Scalapino, R.L. Sugar, and D. Toussaint, Phys. Rev. B 36, 8632 (1987); J.E. Hirsch and R.M. Fye, Phys. Rev. Lett. 56, 2521 (1986).
4. J.E. Hirsch, Phys. Rev. Lett. 51, 1960 (1983), and Phys. Rev. B 31, 4403 (1985).
5. G.M. Eliashberg, JETP 11, 696 (1960).
6. R. Blankenbecler, D.J. Scalapino, and R.L. Sugar, Phys. Rev. D 24, 2278 (1981); D.J. Scalapino and R.L. Sugar, Phys. Rev. B 24, 4295 (1981); and J.E. Gubernatis, D.J. Scalapino, R.L. Sugar, and D. Toussaint, Phys. Rev. B 32, 103 (1985).
7. M. Creutz and B. Freedman, Ann. Phys. 132, 427 (1981).
8. S.R. White and R.L. Sugar (in preparation).

Structure of the Wave Function of Crystalline ^{4}He

S.A. Vitiello, K.J. Runge, and M.H. Kalos

Courant Institute of Mathematical Sciences, New York University,
New York, NY 10012, USA

Abstract: The nature of the wave function for quantum crystals has been an outstanding problem for some time. Although a Jastrow product wave function gives a qualitatively correct description of the liquid, and although such a wave function can describe a crystal, the result is in serious disagreement with experiment. The standard treatment is to construct a lattice and directly couple the particles to its sites. Although this gives fair numerical agreement with some crystal properties, it is at the expense of breaking the Bose symmetry of the crystal. We introduce here a new class of trial wave functions that are symmetric under particle exchange. They give a lower variational energy than, and have properties comparable with those given by trial functions in which atoms are explicitly localized.

If one uses a Jastrow product trial function in a variational calculation of the properties of liquid ^{4}He, one gets qualitatively plausible results for the energy and structure function. Since this problem can be mapped onto a classical system with some pair potential, a solid must be obtained if the correlation between the particles in the wave function is increased sufficiently. Unfortunately, the resulting energy is much too large, and the structure is qualitatively wrong for a quantum crystal. The difficulty is that by increasing the correlation enough to crystalize the system, one departs significantly from satisfying the Schrödinger equation when two particles are very close. Such a trial wave function is no longer a satisfactory description of a quantum crystal.

Good variational energies for the solid phase were reported by HANSEN and LEVESQUE.[1] Following NOSANOW,[2] they introduced Gaussian factors into the wave function coupling the atoms to crystal lattice sites explicitly introduced for that purpose. This destroys both the translational invariance[3] and the Bose symmetry of the trial wave function. We have recently observed[4] that these properties can be restored while retaining a good quantitative description of the crystal.

Our new trial wave function $\Psi_T(R)$, $\quad R \equiv \{\vec{r}_1, \vec{r}_2, ..., \vec{r}_N\}$, uses artificial variables $S \equiv \{\vec{s}_1, \vec{s}_2, ..., \vec{s}_N\}$ which we call "shadow" coordinates. They appear through the definition

$$\Psi_T(R) = \int \Xi(R,S)dS \; ; \tag{1}$$

Ξ is given by

$$\Xi(R,S) = \prod_{i<j} e^{-\frac{1}{2}u(r_{ij})} \quad \prod_{k} e^{-\phi(\vec{r}_k - \vec{s}_k)} \quad \prod_{l<m} e^{-v(s_{lm})} \; , \tag{2}$$

where $r_{ij} = |\vec{r}_i - \vec{r}_j|$ is the distance between particles i and j and $s_{lm} = |\vec{s}_l - \vec{s}_m|$. In (2) the first factor is the usual two-body correlation function of the Jastrow form.

Springer Proceedings in Physics Vol. 33: **Computer Simulation Studies in Condensed Matter Physics**
Editors: D. P. Landau • K. K. Mon • H-B. Schüttler

The last term is a function of the auxiliary variables $\vec{s}_j$ for which a model potential of the form s^{-n} has been taken. In this work ϕ was chosen to be $C(\vec{r}_k - \vec{s}_k)^2$, where C is a variational parameter. The function $\Psi_T(R)$ is both translationally invariant and symmetric under particle interchange.

To understand the motivation that lies behind the shadow wave function it is useful to assign a meaning to the set of variables S in the trial wave function of (1). The shadow particles can be thought of as a set of pseudo-classical particles that interact with each other by a pair potential. The second term in (2) provides a one-to-one coupling between real and shadow particles. The idea here is to incorporate into the quantum domain, in a phenomenological way, the long range crystalline order exhibited at low temperatures by classical particles.

Additional insight into the shadow wave function comes from the path integral picture of a quantum solid.[5] Consider the "center of mass" of all positions of one particle on a Feynman path. This center is likely to behave as a classical particle, here modeled as a shadow particle. The fluctuation of a particle's position with respect to the center of mass may be introduced, at least crudely, by a harmonic interaction.

The important features of the shadow wave function can be seen when one takes into account (1) and (2). Every real particle is more correlated to all orders than a Jastrow product; in previous variational Monte Carlo simulations no more than 3-body correlations have been explicitly introduced.[6] Long-range order is obtained by choosing separate and appropriate strengths of the pseudo-potentials for the shadow and the real particles.

The variational energy is given by

$$E_T = \frac{\int dR \Psi_T H \Psi_T}{\int dR |\Psi_T|^2} , \tag{3}$$

and can be rewritten in terms of the normalized probability density $p(R, S_1, S_2)$ as

$$E_T = \int\int\int dR dS_1 dS_2 \quad p(R, S_1, S_2) \quad \left\{ \frac{H\Xi(R, S_2)}{\Xi(R, S_2)} \right\} \tag{4}$$

where

$$p(R, S_1, S_2) = \frac{\Xi(R, S_1)\Xi(R, S_2)}{\int\int\int dR dS_1 dS_2 \quad \Xi(R, S_1)\Xi(R, S_2)} . \tag{5}$$

The evaluation of the energy in (4) involves $9N$ dimensional integrals. It is evaluated by the standard Metropolis Monte Carlo algorithm.[7] In this method the estimator $H\Xi(R, S_2)/\Xi(R, S_2)$ is averaged with respect to the normalized probability density function $p(R, S_1, S_2)$ by the expression

$$E_T = \frac{1}{M} < \sum_{m=1}^{M} (\mathrm{K} + \mathrm{V})_m > , \tag{6}$$

where M is the number of points in the space $\{R, S_1, S_2\}$ sampled from $p(R, S_1, S_2)$, K is the kinetic energy and V the potential energy:

$$\mathrm{V} = \sum_{i<j} V(r_{ij}) . \tag{7}$$

The interparticle potential $V(r_{ij})$ used is the HFDHE2 of AZIZ et al.[8] For the pseudo-potentials the McMILLAN[9] form has been used:

$$u(r) = \left(\frac{b}{r}\right)^5 \quad \text{and} \quad v(s) = \left(\frac{b_{sh}}{s}\right)^n . \tag{8}$$

Both pseudo-potentials are fitted to a third degree polynomial so that they go smoothly to zero at the sides of the simulation cube. Periodic boundary conditions are applied.

In these computations, the Metropolis random walk involves two types of moves, one for the real particles and another for the shadow particles. Consider first a move of a real particle j. A new coordinate $\vec{r}'_j$ is chosen with uniform probability in a cube of side Δ centered at $\vec{r}_j$. This move is accepted with probability q given by[10]:

$$q = min\left(1, \frac{p(R', S_1, S_2)}{p(R, S_1, S_2)}\right) , \tag{9}$$

where R' contains the new coordinate value. The ratio of the density probability functions can be written explicitly as:

$$\begin{aligned}\frac{p(R', S_1, S_2)}{p(R, S_1, S_2)} &= \exp\{-\sum_{i \neq j} (u(|\vec{r}'_j - \vec{r}_i|) - u(|\vec{r}_j - \vec{r}_i|))\} \\ &\times \exp\{-C\ ((\vec{r}'_j - \vec{s}_{1j})^2 - (\vec{r}_j - \vec{s}_{1j})^2)\} \\ &\times \exp\{-C\ ((\vec{r}'_j - \vec{s}_{2j})^2 - (\vec{r}_j - \vec{s}_{2j})^2)\}. \end{aligned} \tag{10}$$

An attempt to move the next real particle is then made in the same way. In our program, we cycle through all real particles $\{\vec{r}_1, \vec{r}_2, ..., \vec{r}_N\}$ before any shadow moves.

The same procedure is used for the shadow particles except with a cube of side Δ_{sh}. In this case, if an attempt to move particle s_{1j} is made, the probability ratio reads:

$$\begin{aligned}\frac{p(R, S'_1, S_2)}{p(R, S_1, S_2)} &= \exp\{-\sum_{i \neq j} (v(|\vec{s}'_{1j} - \vec{s}_{1i}|) - v(|\vec{s}_{1j} - \vec{s}_{1i}|))\} \\ &\times \exp\{-C\ ((\vec{r}_j - \vec{s}'_{1j})^2 - (\vec{r}_j - \vec{s}_{1j})^2)\}. \end{aligned} \tag{11}$$

By S'_1 we mean the set of of shadow variables $\{\vec{s}_{11}, \ldots, \vec{s}'_{1j}, \ldots, \vec{s}_{1N}\}$. After attempting to move all the shadow particles of set S_1 the same is done for those in set S_2. The parameters Δ and Δ_{sh} were adjusted so that the acceptance was near 50%.

We have computed the energy per particle and the Lindemann ratio (r.m.s. deviation about a lattice site divided by the nearest neighbor distance) in the solid phase of ^{4}He for extensive sets of the variational parameters. Starting from an fcc lattice our best result for the energy together with the Lindemann ratio are shown in Table I. The Lindemann ratios were computed with respect to this starting lattice shifted by the amount the real particles' center of mass had diffused. Long runs were performed to insure that the crystal remained stable. For sake of comparison, we also made runs using a wave function of the Jastrow form localized with a Gaussian. These results are displayed in Table I as well.

Table I. Variational results for the solid phase obtained with the shadow wave function (first two lines) and with a wave function of the Jastrow form localized with Gaussians (third line) at the density $\rho\sigma^3 = 0.55$. The energies are given in K per particle and the unit of length is $\sigma = 2.556$Å.

		Lindemann ratio		Parameters			
	E_T	real	shadow	b	b_{sh}	n	C
Shadow	-3.540 ±0.028	0.23	0.16	1.10	1.68	5	5.7
Shadow	-3.549 ±0.050	0.23	0.13	1.10	1.35	9	5.7
JG	-3.361 ±0.019	0.22	–	1.10	–	–	4.8

As can be seen in Table I the shadow wave function gives an energy that is significanly lower than the results that one gets from a wave function of the Nosanow-Jastrow type. As expected the Lindemann ratio for the real particles are in good agreement for both wave functions. The Lindemann ratio for the shadow particles is comparable with that obtained at melting for a classical Lennard-Jones system.[11]

Table I also shows the behavior of the energy of the crystal as the exponent of the shadow-shadow pair interaction is changed. It is in fact independent of that parameter, when b_{sh} is adjusted to minimize the energy. This insensitivity supports our belief that the reduction in energy is a consequence of the fact that the structure of the trial function has been improved, and not simply that more variational parameters have been introduced.

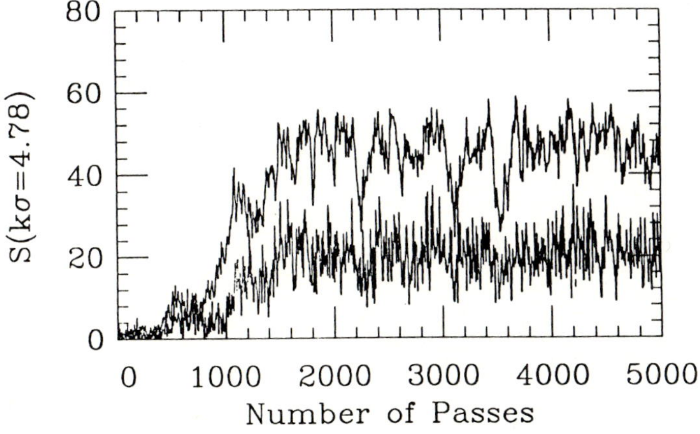

Fig. 1. Monte Carlo evolution of the structure function $S(k)$ with k fixed near the position of the first maximum. The random walk is started with the system in a liquid configuration. The lower trace denotes the time dependence of the real particles' structure function and the upper curve that of the shadow coordinates. Solidification occurs between 1000 and 1500 passes.

The most convincing demonstration that crystal order is described by some trial function is that it appears spontaneously out of a system initialized randomly or as a liquid. We have carried out such simulations in two dimensions. In Fig. 1, we show the structure function S(k), calculated at a momentum in the reciprocal lattice of the triangular lattice expected for a two dimensional solid. The system was initialized as a liquid, and the graph shows S(k) evolving from the disordered state to the crystal for both the shadow and physical particles. Figure 2 shows the same kind of plot for a system initialized in a somewhat compressed square lattice. It too evolves to the triangular order. Finally, Fig. 3 shows the evolution of a system initialized in a triangular lattice. After a very short transient, it shows the same value and qualitatively the same fluctuations as the systems otherwise initialized. This is clear evidence that the shadow wave function describes a system with crystalline order.

For completeness, we investigated also a three-dimensional liquid using a variational trial function having the shadow form. Results are shown in Table II, along

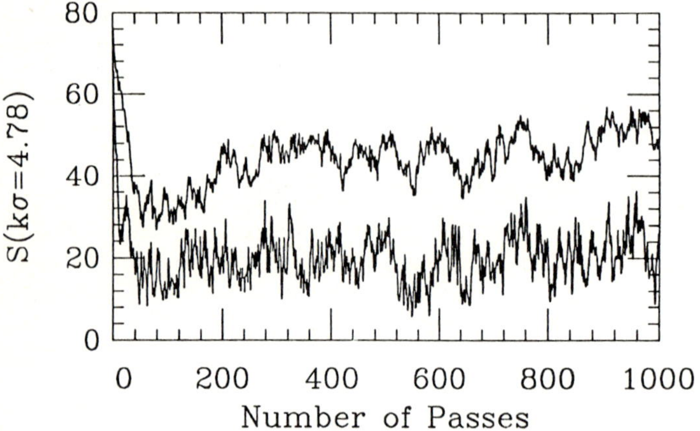

Fig. 2. Monte Carlo evolution of the structure function for the shadow particles (upper curve) and the real particles (lower curve.) The initial configuration is a slightly compressed square lattice placed in a box that accommodates a periodic triangular lattice.

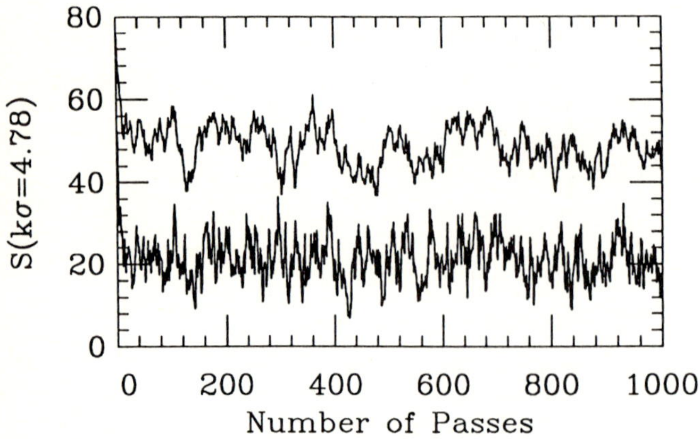

Fig. 3. As in Fig. 2 except the initial configuration is taken as a triangular lattice.

Table II. Variational results determined for ^{4}He liquid at the density $\rho\sigma^3 = 0.365$. Energies are in K per particle and lengths in σ.

	E_T	Parameters b	b_{sh}	n	C
Shadow	-6.062 ±0.036	1.13	1.40	5	4
Jastrow	-5.717 ±0.021	1.20	–	–	–

with those determined from a wave function of the Jastrow form. The shadow wave function gives a lower energy. Note that the same functional form of the wave function has been used in both phases; the parameters b, b_{sh} and C have, of course, been changed. When parameters appropriate to a liquid phase are chosen, the system melted rapidly from an initially crystalline order to an equilibrium liquid state.

The trial functions of the shadow class exhibit translational invariance and symmetry under particle exchange, and suggest a new physical picture of a quantum crystal. We believe that one will be able, with their help, to investigate additional properties of extensive quantum systems. One question in particular may now be accessible, namely that of a condensate in Bose crystals.

Acknowledgements: This work was partially supported by the Condensed Matter Theory Program of the National Science Foundation under Grant DMR-8419083, and by Conselho Nacional de Desenvolvimento Científico e Tecnológico, Brazil. This research was conduct using the Cornell National Supercomputer Facility, a resource of the Center for Theory and Simulation In Science and Engineering at Cornell University, which is funded in part by the National Science Foundation, New York State, and the IBM Corporation.

REFERENCES

1. J. P. Hansen and D. Levesque, *Phys. Rev.* **165**, 293 (1968).

2. L. H. Nosanow, *Phys. Rev. Lett.* **13**, 270 (1964).

3. We are concerned here with systems obeying periodic boundary conditions. In this case it can be shown that the true ground state wave function is translationally invariant even in the solid phase.

4. Silvio Vitiello, Karl Runge and M. H. Kalos, *Phys. Rev. Lett.* **60**, 1970 (1988).

5. R. P. Feynman, *Statistical Mechanics*, W. A. Benjamin, Inc., Reading, 1972.

6. K. Schmidt, M. H. Kalos, M. A. Lee and G. V. Chester, *Phys. Rev. Lett.* **45**, 573 (1980).

7. N. Metropolis, A. W. Rosenbluth, M. N. Rosenbluth, A. H. Teller and E. Teller, *J. Chem. Phys.* **21**, 1087 (1953).

8. R. A. Aziz, V. P. S. Nain, J. S. Carley, W. L. Taylor and G. T. McConville, *J. Chem. Phys.* **70**, 4330 (1970).

9. W. L. McMillan, *Phys. Rev.* **138**, A442 (1965).

10. M. H. Kalos and P. A. Whitlock, *Monte Carlo Methods*, John Wiley & Sons, New York, 1986.

11. J. P. Hansen and I. R. McDonald, *Theory of Simple Liquids*, Academic Press, London, 1976.

Electronic Structure Calculation by Nonlinear Optimization: Application to Metals

R. Benedek[1], *B.I. Min*[2], *C. Woodward*[3], *and J. Garner*[4]

[1]Materials Science Division, Argonne National Laboratory, 9700 S. Cass Ave., Argonne IL 60439, USA
[2]Present address: Pohang Institute of Science and Technology, Pohang 680, Seoul, Korea
[3]Permanent address: Michigan Technological University, Houghton, MI 49931, USA
[4]Permanent address: Bradley University, Peoria, IL 61625, USA

There is considerable interest in the development of novel algorithms for the calculation of electronic structure (e.g., at the level of the local-density approximation of density-functional theory.) In this paper we consider a first-order equation-of-motion method. Two methods of solution are described, one proposed by Williams and Soler, and the other based on a Born-Dyson series expansion. The extension of the approach to metallic systems is outlined and preliminary numerical calculations for Zintl-phase NaTl are presented.

1. INTRODUCTION

One of the general goals of computational physics is to calculate self-consistently the properties of interacting systems of electrons and ions, such as molecules, atomic clusters, solids and liquids. At present, however, this goal is too general to allow a unified method of attack. Let us therefore consider a somewhat more restricted framework in which the ions are regarded as classical and the electrons are adiabatic (i.e. obey the Born-Oppenheimer approximation). This viewpoint is sufficiently general to include a great deal of the physics one wishes to study by computer simulation methods. It was in this latter framework (classical ions and adiabatic electrons) that CAR and PARRINELLO[1] suggested the utility of nonlinear optimization to deal with the electronic degrees of freedom. They introduced the method of "dynamical simulated annealing" (DSA) in which both electrons and ions are formally described by quasi-Newtonian equations of motion. In the case of the electrons, these equations provide an alternative method for solving the Kohn-Sham equations of density functional theory. Compared with conventional methods, DSA has the advantage of avoiding an explicit matrix-diagonalization step, which can be prohibitive for large systems. Another important feature of DSA is the extensive use of fast-Fourier transformation to evaluate convolution sums. In a series of papers CAR, PARRINELLO and co-worker[2-5] applied DSA to study structural properties of atomic clusters, liquids and glasses. Since the method of DSA was introduced, several novel algorithms[6-10] for electronic structure calculation have been explored. In the present contribution, we consider a first order analog to the DSA equations of motion. Two methods of solution are described, one of which was introduced by WILLIAMS and SOLER[7]. Most of the work based on DSA and related methods has focused on semiconductors. We will discuss the modifications required for the treatment of metallic systems. As a first application, we are considering systems with the B32 Zintl-phase structure, such as NaTl.

Springer Proceedings in Physics Vol. 33: **Computer Simulation Studies in Condensed Matter Physics**
Editors: D. P. Landau • K. K. Mon • H-B. Schüttler

2. EQUATIONS OF MOTION

The local-density approximation of density-functional theory[11] has provided a highly robust framework for calculating the ground-state charge density and electronic structure of metals and semiconductors for a fixed arrangement of ions. In the conventional formulation, variational minimization of the total-energy functional leads to the Kohn-Sham equations[12]

$$H\phi_i(r) = \varepsilon_i \phi_i(r), \tag{1}$$

where H is the Kohn-Sham Hamiltonian, and ε_i is the eigenvalue corresponding to orbital $\phi_i(r)$. Normally, the orbitals are expanded in a set of basis functions, $\phi_i = \sum_\alpha C_{i\alpha} \chi_\alpha$. The resultant algebraic eigenvalue problem is usually solved by applying standard matrix diagonalization software. Alternatively, we can start with the optimization principle

$$\delta E/\delta\phi_i^*(r) = 0, \tag{2}$$

subject to the constraint of orthonormality,

$$\langle\phi_i|\phi_j\rangle = \delta_{ij}, \tag{3}$$

where E is the total-energy functional for a fixed arrangement of ions $\{R_\ell\}$. Equation (2) in conjunction with the orthonormality constraint eq. (3) is entirely equivalent mathematically to the Kohn-Sham equations, however, it provides a different point of departure for devising numerical algorithms. It is straightforward to show that eqs. (2) and (3) can be solved by considering the (fictitious) equations of motion

$$\partial\psi_i/\partial t = -(1-P_N)H\,\psi_i, \tag{4}$$

where $P_N \equiv \Sigma|\psi_i\rangle\langle\psi_i|$ projects onto the N occupied orbitals of the system. In applying eq. (4), one constructs an initial set of orbitals $\psi_i(t=0)$ and solves eq. (4) numerically. In the limit $t \to \infty$, $\dot{\psi}_i \to 0$ and P_N corresponds to the Hilbert subspace of occupied Kohn-Sham orbitals. $\psi_i(t\to\infty)$ may, however, differ from the actual Kohn-Sham orbitals ϕ_i by a unitary transformation.

In order to write solutions to eq. (4) in a compact form we specialize to a one-particle system. The generalization of the equations given below to an N particle system is straightforward. For a one-particle system, eq. (4) becomes

$$\partial\psi/\partial t = -(H-\varepsilon)\psi, \tag{5}$$

where $\varepsilon \equiv \langle\psi|H|\psi\rangle$. Equation (5) is identical (apart from the factor ε) to the imaginary-time Schroedinger equation; the equation is also reminiscent of the Bloch equation[13], if we identify t as inverse temperature and ε as the chemical potential. To solve eq. (5), we expand the wavefunction in plane waves (for simplicity, we consider a zone-center state, k=0):

$$\psi(r,t) = \sum_G \psi_G(t)e^{iG\cdot r}. \tag{6}$$

Substituting eq. (6) into eq. (5), we obtain a set of equations for ψ_G. If we treat the terms involving $\psi_{G'}$ with $G' \neq G$ as constants, we can integrate the resultant form analytically:

$$\psi_G(t+\Delta t) = \psi_G(t) + F(G,t)\cdot f(G,t,\Delta t)\cdot \Delta t, \tag{7}$$

where the force is defined $F(G,t) = -\sum_{G'} (H_{GG'} - \varepsilon\delta_{GG'})\,\psi_{G'}(t)$ and f is given by

$$f(x) \equiv (e^x-1)/x, \tag{8}$$

with $x = (H_{GG}-\varepsilon)\Delta t \equiv H_G\Delta t$.

This algorithm, proposed by Williams and Soler[7], is intended to be applied iteratively until satisfactory convergence is achieved. Note that the simplest finite-difference updating scheme based on eq. (5) would correspond to f=1. Equation (7), however, includes some 'nonlinear' terms and therefore enables relatively large values of the time interval Δt to be selected, without resulting in instabilities. Consequently, a faster rate (by about an order of magnitude) of convergence is achieved with eq. (7) than with simple finite-difference updating. A numerical example will be given below, however, we mention first an alternative approach to the numerical solution of eq. (5).

The analogy of eq. (5) to the Bloch equation of statistical mechanics suggests the application of the Dyson series, a standard method of solution to this equation[13]. We obtain

$$\psi_G(t+\Delta t) = \exp(-H_G t) \sum_{n=0}^{\infty} \sum_{G'} B_n(G,t+\Delta t;\ G',t)\ \psi_{G'}(t)(\Delta t)^n. \tag{9}$$

The terms corresponding to a particular value of n are interpreted as n^{th} order scattering processes, with B_n representing the amplitude for scattering from "initial-state" wavevector G' at time t to "final state" wavevector G at time $t+\Delta t$. In particular, we have

$$B_0 = \delta_{G,G'} \tag{10a}$$

$$B_1 = -H_{GG'}\ D_1(H_G,\ H_{G'},\ \Delta t) \tag{10b}$$

$$B_n = (-1)^n \sum H_{GG_1}\ H_{G_1G_2} \cdots H_{G_{n-1}G'} \times D_n(H_G,\ H_{G_1},\ \ldots, H_{G'},\ \Delta t)\ . \tag{10c}$$

In the summation of eq. (10c) and in eq. (10b), only terms in which all matrix elements of H are off-diagonal are included ($G \neq G_1$, $G_1 \neq G_2$, etc.). The functions D_n are derivative-like functions, which we do not reproduce here explicitly. We note, however, that in the limit $\Delta t \rightarrow 0$:

$$\lim_{\Delta t \rightarrow 0} D_n = \left.\frac{d^{n-1}}{dx^{n-1}} f(x)\right|_{x=0} = \frac{1}{n!}, \tag{11}$$

where f(x) was defined in eq.(8). Numerical evaluation of the right-hand side of eq. (9) is greatly facilitated if the approximation $D_n = 1/n$ is adopted. Numerical tests and comparisons of results based on eq. (9) with those based on eq. (7) will be presented elsewhere.

3. EXTENSION TO METALLIC SYSTEMS

As mentioned above, most previous applications of nonlinear optimization in the spirit of DSA[1-6, 8-10] have been to insulating systems. In the case of metals, it is necessary to optimize the Fermi surface as well as the Kohn-Sham orbitals. The total electronic charge density for metallic systems can be written as

$$n(r) = \sum_{\nu} \sum_{k} |\ \phi_{\nu k}(r)|^2\ w(\nu,k)\ \theta(\varepsilon_F - \varepsilon_{\nu k}), \tag{12}$$

where k is the wavevector, ν is a band index, w is a weight factor and ε_F is the Fermi energy. In calculations for insulators a reasonably accurate sampling of the Brillouin zone can usually be accomplished with as few as one or two special points. For metals, the tetrahedron method provides a precise zone integration scheme but may require a large number N_k of sampled k points to achieve satisfactory convergence[14]. We found that the gaussian-broadening method[15] affords a more practical zone integration scheme in the present context.

In order to determine ε_F and the weight factors, it is necessary to calculate a larger number of bands N_ν than the number actually occupied. To construct the charge density, eq. (12), a diagonalization is performed in the subspace of states

$\psi_{\nu},k^{(r)}$, $\nu = 1,\ldots N_{\nu}$, after each iteration step. The eigenvectors obtained in this "subspace diagonalization"[10] are then used to perform the unitary transformation from ψ to ϕ. A common feature of self-consistent electronic structure calculations in metals is the damping of low-frequency, long-wavelength charge-density fluctuations. In the present work this is accomplished by mixing a fraction α of the output charge density in each step with fraction $1-\alpha$ of the input charge density to obtain an input potential for the next iteration. In the calculations described below, $\alpha = 0.2$.

4. APPLICATION TO ZINTL-PHASE COMPOUNDS

As an initial application of the present formalism, we are studying compounds with the (B32) Zintl-phase structure[16], which consists of interpenetrating diamond sublattices. The material of most technological interest with this structure is LiAl, which has a high ionic conductivity. Because of the large concentrations of constitutional defects (vacancies and antisite atoms) typical of this structure, a self-consistent treatment of both atomic and electronic structure will be important for a proper description of properties. This will require larger unit cells than can be accommodated using conventional electronic structure techniques. The test calculations described below, however, pertain to the ideal, defect-free structure.

Figure 1 illustrates the convergence properties of the Williams-Soler algorithm, eq. (7). The ion core potentials were represented by local pseudopotentials of the Heine-Abarenkov-Shaw form[17],

$$v(q) = \frac{4\pi z}{q^2} j_0(\pi q/q_0),$$

where z is the ionic valence, and j_0 is the zeroth order spherical Bessel function. We have employed the parameters q_0=0.75 and 1.45 a.u. to represent monovalent and trivalent ions, respectively, in the compound NaTl. A conventional cubic unit cell was employed with eight atoms of each type, which would correspond to sixteen doubly occupied bands at each k-point if NaTl were an insulator. In the present calculations, $N_{\nu} = 28$. $N_k = 10$ points were sampled in the irreducible wedge; very similar results were obtained in tests with $N_k = 20$ points. The basis set included 251 plane waves.

In order to apply eq. (7), initial states $\psi_{\nu kG}(t=0)$ must be constructed. In our illustrative example, initial states were generated using random numbers, and

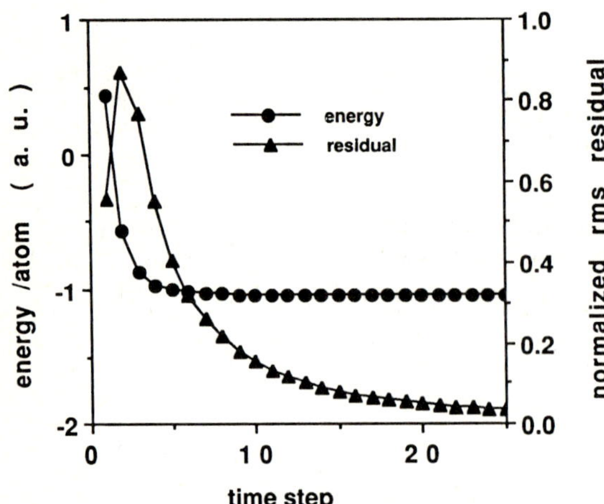

Fig.1. Numerical calculations for NaTl based on Williams-Soler algorithm, eq. (7). See text for discussion.

orthornormalization was accomplished by the Gram-Schmidt procedure. We observe in Fig. 1 that convergence of the total energy per atom is rapid, in spite of the arbitrary initial states. The time step was $\Delta t=1.6$ a.u. As a measure of wave function convergence we plot

$$R = \sqrt{\langle F^2 \rangle / \langle \varepsilon^2 \rangle}, \tag{13}$$

where $F \equiv (H-\varepsilon)\psi$ represents the deviation ("residual") from the Kohn-Sham eq. and we have normalized by the rms eigenvalue. The results for R plotted in Fig. 1 are based on a simple average over N_k k points and $N_\nu=28$ bands; if we had weighted the average by the factor $w(\nu,k)$, for example, the wavefunction convergence factor R would decrease somewhat faster. Note that in Fig. 1, the increase in R during the first time step is due to the unphysically large eigenvalues ε corresponding to the (random) initial states.

Fig. 2 shows density of states and integrated density of states obtained after 25 time steps. The results are similar to previous self-consistent calculations for NaTl[16], however, certain differences are found. The Fermi energy should lie just to the right-hand side of a DOS minimum and the minimum should be somewhat deeper than that shown in Fig. 2. We believe that first-principles nonlocal pseudopotentials will be required to properly reproduce such detailed features in the DOS. Computationally convenient in the present context is the form proposed by Bylander and Kleinman (see [9]), and we are presently incorporating this pseudopotential into our methodology.

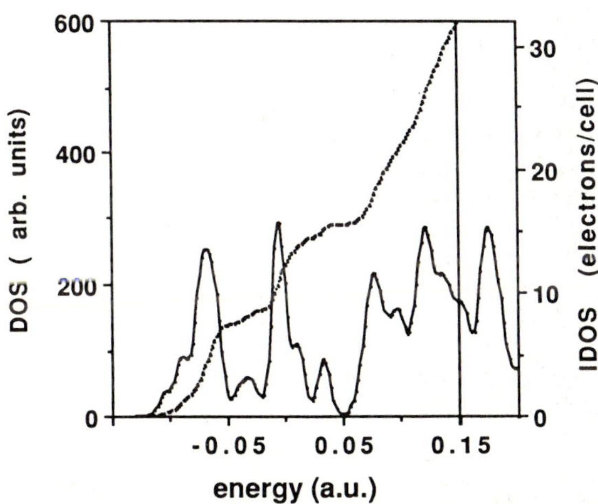

Fig. 2. Density of states (DOS) and integrated density of state (IDOS) for NaTl. Vertical line corresponding to IDOS-32 indicates Fermi energy.

SUMMARY AND CONCLUSION

Methods of nonlinear optimization are being developed with the objective of calculating the electronic structure of systems too large to be treated by conventional techniques. We have discussed two particular algorithms for solving a first order equation of motion for the Kohn-Sham orbitals. The modifications of these procedures required in the case of metallic systems were outlined; the most significant point is the sampling of the zone on a finer scale than is required for insulators. A preliminary application to the compound NaTl was described.

ACKNOWLEDGEMENTS

This work was supported by the U.S. Department of Energy, BES-Materials Sciences under contract #W-31-109-ENG-38. It benefited from an allocation of time on the ER-CRAY at the MFE Computer Center.

REFERENCES

1. R. Car and M. Parrinello, Phys, Rev. Lett. 55, 2471 (1985)

2. R. Car, M. Parrinello, and W. Andreoni, in Proc. 1st NEC Symp. on Fundamental Approach to New Material Phases, ed. S. Sugano and S. Ohnishi (Springer, Berlin, 1987) p. 134.

3. D. Hohl, R. O. Jones, R. Car, and M. Parrinello, Chem. Phys. Lett. 139, 540 (1987), and J. Chem. Phys., in press.

4. R. Car and M. Parrinello, Phys. Rev. Lett. 60, 204 (1988).

5. R. Car and M. Parrinello in The Physics of Semiconductors, edited by O. Engström (World Scientific, Singapore, 1986) p. 1165.

6. M. C. Payne, J. D. Joannopoulos, D. C. Allan, M. P. Teter, and D. H. Vanderbilt, Phys, Rev. Lett. 56, 2656 (1986); M. C. Payne, P. D. Bristowe, and J. D. Joannopoulos, ibid, 58, 1348 (1987); M. Needels, M. C. Payne, and J. D. Joannopoulos, ibid, 58, 1765 (1987)

7. A. Williams and J. Soler, Bull. Am. Phys. Soc. 32, 562 (1987).

8. J. L. Martins and M. L. Cohen, Phys. Rev. B., in press (1988).

9. D. C. Allan and M. Teter, Phys. Rev. Lett. 59, 1136 (1987).

10. M. P. Teter and D. C. Allan, preprint.

11. See, e.g., W. Kohn and P. Vashishta in Theory of the Inhomogeneous Electron Gas, ed. S. Lundquist and N. H. March (Plenum, New York, 1983).

12. W. Kohn and L. J. Sham, Phys. Rev. 140, A1133 (1965).

13. See, e. g., R. D. Mattuck, A. Guide to Feynman Diagrams in the Many-Body Problem, (McGraw-Hill, New York, 1967).

14. H. J. F. Jansen and A. J. Freeman, Phys. Rev. B. 30, 561 (1984).

15. C. L. Fu and K. -M. Ho, Phys. Rev. B. 28, 5480 (1983).

16. P. C. Schmidt, in Structure and Bonding, Vol. 65, Springer Ser. Solid-State Chemistry (Springer, Berlin, Heidelberg 1987) p 91.

17. D. D. Ling and C. D. Gelatt, Jr., Phys. Rev. B 22, 557 (1980).

A Method for Determining Many Body Wavefunctions

C.J. Umrigar, K.G. Wilson[†]*, and J.W. Wilkins*[*]

Theory Center and Laboratory of Atomic and Solid State Physics, Cornell University, Ithaca, NY 14853, USA

We present a procedure for obtaining trial wavefunctions for use in Quantum Monte Carlo simulations that have both smaller statistical errors and improved expectation values, compared to commonly used functions. Results are presented for the ground states of H^-, Li^-, Be^{2+}, Be and Ne, and ground and excited states of He.

1. Introduction

It is well known [1] that cusps in wavefunctions can control the rate of convergence of wavefunctions. The reason why the wavefunctions used in quantum chemistry calculations, e.g. Configuration Interaction (CI) wavefunctions, converge slowly is that the expansion in determinants of single-particle orbitals cannot reproduce the electron-electron cusp. The advantage of Monte Carlo(MC) methods is that no such limitation is placed on the form of the wavefunction. However, so far MC methods have typically exploited this fact only to the extent of introducing the correct cusp in the wavefunctions by multiplying a Hartree Fock(HF) or a small Multiconfiguration Self-Consistent Field (MCSCF) wavefunction, produced by a conventional quantum chemical calculation, by a Jastrow factor that has the correct electron-electron cusps. These HF-J and MCSCF-J wavefunctions are improvements to the HF and MCSCF wavefunctions for small atoms, but the improvements decrease for larger atoms. Thus, for example, Variational MC calculations using HF-J wavefunctions recover 62% of the correlation energy for He, but at most a few percent of the correlation energy for Ne. The Jastrow factor does not have sufficient variational freedom to capture the major part of the correlation; a more general functional form is required. Even if this restricted form is used, the optimal set of orbitals are not the HF or MCSCF orbitals but ones that are somewhat contracted toward the nucleus because the Jastrow factor tends to keep electrons apart, thereby increasing the size of the atoms.

Attempts to take this into account have been made by either introducing an overall scaling factor or by multiplying the wavefunction with an electron-nuclear Jastrow factor [2]. For the most part however, instead of searching for better wavefunctions, the approach that has so far been taken is to perform Diffusion MC or Green Function MC (GFMC) calculations [3] with existing wavefunctions. The expectation value of the energy for these methods is the lowest energy possible for any wavefunction that has the same nodes as the trial wavefunction; hence they often recover a much larger percentage of the correlation energy than Variational MC. (Since Diffusion MC, in the limit of sufficiently small time steps, and GFMC calculations yield energies of comparable accuracy, we will refer to both

[†]Address after August 1988: Department of Physics, The Ohio State University, Columbus, OH 43210.

[*]Present address: Department of Physics, The Ohio State University, Columbus, OH 43210.

Springer Proceedings in Physics Vol. 33: **Computer Simulation Studies in Condensed Matter Physics**
Editors: D. P. Landau · K. K. Mon · H-B. Schüttler

of these as GFMC in what follows.) There are two problems with this approach. First, whereas GFMC calculations recover most of the correlation energy for small atoms and molecules, e.g., 102.5($\pm$0.7)% [4], 96($\pm$2)% [5], 98($\pm$3)% [6], 99.1($\pm$0.3)% [7] for LiH [4], they do less well for somewhat larger systems, e.g., 84($\pm$2)% [5], 91($\pm$4)% [6] for water. For the larger systems, the nodes of the simple wavefunctions used are not sufficiently accurate. Second, for comparable statistical accuracy, GFMC calculations are typically a couple of orders of magnitude more time consuming than Variational MC calculations because of the very long auto-correlation times. Clearly, it is desirable to develop ways of finding wavefunctions which are compact, but which nevertheless recover most of the correlation energy even in a Variational MC calculation.

2. Procedure for Finding Wavefunctions

Past attempts at finding improved trial wavefunctions have mostly consisted of adjusting the parameters of the wavefunction to minimize the expectation value of the energy $\bar{E}$. Using this method it has been possible to optimize at most a few parameters since a single calculation of $\bar{E}$ takes a significant amount of computer time, and this must be repeated several times before an optimal set of parameters is found. Instead, we observe that if ψ is an eigenfunction, the local energy $H\psi/\psi$ equals the energy of the state, for any instantaneous position of the electrons. Hence, instead of varying the parameters in the wavefunction to minimize the energy, in common with some earlier work, we vary them to minimize the variance of the local energy [8]

$$\sigma_{opt}^2 = \frac{\sum_{i=1}^{N_{opt}} \left(H\psi(i)/\psi(i) - E_g\right)^2 w(i)}{\sum_{i=1}^{N_{opt}} w(i)} \tag{1}$$

where $w(i) = |\psi(i)/\psi_o(i)|^2$, E_g is a guess for the energy of the state we are interested in and the sum is over a fixed set of configurations of the electrons sampled from $|\psi_o|^2$. ψ_o is taken to be the best wavefunction available before we start the optimization procedure, usually HF-J.

How large does N_{opt} need to be to determine the parameters? One might expect that N_{opt} would have to be exceedingly large, since one needs to have a representative sample of configurations that live in a large dimensional space, and since the number of parameters that one would like to optimize is large. The remarkable fact is that about a 1000 configurations are sufficient, for a 30 dimensional space (Ne atom) when as many as 50 parameters are being optimized. There are two reasons for this. First, the configurations over which the optimization is performed are fixed, so we are using correlated sampling to arrive at an optimal set of parameters. Hence the difference in the σ_{opt} 's, for 2 sets of values of the parameters being optimized, is much more accurately determined than the values of the σ_{opt}'s themselves. Second, we are performing a fit, not an integral. The better the form of a n-parameter trial wavefunction, the smaller is the number of configurations needed to pin down the n parameters. If the true wavefunction were representable by an n-parameter trial wavefunction then only n configurations would be necessary to determine the n parameters exactly.

There are other important advantages to our procedure:

1) The quantity being optimized has a known lower bound (namely zero).
2) Since any eigenstate has zero variance of the local energy, it is possible to calculate true excited states, i.e., excited states that are not the lowest lying state of that symmetry.
3) It is helpful in performing the optimization to know not only the gradient of the quantity being optimized, but also the second derivative matrix. When the quantity being optimized can be written as a sum of squares, as is the case in (1), then it is possible to calculate an approximate second derivative matrix in linear time, making the optimization procedure more efficient [9].

4) The wavefunctions found by the normal energy minimization methods are often not very accurate, since it is possible to have an approximate cancellation of errors in the energy from regions where the local energy is too low and where it is too high. On the other hand, the only way that σ can be small is to have a good local energy and therefore a good wavefunction in all regions of configuration space that have an appreciable probability of being sampled. In fact, this notion has in the past been used to test the quality of wavefunctions generated by other methods [10]. Hence our wavefunctions are likely to be more accurate than those generated by energy minimization. Further, if we are interested in obtaining particularly accurate wavefunctions in a certain region of configuration space, then we can do so by concentrating additional configurations in that region.
5) Since our wavefunctions are relatively compact, it may be easier to obtain a physical feel for them.

As with all non-linear optimization problems, it is possible to get stuck in local minima. Hence the wavefunctions that we have determined are not necessarily the best possible, for the form of the function used. This is not as serious a difficulty as in a general minimization problem because the absolute minimum is known to be greater than zero. Although we do not have any guarantee of having found the absolute minimum, we demonstrate that it is relatively easy to find wavefunctions that are much better than those commonly used.

The initial value of E_g in (1) can be taken to be the expectation value of the energy for the initial trial wavefunction (HF-J or MCSCF-J). A better guess for E_g can be obtained by estimating the correlation energy, which varies only between 0.4 eV and 1.2 eV per electron, and in a systematic way, for the ground states of all atoms with $Z \leq 18$ [11]. Once an improved trial wavefunction is found, another MC run is performed using this improved function and the expectation value of the energy from this MC run is used as the new E_g. Typically only 2 iterations are required to obtain a self-consistent value of E_g. Taking E_g to be lower than the true energy is equivalent to minimizing a linear combination of the variance and the expectation value of the energy.

The weights $w(i)$ in (1) are the correct weights needed to calculate the expectation value of the energy for ψ, when the sampling is done not from $|\psi|^2$ but from $|\psi_o|^2$. Since we are not calculating an expectation value, but rather performing a fit, it is from this point of view permissible to replace the $w(i)$ by unity. However, there is another reason why it is at times important to have the $w(i)$. As the wavefunction is optimized, its nodes will in general move. Now, if one of the N_{opt} configurations in (1) happens to lie between the nodes of the initial wavefunction and the final one, then it can block the motion of the nodes if the $w(i)$ are set to unity, but not if we use the values of $w(i)$ in (1).

3. Functional Form of Trial Wavefunctions

The most commonly used trial wavefunctions have the form

$$\psi = \sum_{n=1}^{N_{det}} (d_n D_n^{\uparrow} D_n^{\downarrow}) \prod_{i<j} \exp\left(a_{ij} r_{ij}/(1 + b r_{ij})\right)$$

where $D^{\uparrow}(D^{\downarrow})$ are the up(down) spin Slater determinants and r_{ij} is the inter-electron distance of the ith and j^{th} electrons. This form consists of a MCSCF-like part (consisting of a sum of determinants) and a Jastrow part (involving a product over all pairs of electrons). The orbitals in the determinants are themselves linear combinations of products of Slater functions and spherical harmonics

$$\phi_i(\vec{r}) = \sum_{j=1}^{N_{basis}} C_{ij} r^{n_j - 1} e^{-\varsigma_j r} Y_{l_j m_j}(\hat{r}). \tag{2}$$

The electron-electron cusp condition [12] implies that the Jastrow a_{ij} equal 1/2 for antiparallel spin electrons and 1/4 for parallel spin electrons. Having a different Jastrow b for antiparallel and parallel spin electrons does not result in a significantly better wavefunction; hence a single b is used.

As a first step in finding better wavefunctions we used the same form for the wavefunctions but optimized the d_n, C_{ij}, ς_j, and the Jastrow b. This leads to significant but not large improvements in the wavefunction. Next we altered the form of the wavefunctions. Several functional forms were tried with varying degrees of success. Here, we present results for one of the more successful forms. The Jastrow factor correlates 2 electrons. The correlation of the 2 electrons is dependent on where in the atom they lie, but this dependence is not described by the Jastrow factor.

Here, we generalize the Jastrow correlation function to include correlations between 2 electrons *and the nucleus*[13]

$$\prod_{i<j} \exp\left(\frac{P(\{a\}, r_i, r_j, r_{ij})}{1 + P(\{b\}, r_i, r_j, r_{ij})}\right)$$

where P is a polynomial in r_i, r_j, r_{ij} [14] with coefficients $\{a\}$ or $\{b\}$. Antisymmetry of the wavefunction under the interchange of two electrons implies that P is symmetric in r_i and r_j. For large r_i, r_j, r_{ij} the correlation function tends to a constant that depends on the relative ratios of r_i, r_j and r_{ij}. There are two points we wish to make about this form for the wavefunction.

1) For a 2-electron atom or ion in an S state, after one takes into account the symmetries, there are only 3 independent variables. So, in the limit that the order of P is infinite, it fully describes the correlation between 2 electrons and a nucleus in a 2-electron atom or ion. Hence, we expect the wavefunctions to be very accurate for 2-electron atoms and ions. We will see later (Table 1) that more than 99.99% of the correlation energy is recovered for these systems.

2) The question arises, whether this form of the wavefunction includes most of the correlation even in systems with more than 2 electrons. To answer this, we observe that in the language of cluster expansions, we have included clusters of all sizes, but correctly only those that can be decomposed into triples consisting of pairs of electrons and a nucleus, such that the electrons in one triplet are not connected to the electrons in any other triplet. However, due to the exclusion principle, it is rare for 3 or more electrons to be close, since at least 2 electrons must necessarily have the same spin, so the terms we have included are the most important ones for multi-electron systems.

The local energy $H\psi/\psi$ in (1) diverges when an electron and a nucleus overlap ($r_i = 0$) or when 2 electrons overlap ($r_{ij} = 0$) unless the wavefunction obeys the cusp-conditions [12]. The cusp conditions can be deduced from Schrodinger's equation by equating the coefficients of the divergent terms to zero when an electron and a nucleus overlap ($r_i = 0$) or when 2 electrons overlap ($r_{ij} = 0$).

The cusp condition for the electron-nucleus cusp is: $\left.\frac{\partial\psi}{\partial r_i}\right|_{r_i=0} = -Z\psi$ and that for the electron-electron cusp is $\left.\frac{\partial\psi}{\partial r_{ij}}\right|_{r_{ij}=0} = \psi/2$ or $\psi/4$ for antiparallel and parallel spins, respectively. Imposing these conditions on our ψ results in $17 + n_{orb}$ coupled quadratic equations where n_{orb} is the number of orbitals in the determinantal part of ψ. For maximum variational flexibility we imposed the cusp conditions approximately by including in (1) additional terms proportional to the square of the errors in satisfying the cusp conditions. By scaling up or down the prefactor of these terms, the cusp-conditions can be imposed more or less exactly.

Once an optimized wavefunction is obtained, a long MC calculation is carried out to calculate the expectation value of the energy $\bar{E}$ and the fluctuation in the local energy σ.

$$\sigma = \sqrt{(1/N_{MC}) \sum_{i=1}^{N_{MC}} \left(H\psi(i)/\psi(i) - \bar{E}\right)^2} \quad . \tag{3}$$

The sums are over N_{MC} configurations of electrons sampled from $|\psi|^2$. If N_{opt} in (1) is not sufficiently large or the cusp conditions are ignored, then σ from (3) will be large even though σ_{opt} from (1) is small.

4. Results for Ground and Excited States of Atoms and Ions

We will use two quantities, the error in the expectation value of the energy $\Delta E = \bar{E} - E_{true}$ and σ to evaluate the quality of the optimized wavefunctions. The number of MC updates required to reduce the statistical error to any given value is proportional to σ^2. In Table 1 we present results for the ground state of H^- and Be^{2+} and the ground and two excited states of He, each for four different wavefunctions that get progressively better. For each of the wavefunctions, the first line is the error in the expectation value of the energy ΔE the second line is σ. The number in parentheses is the expected statistical error in the last digit, and can be made arbitrarily small by increasing the number of MC updates. The value of N_{MC} was in every case between 10^6 and 10^7.

For H^- , the HF prediction is that it is not bound. Multiplying the HF wavefunction by an optimized Jastrow (wavefunction 2) reduces ΔE by about a factor of 3 and σ by about a factor of 4. The HF determinant yields the lowest possible energy of all single determinantal wavefunctions, but is not the optimal determinant when multiplied by a Jastrow. Hence, further improvements are obtained upon simultaneous optimization of the determinant and the Jastrow (wavefunction 3). However, the really large improvements are only obtained upon replacing the Jastrow by a 4th order exponential Padé (wavefunction 4).

To illustrate the fact that the optimization method works for excited states as well as ground states, we have calculated those states of the excitation spectrum of He that have been circled in Fig. 1. The $2\,{}^3S$ state is the lowest state of ${}^3S$ symmetry and therefore no more difficult than the ground state, but the $3\,{}^3S$ is a true exited state. As shown in Table 1 large improvements are obtained for all the states.

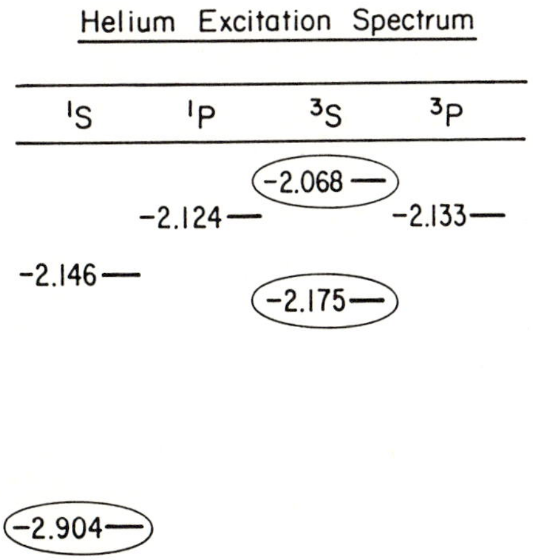

Helium Excitation Spectrum

^1S	^1P	^3S	^3P
		−2.068	
	−2.124		−2.133
−2.146			
		−2.175	
−2.904			

Fig. 1.
Excitation spectrum of He. The calculations for the circled states are shown in Table 1.

Table 1. Performance of various wavefunctions for ground and excited states of 2-electron ions. For each wavefunction, the first line gives the error in the expectation value of the energy and the second line gives the standard deviation of the local energy, in Hartrees. The expected statistical error in the last digit is in parentheses.

Wavefunction	$1\ ^1S$ H^-	$1\ ^1S$ He	$2\ ^3S$ He	$3\ ^3S$ He	$1\ ^1S$ Be^{2+}
"Exact" Energy (non-relativistic, ∞ mass)	0.527 751[a]	2.903 724[a]	2.175 229[b]	2.068 689[b]	13.655 566[a]
1) HF	0.039 821	0.042 044	0.000 978	0.000 21	0.044 267
0 free params.	0.31	0.65	0.054	0.024	1.4
2) HF + Optim(Jastrow)	0.012 3(3)	0.015 9(6)	0.000 5(1)	0.000 10(3)	0.016(1)
1 free param.	0.082	0.23	0.028	0.010	0.48
3) Optim(Det + Jastrow)	0.007 9(2)	0.004 1(3)	0.000 05(5)	-0.000 04(3)	0.003 5(2)
3,4,7 free params. for $1\ ^1S$, $2\ ^3S$, $3\ ^3S$ states resp.	0.070	0.14	0.020	0.009 1	0.30
4) Optim(Det + Expon Pade)	0.000 005(3)	-0.000 002(4)	-0.000 003(2)	0.000 001(1)	0.000 001(6)
41,44,47 free parms. for $1\ ^1S$, $2\ ^3S$, $3\ ^3S$ states resp.	0.002 1	0.001 1	0.000 71	0.001 3	0.003 4

[a] Ref. 15; [b] Ref. 16

It may be noticed that the HF wavefunction is more accurate for the excited states than for the ground state. This is because the two HF orbitals have peaks in rather different regions of space for the excited states and therefore the two electrons are not highly correlated. This situation does not occur for the excited states of larger systems.

In Table 2 we present our results for two 4-electron systems, Li^- and Be and for a 10-electron system Ne. As expected, the improvements are not as large as for the 2-electron systems but they are nevertheless considerable. As pointed out by Watson [17], there are two types of correlation. The first is due to the inter-electron repulsion and is always present. The second, is due to a near degeneracy of occupied and unoccupied HF orbitals and is of course only present in those systems where such a near-degeneracy exists, e.g. Li^- and Be. These are referred to in the quantum chemistry literature [18] as dynamic and non-dynamic correlation respectively. Since for Li^- and Be there is a near degeneracy of the occupied HF 2s and the unoccupied HF 2p orbitals, we include in Table 2 two wavefunctions that include the $1s^2 2p^2$ configuration in addition to HF ground state $1s^2 2s^2$ configuration. This second configuration must be written as the sum of 3 determinants in order to have the correct ^1S symmetry. Our best wavefunction recovers 99.2($\pm$0.2)% of the correlation energy for Li^- and 99.0($\pm$0.3)% of the correlation energy for Be. These values are comparable to the best CI, 98.6%, and CI-Hylleraas, 99.7%, calculations for Li^- [19] and to the best CI, 99.5%, calculations for Be [20]. Our best results for Be are not only much better than the best existing [21] Variational MC results 79($\pm$0.8)% [7], but also somewhat better than the best GFMC calculation 98.0($\pm$0.4)% [7] that has sufficiently small statistical errors to make a comparison possible. We observe from Table 2 that these excellent results are only obtained if one includes both the exponential Padé and the

Table 2. Same as Table 1, but for Li^-, Be and Ne

Wavefunction	^1S Li^-	^1S Be	^1S Ne
"Exact" Energy (non-relativistic, ∞ mass)	7.500 4[a]	14.667 33[b]	128.937[c]
1) HF	0.072 2 1.0	0.094 31 1.5	0.390 4.9
2) HF + Optim.(Jastrow)	0.043(1) 0.37	0.060(2) 0.53	0.39(2) 2.7
3) Optim.(Det. + Jastrow)	0.031(1) 0.25	0.040(2) 0.35	0.30(1) 1.8
4) Optim.(4 Det. + Jastrow)	0.008 2(8) 0.24	0.009 4(9) 0.34	
5) Optim.(Det. + Expon Pade)	0.008 4(5) 0.065	0.022 6(9) 0.15	0.053(4) 1.0
6) Optim.(4 Det. + Expon Pade)	0.000 6(1) 0.045	0.000 9(3) 0.092	

[a]Extrapolated value from Ref. 19 The estimated error is 2 in the last digit.
[b]Extrapolated value from Ref. 20 The estimated error is 3 in the last digit.
[c]From Ref. 22. It should be noted that the Lamb shift has been added in with the wrong sign in Table X of Ref. 22.

two configurations in the wavefunction. It appears that the exponential Padé is effective at describing the dynamic correlation, but that the near-degeneracy correlation is more effectively described by including the additional configuration in the wavefunction. The slow convergence of CI calculations is due to the dynamic correlation, but that is precisely the part of the correlation that is easily described by the exponential Padé . Hence it is plausible, even for systems more complicated than those discussed here, that compact and accurate wavefunctions can be found using wavefunctions that are products of a small linear combination of configurations and an exponential Padé . It is interesting to observe that whereas wavefunction 4 has a smaller ΔE , wavefunction 5 has a smaller σ. The second configuration is more effective in reducing ΔE whereas the exponential Padé is more effective in reducing σ.

Our best wavefunction for Ne recovers about 86% of the correlation energy. It should be noted that the equivalent wavefunction tor Be recovers only about 76% of the correlation energy. In order to study the question whether the optimization works well for heavier atoms, one should compare atoms in the same column of the periodic table; something we have not yet done, except for He and Ne. It is clear however, that due to the special form of the trial wavefunction the optimization works much better for 2-electron systems than those with more than 2 electrons, but we hope that the performance as measured by the percentage of the correlation energy recovered will not deteriorate sharply with increasing number of electrons beyond 3 electrons. The reason for our optimism is that we are including clusters of all sizes. Hence, in common with the Coupled-Cluster Singles Doubles[23] approach our energy ought to be size-extensive [24].

It should be stressed that all the results presented here are for Variational MC. In those cases where even the optimized wavefunctions miss a significant fraction of the correlation energy, e.g. Ne, it is possible to perform GFMC calculations to obtain yet better expectation values of the energy. Our optimized wavefunctions yield smaller statistical errors and hopefully improved expectation values compared to conventional wavefunctions when used in GFMC calculations as well. A GFMC calculation for Ne, using our optimized wavefunction recovered 96($\pm$10)% of the correlation energy. Longer simulations are required to reduce the statistical error.

For some of the larger systems, even GFMC calculations using simple conventional wavefunctions miss a significant fraction of the correlation energy, e.g., 84($\pm$2)% of the correlation energy is recovered in GFMC calculations of water using HF-J wavefunctions [5]. The same wavefunction recovered only 17% of the correlation energy in a Variational MC calculation [5]. Since our wavefunctions, even at the variational level recover a large percentage of the correlation energy, and since the nodes of the wavefunction are improved in the optimization process, they will hopefully recover almost all of the correlation energy in a GFMC calculation.

5. Conclusions

By minimizing the deviations of the local energy, it is possible to find compact wavefunctions that have significantly better expectation values and smaller statistical errors than those in common use. The method works for excited states as well as ground states. Although only the fluctuations are explicitly minimized, the error in the expectation value of the energy is often reduced by an even larger amount. The results for Ne show that N_{opt} does not increase rapidly with the number of electrons and that the form of the wavefunction is capable of describing the correlations in atoms with more than just a few electrons moderately well. No doubt, with some chemical intuition, yet better functional forms will be found. The wavefunctions used in conventional electronic structure methods are constrained to be constructed from single-electron orbitals. MC methods, combined with the optimization scheme described here open up the possibility of using new classes of functions involving multi-electron coordinates that are capable of describing the wavefunction more compactly and/or more exactly.

The feasibility of the optimization scheme has been demonstrated here for atoms and ions, but the scheme can be applied to any quantum MC calculation. In fact, as pointed out by Nightingale [25], since classical MC calculations using the transfer matrix technique require the determination of the eigenvector of the transfer matrix, this technique can be used there as well.

This work was supported by the NSF. JWW is supported by the DOE - Basic Energy Sciences, Division of Materials Research. We are grateful to Kevin Schmidt for giving us his MC program. CJU is grateful to Zachary Levine, John Rehr, Karl Runge, Jim Sethna and Steve White for many useful discussions and the Aspen Center for Physics for its hospitality. The computations were performed on a Gould PN-9050 minicomputer and on Sun-3/50 workstations.

References

1. C. Schwartz: *Meth. Comp. Phys.* **2**, 241 (1963), W. Lakin: *J. Chem. Phys.* **43**, 2954 (1965), B. Klahn and J.D. Morgan: *ibid.*,**81**, 410 (1984), R.N. Hill: *ibid.*,**83**, 1173 (1985).
2. P.J. Reynolds, M. Dupuis and W.A. Lester: *J. Chem. Phys.* **82**, 1983 (1985).
3. D.M. Ceperley, M.H. Kalos: In *Monte Carlo Methods in Statistical Physics*, ed. by K. Binder, Topics Current Phys., Vol.7 (Springer, Berlin, Heidelberg 1979) Chap.4; K.E. Schmidt, M.H. Kalos: In *Applications of the Monte Carlo Methods in Statistical Physics*, ed. by K. Binder, Topics Current Phys., Vol.36 (Springer, Berlin, Heidelberg 1984) Chap.4
4. J.W. Moskowitz, K.E. Schmidt, M.A. Lee and M.H. Kalos: *J. Chem. Phys.* **77**, 349 (1982). The statistical error was probably underestimated since the GFMC energy should be an upper bound to the true energy.
5. P.J. Reynolds, D.M. Ceperley, B.J. Alder and W.A. Lester: *J. Chem. Phys.* **77**, 5593 (1982).
6. J.W. Moskowitz, K.E. Schmidt: *J. Chem. Phys.* **85**, 2868 (1985).
7. R.J. Harrison and N.C. Handy: *Chem. Phys. Lett.* **113**, 257 (1985).
8. The idea of minimizing the variance of the local energy goes back to at least 1935. J.H. Bartlett, J.J. Gibbons and C.G. Dunn: *Phys. Rev.* **47**, 679 (1935) credit it to J. Boussinesq: *Theorie de la chaleur* (I). In the context of Monte Carlo simulations, a procedure similar to ours was developed by R.L. Coldwell: *Int. J. Quant. Chem. Symp.* **11**, 215 (1977), and applied to the Li atom. However only 36% of the correlation energy was recovered. The method was also used to calculate an accurate wavefunction and interaction potential for two helium atoms, starting from an accurate 189-term Hylleraas-type atomic wavefunction, R.E. Lowther and R.L. Coldwell: *Phys. Rev. A* **22**, 14 (1980).
9. W.H. Press, B.P. Flannery, S.A. Teukolsky and W.T. Vettering: *Numerical Recipes*, Cambridge University Press (Cambridge, 1986).
10. J.H. Bartlett: *Phys. Rev.* **98**, 1067 (1955).
11. E. Clementi: *J. Chem. Phys.* **39**, 175 (1963).
12. T. Kato: *Comm. Pure Appl. Math.* **10** , 151 (1957).
13. Using a Jastrow function that is a more general form of r_{ij} than the usual Jastrow function, but not a function of the r_i and the r_j, does not result in a significant improvement over the usual Jastrow function.

14. We have included all terms upto 4th order. In each polynomial, symmetry of P under interchange of the two electrons implies that only 21 of the 34 coefficients are independent. In principle, a different correlation function ought to be used for electrons with parallel and anti-parallel spins. However, we find that optimizations performed with the constraints, $\{a\}_{\uparrow\uparrow} = 0.5\ \{a\}_{\uparrow\downarrow}$, $\{b\}_{\uparrow\uparrow} = \{b\}_{\uparrow\downarrow}$, result in wavefunctions that are almost as good.

15. David E. Freund, Barton D. Huxtable and John D. Morgan: *Phys. Rev. A* **29**, 980 (1984).

16. Y. Accad, C.L. Pekeris and B. Schiff: *Phys. Rev. A* **4**, 516 (1971).

17. R.E. Watson: *Ann. Phys.* **13**, 250 (1961).

18. I. Shavitt: In *Advanced Theories and Computational Approaches to the Electronic Structure of Molecules*, ed. by C.E. Dykstra (Reidel, Dordrecht 1984).

19. J.S. Sims, S.A. Hagstrom, D. Munch and C.F. Bunge: *Phys. Rev. A* **13**, 560 (1976).

20. Carlos F. Bunge: *Phys. Rev. A* **14**, 1965 (1976).

21. J.W. Moskowitz, K.E. Schmidt, M.A. Lee and M.H. Kalos: *J. Chem. Phys.* **76**, 1064 (1982) have studied a variety of correlated wavefunctions for Be. We are unable to compare with their results because of several typographical inconsistencies in their Table Ib.

22. A. Veillard and E. Clementi: *J. Chem. Phys.* **49**, 2415 (1968).

23. R.J. Bartlett, C.E. Dykstra and J. Paldus: In *Advanced Theories and Computational Approaches to the Electronic Structure of Molecules*, ed. by C.E. Dykstra (Reidel, Dordrecht 1984).

24. The term *size-extensive* means that the percent correlation energy recovered does not depend on the size of the system. The most commonly used technique for large-scale quantum chemistry calculations, namely Configuration Interaction carried out to any fixed order, is not size-extensive. i.e. the percent correlation energy goes to zero as the size of the system is increased.

25. Peter Nightingale: personal communication; contribution by Peter Nightingale and Robert Caflisch in these proceedings.

Part III

Computer Graphics

Computer Graphics for Scientists and Engineers

S. Follin

ETA Systems, Inc., Computer Services Annex, University of Georgia, Athens, GA 30602, USA

Computer graphics is a powerful tool for interpreting scientific data and the results of computer simulations. This paper provides introductory information about computer graphics for scientists and engineers. Computer graphics terminology is introduced, basic graphics capabilities are described and references to computer graphics literature are provided. Finally, examples demonstrating the appropriateness of supercomputers for certain graphics applications are presented.

1. INTRODUCTION

There is currently a great deal of discussion in the computer graphics community about computer graphics for scientific applications. This discussion was prompted by a report, sponsored by the National Science Foundation, entitled "Visualization in Scientific Computing" [1]. This report points out how current computer graphics technology could revolutionize scientists' abilities to interpret experimental data and the output of simulations. It also requests funding agencies to consider sponsoring research and development projects that would provide scientists with high performance, high quality computer graphics software tools.

The NSF report was released at the 1987 ACM SIGGRAPH conference, and several of the conference sessions were devoted to discussions of computer graphics for science. The article "The Art and Science of Visualizing Data" in [2] is a synopsis of those sessions.

The visualization report and the ensuing discussions are worthy of notice by scientists, such as the participants of this conference on simulational physics. There are very sophisticated computer graphics tools and techniques, commonly used in the television and motion picture industries, that should be used everyday by scientists and engineers. Unfortunately, these tools are neither readily available nor convenient to use. In addition, graphics hardware must be purchased. To date, these two factors have combined to keep computer graphics, especially interactive color computer graphics, from being commonly used in the scientific community.

The cost of computer graphics equipment has decreased dramatically over the last ten years. As a result, hardware acquisition costs are becoming less of a barrier for scientists. Sophisticated, color graphics images can be generated and displayed on systems costing under ten thousand dollars. Forty thousand dollars will buy a reasonably powerful workstation with excellent graphics capabilities.

This paper explains some commonly used computer graphics terms, describes several graphics capabilities that are available, and

Springer Proceedings in Physics Vol. 33: **Computer Simulation Studies in Condensed Matter Physics**
Editors: D. P. Landau • K. K. Mon • H-B. Schüttler

presents examples of using high quality computer graphics for scientific visualization. In particular, since many conference participants are supercomputer users, it presents two examples demonstrating the usefulness of supercomputers for certain graphics applications.

The contents of this paper are unfortunately constrained by the medium, paper, used to convey it. The oral presentation that this paper documents used color videotapes and slides extensively. Although the impact of the images included in this paper are considerably decreased by their reproduction as half-tone photographs, these images are still useful for demonstrating certain graphics capabilities. Readers interested in seeing more visually powerful results of data visualization are encouraged to view demonstration tapes of color animations of scientific data, such as [11].

2. REFERENCES

Two textbooks that are highly recommended for readers desiring an introduction to computer graphics are [3] and [4]. Both books include discussions of graphics algorithms and hardware. In many cases, the hardware described is now obsolete, but even these descriptions are valuable since they make it easier to understand the more extensive capabilities in modern graphics devices. Reference [5] is a later publication that includes some recently developed methods, such as fractal techniques, not included in [3] and [4].

Computer graphics is a young and rapidly changing discipline, so many of the techniques in common use have not yet been documented in textbooks. Several journals that publish articles on graphics hardware and software developments are [6], [7], [8] and [9]. Two of the major annual graphics conferences are the NCGA (National Computer Graphics Association) Conference and the ACM (Association for Computing Machinery) SIGGRAPH (Special Interest Group on Computer Graphics) Conference. These provide excellent opportunities to see the graphics hardware and software commercially available, to learn about the current research on graphics algorithms and architectures, and to view the images and films produced by leading computer graphics organizations and individuals. The proceedings of each SIGGRAPH meeting (e.g. [10]) is a compendium of the technical papers that is a useful reference: many state-of-the art techniques are first presented at SIGGRAPH meetings.

3. TECHNICAL ISSUES: TERMINOLOGY AND PERFORMANCE CONSTRAINTS

This section introduces several technical terms that are frequently used in computer graphics. More importantly, performance constraints related to some of the concepts introduced are discussed. Most of these constraints are critical: how they are resolved determines the graphics quality or capabilities available to the user.

The following terms will be introduced and discussed: vector graphics device, raster graphics device, pixel, scan line, RGB values, device resolution, aliasing, antialiasing, and bit planes.

Computer graphics devices are generally either vector (i.e. line drawing) or raster devices. Figure 1 shows an image of a jet fighter generated on a vector device, a color CRT (cathode ray tube). Figure 2 is the same (geometrical) object, displayed on a raster CRT.

The vector image (Fig. 1) consists entirely of lines. This is because the device itself is only capable of generating lines. The input data (which is supplied by the user or the user's program) for such an image is just a series of MOVE (i.e. move, without drawing a line, to a specified coordinate on the screen) and DRAW (i.e. draw a line from the current position on the screen to a specified coordinate on the screen) commands. Clearly, the amount of data required by the device to display a drawing is proportional to the number of lines drawn.

The following hypothetical calculation may provide a better appreciation for the data requirements. Assume that the user has a screen coordinate space of 1024 by 1024. That is, the x and y (integer) coordinates specified for a MOVE or DRAW lie in the range 0 to 1023. Assume also that the command for a MOVE or DRAW is one character (8 bits). Then each command requires 28 bits: 8 bits for the command, 10 bits for the x-coordinate and 10-bits for the y-coordinate. The data required to draw 100 lines for the worst case (none of the starting and ending points coincide) is then 5600 bits (100*28*2). At 9600 bits per second (bps), a typical terminal connection speed, 0.58 seconds are required to transmit this data.

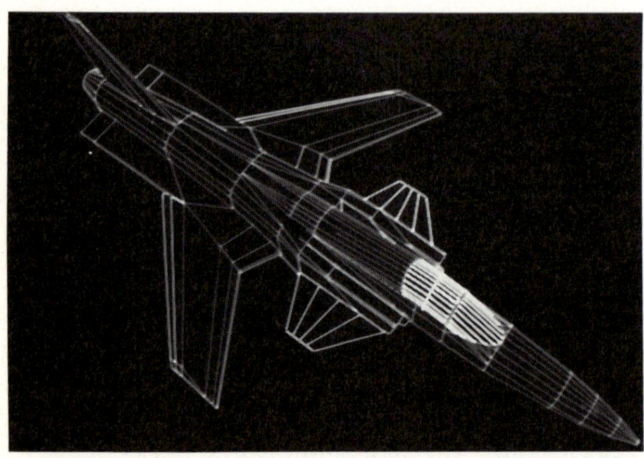

Fig. 1 Image photographed from vector CRT (Courtesy of Grumman Aerospace and Evans and Sutherland)

Fig. 2 Image photographed from raster CRT (Courtesy of Grumman Aerospace and Evans and Sutherland)

The hypothetical example above is a very conservative one; better efficiency (i.e. fewer bits per line) is actually obtained. Thus, this example points out that using a vector CRT to display data residing on a remote computer is quite feasible: images can be generated sufficiently rapidly provided that there are not too many lines. Animations are even feasible (An animation is a sequence of images generated so quickly that the viewer perceives a scene with objects in motion, rather than a sequence of images. Generally, at least 24 images per second are required). This is especially true if only new or changed coordinates need to be transmitted, which is possible with "smart" vector devices. Moreover, higher data transmission speeds are now being implemented at many sites, so data transmission rates above 9600 bps are often available. For example, the Ethernet (hardware) data rate is 10 million bits per second (10 Mbits) and 100 thousand bits per second (100 Kbits) is often achieved for data transfers on an operational Ethernet network.

Raster images, such as Fig. 2, require considerably more data to produce. Raster devices, such as the CRT from which Fig. 2 was photographed, create images by coloring individual elements of the screen. These elements, uniformly sized rectangles (or squares), are called pixels (for picture elements). The resolution of the raster device is determined by the number of pixels on the screen: it is normally specified by giving the number of pixels in the x-direction followed by the number of pixels in the y-direction. For example, the resolution of the CRT used for Fig. 2 is 640 by 480. That is, there are 640 pixels in the x-direction and 480 pixels in the y-direction, for a total of 307,200 pixels.

Pixels are normally considered to be horizontally contiguous since most devices generate the pixel colors by rows. Each row is called a scan line. A 512 by 512 device is said to have 512 scan lines, each of 512 pixels.

Devices, especially CRT devices, with a resolution of 512 by 512 or greater are often considered high resolution devices. Other common high resolution specifications, besides the two already mentioned, are 1024 by 780, 1024 by 1024 and 1280 by 1024. Even higher resolutions are commonly available in film recorders, which put digital images on film. These attain resolutions of 2048 by 2048 (2K by 2K) and 4096 by 4096 (4K by 4K). A resolution of 512 by 512 is normally adequate for displaying images of scientific data. In some cases, such as computational fluid dynamics or aerodynamics, a resolution of 1024 by 1024 is needed to distinguish fine details. Higher resolutions, commonly used in producing images for the television and movie industry, are generally not required for technical applications.

The data requirements for a color raster image depend on the device resolution and its color capabilities. The color for each pixel is determined by specifying the red, green and blue components, called the RGB values, of the color. All colors are produced by combining intensities of these three primary colors. The color intensities are specified on a scale of 0 to (N-1) for each primary color, where N is the number of intensities available (and is device dependent). "Full" color capability normally means that there are 256 intensities available for each primary color. Therefore, a total of 16,777,216 (=256**3) colors are available. Thus, for a full color device with a resolution of 512 by 512, each image requires 6,291,456 bits: 512**2 pixels times 24 bits/pixel (8 bits each for the red, green and blue intensities). At 9600 bps, approximately 655 seconds are required to transmit this data. At 100 Kbps, about 63 seconds are required.

The amount of data required for a full color, high resolution image is about three orders of magnitude greater than for a vector image. A full-color animation (24 frames/second) at 512 by 512 requires a transmission rate of 151 Mbits/sec.

Considerable research has been done (and is in progress) on methods for reducing the storage and transmission requirements for raster images. These techniques have produced noteworthy results: image compression ratios of 400 have been achieved in special cases. One algorithm developed reduces the data transmission requirement to two (instead of 24) bits per pixel for 24-bit color images, although some extra overhead (image sampling) is required [12]. Unfortunately, image compression techniques are not currently available in commercial hardware or software, so users must implement them. This is a significant barrier to the adoption of these methods, since most users do not want to invest the time required to learn, implement, test and debug image compression algorithms.

An additional method for reducing the amount of data per image is to restrict the number of colors available to the user. For example, if only 16 intensities are allowed for each of the primary colors (instead of 256), then only 12 bits (four for each of the three primary colors) are needed per pixel. The data required is reduced by one half and the number of colors is reduced to 4096 (= 16**3).

Restricting the number of colors available on a raster CRT has the added advantage that the device becomes less expensive since less memory is required. In a raster CRT device, there is physical memory that corresponds to each pixel on the screen. If the device is designed to display 4096 colors, only 12 bits per pixel are required. If 16.7 million colors (full color) is to be supported, then 24 bits per pixel must be installed. Display memory is often described in terms of bit planes. A CRT with 12 bits per pixel will be described as having 12 bit planes. This terminology results from the manner in which physical memory is actually installed. The first bit plane (designated plane zero) contains one bit (designated bit zero) of memory for each pixel on the screen. The second bit plane contains the second bit (designated bit 1) for each pixel, etc. Thus, the number of bit planes in a device determines the number of colors that can actually be displayed (simultaneously).

The number of colors available is an important concern for scientists and engineers. If too few colors can be displayed, then objects will have stripes or bands of color, rather than being smoothly shaded. This can easily lead to misconceptions about the data, especially if color values represent physical properties, such as temperature or pressure. Full color devices (24 bit-planes) can provide high quality, smoothly shaded images. Devices with fewer than 4096 colors (12 bit planes) should be avoided for technical applications since a wide range of colors (red, green and blue) often needs to be displayed, due to the color-coding of data values.

4. USING COMPUTER GRAPHICS

The basic processes involved in using computer graphics are outlined in this section. The fundamental elements are the same as for any computational application: a computer graphics program (commercially distributed or otherwise) is selected to run on hardware available to the user. Input data acceptable to the program must be prepared and some means for preserving and interpreting the output must be arranged. In the case of computer graphics, this includes having a means for viewing the computed images.

For the remainder of this paper, it will be assumed that the graphics display available to the user is a high resolution (at least 512 by 512), full color raster device. In particular, the subsequent discussions do not pertain to vector graphics devices, low resolution devices or devices with severely limited color capabilities. These devices are not excluded because they are useless: their limitations simply introduce considerations that this paper will not address.

4.1 Input for Graphics Programs

Computer graphics programs require several types of input. Minimally, geometry data for the objects in the scene and input specifying the viewing conditions are required. Other data, such as function values, may also be input to many programs. These input data types are briefly discussed below.

Geometry data describes the geometry of the objects that will be rendered. One commonly used method of describing objects is to build their surfaces from polygons, such as triangles and rectangles. The actual geometry data file consists of polygon data: nodal positions and lists of the nodes (in clockwise or counter clockwise order) for each polygon. Another common method for describing object geometry, called constructive solid geometry (CSG), is in terms of primitives such as spheres, cubes, etc. Objects are specified as intersections and unions of scaled, translated and rotated primitives.

Function value data is input when there is a functional value, such as pressure or temperature, associated with positions on an object or in a scene. These data are input in such a manner that the graphics program is able to associate the value with a location, such as a node.

Scene specification data provides the graphics program with information about the objects and the conditions under which they are being viewed. The user specifies where light sources are located and their brightness. Surface properties, such as color, transparency reflectance, must also be specified so that surface shading calculations can be done. The location of the objects and the viewer, plus viewing parameters such as the viewers field of vision, are also input. If the graphics program supports animation, then animation specifications (e.g. how much each object should move between images) can be entered so that a sequence of images is produced. Finally, the attributes of the rendering are specified. These select the algorithms used to compute the image and, hence, determine which rendering capabilities are employed. The next section describes the basic rendering capabilities that are often available to users.

4.2 Basic Graphics Capabilities

Listed below are graphics capabilities that are often available for rendering (i.e. computing) color images of three dimensional objects and scenes. These are the primary tools that the scientist or engineer has available for visualizing data. Five years ago, these tools were implemented only in software, which was often run on a minicomputer. There are now many workstations that have these capabilities implemented in hardware and accessible to the user via system subroutine calls. More recently, add-on boards with some of these capabilities have become available for personal computers [13].

Shaded raster image: This is the most basic capability: the program must generate raster images of three dimensional scenes.

This includes, by assumption, the removal of surface patches hidden by other surfaces closer to the viewer. Shading refers to computing the color intensities (i.e. RGB values) for the surfaces in the scene. There are several shading algorithms that are commonly used: flat, Gouraud and Phong shading are three of these. Reference [3] has a detailed discussion of these models and their differences. Some methods, such as Phong shading, provide smoother, more realistic objects. The tradeoff is that more computation is required. Graphics programs often have several shading algorithms included. The user can then select a fast algorithm for initial renderings, when lighting or positioning is being adjusted. A higher quality, more expensive algorithm can be used for later scene renderings.

Object and Viewer Positioning: Graphics programs allow the objects in a scene to be relocated from their original (coordinate) positions. The position of the viewer is also often controllable. The relationship between the fundamental primitives, such as polygons, that constitute an object and the higher level primitives, such as objects, is normally input as part of the geometry data. Thus, the user needs only to position the higher level primitives, not every constituent piece. Some programs support hierarchical geometric structures, so that if one object is moved (e.g. an arm) then parts logically attached (e.g. a hand at the end of the arm) move properly.

Interactivity: Some of the input parameters, such as object and viewer positions, are more conveniently input interactively (if the resulting image can be seen relatively soon). Interactivity makes objects and viewer positioning particularly simple since the user quickly learns how a particular operation, such as translation or rotation, affects the scene. This is especially true if the program and hardware allow the use of dials, a joystick or a mouse for input.

The incorporation of graphics algorithms into hardware has decreased response time to the point that many raster images can be rendered in real time. Thus, it is now possible to "move" through a scene or around an object by moving a mouse or turning a dial. Other input functions are also usually available for interactive control. For example, entering a sequence of keystrokes might cause a dial or mouse to control the position of a light source. The user can then watch the image change as the light source is moved with the dial or mouse. Interactivity of this nature greatly simplifies the control of objects, light sources and the viewpoint for the scene. It is this capability (to easily move through a scene) that promises to revolutionize the analysis and interputation of scientific data.

Transparency: Scenes containing surfaces that are transparent or semi-transparent can often rendered by graphics programs. The surface transparency is specified with the scene specification input data. Opacity (no transparency) is normally the default. Transparency is a very important tool for technical graphics. In many applications (e.g. computational fluid dynamics) bounding surfaces (e.g. container walls) often obscure the phenomena of interest, such as a flow field. Making the surface semi-transparent allows the flow to be seen, but still keeps the bounding surface visible. The automobile windows in Fig. 3 are examples of transparent surfaces that have been rendered.

Highlights: Many graphics programs are able to render highlights on objects. These are brightly lit areas on objects that are caused by reflectance from light sources illuminating the scene. The jet fighter in Fig. 2 has several highlights, most notably near the front on the fuselage. Highlights can provide the viewer with important visual clues about the reflectance properties, the texture and the

Fig. 3 Scene with 70,000 polygons computed using MOVIE.BYU on a CDC Cyber 205 supercomputer (Automobile geometry data courtesy of The Chrysler Corporation)

Fig. 4 Locomotive rendered using highly vectorized ray tracing program on CDC Cyper 205 Supercomputer (Image courtesy of Purdue University CADLAB)

shape of a surface. They are very useful when a scene contains transparent surfaces: highlights on those surfaces can help make the shape and extent of the surface more obvious.

Shadows: The shadows cast by objects can also be computed and displayed. The shadow of the locomotive in Fig. 4 is an example. Shadows are useful in helping to convey the three dimensional shape of object or flow: they provide powerful visual cues. The primary drawback of shadows is that they are expensive to compute: having shadows computed can easily double the execution time of a program. Increases by a factor of five or more are common for scenes containing many polygons or geometric primitives.

Antialiasing: Many difficulties in developing computer graphics rendering algorithms arise from the discrete nature of a computer graphics image: it consists of a finite number of rectangles, each of which is a single color. The result of this finite resolution is that, in some cases, unrealistic images are rendered. A particularly noticeable example of this occurs with the edges of objects; straight edges appear to be jagged if the angle of the edge (in the image) is close to horizontal or vertical. This is called aliasing, and is often referred to as "the jaggies" in computer graphics. The railroad tracks in Fig. 4 show this effect. The solution to this problem is to blur the edge by putting lighter shades of the same color nearby. This is called antialiasing. The human mind will perceive the edge as being straight if this is properly done. Antialiasing does increase computational cost, but not nearly as dramatically as shadows.

Animation: Many programs allow the user to specify an animation sequence. The program then renders a sequence of images in which objects (or the viewer) move according to the animation specification. Devices that allow object control using dials (or a mouse, etc.) and can render images in real-time are especially powerful animation systems. The user specifies the animation by merely turning a dial and the animation occurs immediately. This is the ideal situation for interpreting scientific data: the viewer can easily move around and view the data from different positions.

Function value interpretation: Graphics programs oriented toward technical applications often provide the ability to color objects according to function values (input by the user) associated with positions on the objects. The user specifies the colors corresponding to certain function values and the program then assigns colors by interpolating from the RGB values assigned. This feature is particularly useful for visualizing stresses and pressures on three dimensional objects; the colors make it apparent where minimal and maximal function values occur.

4.3 Graphics Output

Raster output from computer graphics programs occurs in two forms: an image displayed on a display device or a file containing data describing the image. It is desirable to have both forms of output available since each one provides advantages for certain analysis and interpretation tasks. Output in the form of an interactively displayed raster image is generated by having the computer graphics program generating the image transmit the raster output data to the desired display device. Inherent in this process is putting the data in a format the display device expects and sending any necessary control characters. Most commercially licensed graphics programs have device-specific modules, called device drivers, that "package" the image data for specific display devices. Thus, users need only have device drivers for the devices available to them. Furthermore, each driver can utilize any special display hardware features available since it is intended for a particular device.

The capabilities available for image display depend critically on the data transmission speed. The capability of displaying images is limited by the time required to convert computed pixel RGB values into a colored pixel on the display screen. In particular, as pointed out in Section 3, animation requires an especially high bandwidth.

Computer graphics image files are often called metafiles. In the past, different commercial packages have used their own metafile format. There is now a movement among graphics program vendors and device manufacturers to support a standard format (for both raster and vector images), called the Computer Graphics Metafile (CGM) standard. This will greatly facilitate the transfer of computed images between devices and sites. Metafile interpreters (programs) for graphics devices will allow computed images to be displayed without any conversion as long as the CGM format was used for the metafile.

4.4 Interactive Graphics

The environment envisioned by the NSF report on visualization is one in which the scientist or engineer can interact with high resolution, full color graphics images of data. This includes the ability to view output data and enter input data quickly enough that the program generating the data could be directed (e.g. steered) by the user.

The desired interactive graphics environment can only occur if certain functional requirements are met. Two critical ones are that the program must be capable of executing the application calculations and generating images sufficiently quickly to make interactivity desirable and the transmission speed between the memory containing the RGB values and the display device must be adequate.

For a certain class of problems the interactive graphics environment is already available. These problems are ones for which the scientific and image calculations can be done in real time on graphics workstations such as those available from Apollo, Ardent, Hewlett-Packard, Prime, Silicon Graphics, Stellar, Sun, etc. As these workstations continue to get more powerful this class of problems will be enlarged. Note that the bandwidth issue is overcome more easily since internal busses are used for data transmission.

The problems currently attracting the most attention are those where a supercomputer is necessary or desirable for the application computations and immense amounts of output data are often created. There are three options in this situation: (1) render images on the supercomputer and transmit them to a display device; (2) download geometry and other environmental data for each scene to a graphics workstation and have it render each of the images; and, (3) download initial geometry and scene specification data to a graphics workstation, then update this data as it becomes available.

A detailed discussion of the three options above is beyond the scope of this paper, but several general observations can be made. Option 1 is attractive when rendering the images on a workstation is impractical due to scene complexity or the need for special graphics effects. However, Option 1 requires adequate bandwidth for the graphics capability desired (number of image per second). Option 2 is desirable when the amount of geometry and scene data is less than the amount of image data that would be generated. Option 3 is ideal when extensive calculations are required to compute the updated scene data, but a relatively small amount of data is generated. The distributed graphics system used in [14] and [15] is a good example of this.

5. SUPERCOMPUTER GRAPHICS

This section presents two examples illustrating the utility of using supercomputers to render images. In both cases, metafiles were created and then transferred to other systems where they were viewed. These are examples of situations, such as Option 1 described in Section 4.4, where the amount of calculation and memory required make the computations impractical on anything except a supercomputer.

Figure 4 is an image computed on a Control Data Corporation (CDC) Cyber 205 using a highly vectorized graphics program developed at Purdue University. It is one image of 864 computed for a 36 second animation of the locomotive moving along the track. The rendering is done using a computationally intensive method (ray tracing) that is capable of producing many special effects. The scene in Fig. 4 has three light sources and 263 objects; it was computed at a resolution of 2080 by 1520. This image required 2,008 CPU seconds on the Cyber 205, but it is estimated that the same image, without shadows, would require over 83,000 CPU seconds on a minicomputer, such as a CDC Cyber 720. The entire animation required 232 Cyber 205 CPU hours and 34 tapes were needed to store image metafiles (at 1600 bpi). Approximately 14 years of CPU time would be required to compute this animation on a dedicated minicomputer.

Figure 3 is an image also computed on a CDC Cyber 205. The image was rendered by MOVIE.BYU a general purpose, solid modeling graphics package available from Brigham Young University. This program was modified so that more complex scenes (i.e. scenes with more polygons) could be rendered. The scene in Fig. 3 contains approximately 70,000 polygons and was rendered in minutes.

6. CONCLUSION

An introduction to computer graphics terminology and a description of the basic features available in computer graphics hardware and software have been presented. Issues of critical importance to scientists and engineers wishing to use high resolution, full color computer graphics devices to interpret and analyze data have been identified and briefly discussed. References to sources of more detailed information were given. Sample images illustrated graphics rendering features. Two of the images illustrate graphics computations appropriate for supercomputers.

REFERENCES

1. B.H. McCormick, T.A. DeFanti, M.D. Brown, Eds, Visualization in Scientific Computing, <u>Computer Graphics 21</u>, 6(Nov. 1987), ACM SIGGRAPH, New York.
2. K.A. Frenkel, <u>Communications of the ACM</u>,<u>31</u>,2(Feb. 1988), p.110.
3. J.D. Foley, A. Van Dam, <u>Fundamentals of Computer Graphics</u>, Addison-Wesley Publishing Co., 1982.
4. W.M. Newman, R.F. Sproull, <u>Principles of Computer Graphics</u>, 2nd ed., McGraw-Hill Book Co., 1979.
5. D. Hearn, M.P. Baker, <u>Computer Graphics</u>, Prentice-Hall Inc., 1986.
6. <u>Transactions on Graphics</u>, Association for Computing Machinery, Inc., 11 W. 42nd St., New York, NY 10036.
7. <u>Computer Graphics</u>, publ. by Special Interest Group on Computer Graphics of the Association for Computing Machinery, Association for Computing Machinery, Inc., 11 W. 42nd St., New York, NY 10036.
8. <u>Communications of the ACM</u>, Association for Computing Machinery, Inc., 11 W. 42nd St., New York, NY 10036.
9. <u>IEEE Computer Graphics and Applications</u>, The Computer Society, 10662 Los Vaqueros Circle, Los Alamitos, CA 90720.
10. <u>Computer Graphics</u> <u>21</u>, 4(July 1987).
11. <u>ACM SIGGRAPH Video Review</u> <u>2</u>, 28, ACM Order Dept., P.O. Box 64145, Baltimore, MD 21264, (301)528-4261
12. G. Campbell et al, <u>Computer Graphics</u> <u>20</u>, 4(Aug. 1986), p.215
13. S. Jelovcich, <u>S.Klein Computer Graphics Review</u> <u>3</u>, 2(March/April 1988), p.59
14. D. Choi, C. Levit, <u>International Jour. of Supercomputer Applications</u> <u>1</u>, 4(Winter 1987), p.82
15. S.E. Rogers, P.G. Buning, F.J. Merritt, <u>International Journal of Supercomputer Applications</u> <u>1</u>, 4(Winter 1987), p.96

Part IV

Contributed Papers

Monte Carlo Calculation of Transfer Matrix Eigenvalues

M.P. Nightingale and R.G. Caflisch

Department of Physics, University of Rhode Island, Kingston, RI02881, USA

ABSTRACT

Results of various variance reduction schemes are reported for the Monte Carlo calculation of eigenvalues of the transfer matrices of Z(n) models in two dimensions and the Ising model in three dimensions. Numerical evidence is presented for absence of critical slowing down in a weak sense. The method is applied to confirm the conjectured value of the conformal anomaly number $c=8/7\approx 1.14286$ for the end-point of the critical fan of the Z(5) model from the finite-size behavior of strips up to width 50. For strips up to size 11 exact numerical calculation of eigenvalues is used, this yields an estimate $c=1.1432\pm 0.0014$.

I. The Z(n) model

We consider a Z(n) model defined on a simple quadratic lattice in the form of a strip L sites wide and with helical boundary conditions, as defined more precisely below. At each lattice site $i=1,2,\ldots,L(M+1)+1$ there is a spin s_i, assuming integral values from 0 to n−1. The Boltzmann weight of a configuration is given by $\prod_{i=1}^{LM} w(s_i,s_{i+1}|s_{i+L},s_{i+L+1})$, where the weight $w(s,t|u,v)$ is nonnegative and incorporates all possible interactions of the spins s, t, u, and v around a lattice face. The w are normalized such that w=1 for every face in a ferromagnetic ground state.

Denoting by $\underline{s}=(s_0,\ldots,s_L)$ and $\underline{t}=(t_0,\ldots,t_L)$ arbitrary configurations of L+1 spins, a transfer matrix **T** of order n^{L+1} is defined

$$T(\underline{s}|\underline{t})=\prod_{i=1}^{L}\delta_{s_i,t_{i-1}}w(s_0,s_1|t_{L-1},t_L), \qquad (1)$$

using the Kronecker δ. The partition function per site for large M is the largest eigenvalue (μ_o) of the transfer matrix, and the dimensionless free energy per site is $f_h=-\ln\mu_o$.

Z(n) models have special points where the free energy can be calculated exactly [1]. For large systems finite-size scaling and conformal invariance predict the size dependence of f_L, the dimensionless free energy per site at criticality, to be given by [2] $f_L=f_\infty+c\pi/6L^2$. The number c, the conformal anomaly number, characterizes universality classes [3], and it is conjectured that $c=c_n=2(n-1)/(n+2)$ at the special point [4] of the Z(n) model. Using system sizes up to L=8 for the one-dimensional quantum mechanical version of the Z(n) model, Alcaraz [5] estimates $c_5=1.142\pm 0.009$, in agreement with the theoretical value $8/7\approx 1.1429$. Our result is $c_5=1.1432\pm 0.0014$, where the error is twice the

Springer Proceedings in Physics Vol. 33: **Computer Simulation Studies in Condensed Matter Physics**
Editors: D. P. Landau • K. K. Mon • H-B. Schüttler

difference of the results for the two sets of largest systems. This result is based on data obtained by exact numerical calculations for systems with periodic boundary conditions up to width 11 (see Table 1). The infinite system free energy was subtracted and the results multiplied by $6L^2/\pi$ and extrapolated twice using 3-point fits [6].

TABLE 1. Free energy $-f_L$ of Z(5) model at the special point as given in the text. The third column (It.) lists the number of iterations necessary for the conjugate gradient method to converge to an estimated relative accuracy of 5×10^{-16}. The calculation of the eigenvalue for L=11 took roughly one hour of CPU time on the CNSF IBM 3090.

L	$-f_L$	It.	L	$-f_L$	It.
2	0.2493490141895	3	6	0.0838797415800	10
3	0.1412830950983	6	7	0.0792405661694	11
4	0.1067196437860	8	8	0.0762658547985	13
5	0.0917310473509	9	9	0.0742423880320	14
6	0.0838797415800	10	10	0.0728028906334	14
7	0.0792405661694	11	11	0.0717420540140	15

II. Variance Reduction

An approximate right (left) eigenvector, say $\boldsymbol{\gamma}$ ($\hat{\boldsymbol{\gamma}}$), can be used to reduce the variance of the eigenvalue estimates obtained with the Monte Carlo transfer matrix algorithm as given in [7]. Two well-known ways of doing this [8] will be discussed. One method consists in applying the method described above to a similarity transform of the original transfer matrix, say **T**. This new matrix **T'** is defined to have elements

$$T'(\underline{s}|\underline{t})=\hat{\gamma}(\underline{s})T(\underline{s}|\underline{t})/\hat{\gamma}(\underline{t}). \qquad (2)$$

An alternative use of the approximate eigenvector $\boldsymbol{\gamma}$ relies on the relation

$$\mu_0\approx\hat{\boldsymbol{\gamma}}\cdot\mathrm{T}\cdot\boldsymbol{\Phi}_0/\hat{\boldsymbol{\gamma}}\cdot\boldsymbol{\Phi}_0, \qquad (3)$$

where inner products involving $\boldsymbol{\Phi}_0$, the Monte Carlo estimate of the right eigenvector of the transfer matrix, are calculated with the stochastic process defined with the original matrix **T**. Again, for an exact left eigenvector $\hat{\boldsymbol{\gamma}}$, the variance vanishes.

The variational eigenvectors that will be discussed first are generalizations of a hierarchy proposed by Baxter [9] generalized to the geometry of a helix [10]:

$$\gamma(\underline{s})=\mathrm{Tr}\,[\mathbf{F}(s_0,s_1)\cdot\ldots\cdot\mathbf{F}(s_{L-1},s_L)\cdot\mathbf{C}(s_L,s_0)]. \qquad (4)$$

The **F** (and **C**) are n^2 matrices of order n^b, where n^b is a natural number and determines the level of the approximation; in the special case b=0 the matrices reduce to real numbers. We also consider iterates of $\hat{\boldsymbol{\gamma}}$ of the form

$$\hat{\boldsymbol{\gamma}}^{(\ell)}=\hat{\boldsymbol{\gamma}}\cdot\mathrm{T}^{\ell}, \qquad (5)$$

which for small values of ℓ are easy to calculate. They yield very convenient, improved eigenvectors, as can be seen by considering them as the low-order iterates of the power method applied to initial vector $\hat{\boldsymbol{\gamma}}$. For ℓ=L+1 the vector in eq.(5) can be shown to be again of the form (4) with the order b of the approximation increased by one, which is precisely the rationale of the form (4) in the first place.

Implementation of the Monte Carlo process with modified matrices T' defined in terms of vectors of level b requires two or three multiplications of matrices of order n^b for every single spin update. For the Z(5) model, which has served as a test case, this slows down the calculation by an estimated factor of 500, which the variance reduction would most likely make up for only for very large systems. For models with fewer states per site the prospects are far better. In the calculations reported here, only b=0 vectors were employed to guide the stochastic process itself. In this case most of the factors, which in this case are real numbers, in the numerator and denominator in eq.(2) cancel. Note that for $\ell=0$ all one actually does is make the best possible use of the non-uniqueness of associating the nearest neighbor contributions to the Boltzmann weights ω with the faces of the lattice.

We also used a different scheme of calculating optimal trial vectors as follows:

$$\eta(\underline{s})=\prod_{i>j}\omega_{ij}(s_i-s_j), \tag{6}$$

with the normalization $\omega_{ij}(0)=1$. The ω_{ij} are variational parameters determined by the Monte Carlo scheme introduced by Umrigar et al. [11], i.e. by minimization of the variance of

$$\hat{D}(\underline{s})=\sum_{\underline{t}}\hat{\eta}(\underline{t})T(\underline{t}|\underline{s})/\hat{\eta}(\underline{s}). \tag{7}$$

Adequate estimates of the variational parameters were in all cases obtained by minimizing this expression over a single generation of random walkers. Guided by the geometry of the surface, the ω_{ij} were chosen as follows. At all times ω_{01} and ω_{0L} were fixed at the Boltzmann weights associated with the corresponding nearest neighbor bonds. Define the distance d_{ij} between sites i and j at the surface of the helix: $d_{ij}=1$ for nearest neighbor sites; otherwise, d_{ij} is defined to be the length of the shortest path between i and j that visits no (other) surface sites. The ω_{ij} were taken to be a function only of this distance. Exceptions to account for the lack of translational invariance of the surface were made for paths that involved sites 1 or L, or straddled the surface defect. For paths of length greater than 5 we took $\omega_{ij}=1$.

Monte Carlo calculation of eigenvalues was performed, again for the special point of the Z(5) model. For L=8, the estimate $\mu_0(8)= 1.07907\pm0.00002$ agrees with the value 1.079075 obtained by numerically exact diagonalization. For L=14, 20, 30, and 40, where no such results are available, the respective estimates $\mu_0(14)=1.07231\pm0.00002$, $\mu_0(20)= 1.07060\pm0.00002$, $\mu_0(30)=1.06973\pm0.00002$, $\mu_0(40)=1.06942\pm0.00002$ agree with the prediction based on conformal invariance, viz. $\mu_0(L)\approx \mu_0(\infty)(1++\pi c_5/6L^2)\approx1.069020+0.639700/L^2$, using the exact results of Fateev and Zamolodchikov [1] for the free energy in the thermodynamic limit. This gives $\mu_0(14)=1.072289$, $\mu_0(20)=1.070621$, $\mu_0(30)=1.069731$, and $\mu_0(40)=$ 1.069420 respectively.

In Table 2 results for the various variance reduction schemes are shown. The true measure of the relative efficiencies is the ratio of the product of the one-flip error and the square root of the time per flip.

We define the one-flip error e_2 as follows: $e_2=\sigma\surd(M_0 r_0 L^{-1})$, where M_0 is the length of a Monte Carlo run with standard error σ, and r_0 the number of random walkers. The factor L^{-1} is inserted for compatibility with standard Monte Carlo methods to account for the reduction of the effective dimension by one in the

TABLE 2: Results of variance reduction for the Z(5) model. Listed are: (1) e_1 defined as the estimated percentage error in the free energy relative to its the singular (i.e. size-dependent) part as given in the text; (2) e_2 the one-flip percentage error as defined in the text; (3) the order n^b of the matrices used for the approximate eigenvector [n^b=0 indicates use of the original matrix of eq.(3)]; (4) ℓ the number of transfer matrix pre-multiplications as defined in eq.(5). Results obtained with trial functions of the form (6) are indicated as MC; (5) t the time per flip per random walker in μsec for L=20. For the times indicated by asterisks these numbers are estimates. The target number of random walkers was 2500 for the results in the first part of the table, and 5000 for the second part.

L=8		L=15		L=20		L=30		L=40				
e_1	e_2	e_1	e_2	e_1	e_2	e_1	e_2	e_1	e_2	n^b	ℓ	t
6	100	100	400	70	100					0	0	5
0.3	5	1	4	3	6					1	0	–
0.5	9	3	10	5	10					2	0	7 *
0.2	3	1	4	3	4					5	0	15 *
0.2	3	0.8	3	1	3					1	1	3
0.1	2	0.8	3	1	3	3	3	16	5	1	2	3
				0.7	2	2	2	5	2	MC		7 *

TABLE 3: Results of variance reduction for the 2-dimensional Ising model (same notations as Table 2); 5000 random walkers were used.

L=20		L=40		L=60				
e_1	e_2	e_1	e_2	e_1	e_2	n^b	ℓ	t
20	10	100	20	–	–	0	0	3
2	1	6	0.9	10	0.7	1	0	2
1	0.6	2	0.4	7	0.4	1	2	2
1	0.6	2	0.4	6	0.3	MC		2

states defining the transfer matrix. For the cases n^b=2 and n^b=5, the estimates were obtained using eq.(3); the required traces were calculated once out of every two and three flips per spin per random walker, respectively. In these cases the transfer matrix was transformed using the n^b=1, ℓ=0 vector. Results for the accuracy of the calculated free energy of the Ising model (n=2) at its critical point are shown in Table 3.

From the results in Tables 2 and 3 we conclude the following. First, the simplest trial vectors, of the form (5) with n^b=1 and ℓ=0, derived from Baxter's optimization scheme, lead to a substantial variance reduction compared to the process defined in terms of the original matrix (1).
Second, the most flexible scheme, using the parameters optimized by Monte Carlo, gives the best results. Third, the one-flip error does not depend on system size over the range of systems studied. This can be explained as follows, if the assumption is made that all random walkers within one generation are independent. In that case the correlation time of the Monte Carlo process is proportional to the correlation length of the system described by the transfer matrix, and therefore at criticality proportional to system width L. In other words,

starting from a given initial configuration it takes on the order of L^d spin flips per random walker in d dimensions to generate a statistically independent configuration. We do not expect this behavior to continue indefinitely. In particular, if the system gets so big that a random walker is likely to die within L^d steps the assumption of statistical independence of the random walkers becomes invalid, at which point the method probably will become intractable. An order of magnitude estimate of the size at which this will occur can be derived in terms of the rms deviation σ_D of $\hat{D}$ defined in eq. (7). A random walker will die when its weight changes by an amount of order unity: $L^{d/2}\sigma_D \approx 1$.

Finally we discuss some results for the 3-dimensional Ising model on a helix. The transfer matrix is defined for a stack of planes of $N=n_x \times n_y$ sites each:

$$T(\underline{s}|\underline{t}) = \prod_{i=2}^{N-1} \delta_{s_i,t_{i-1}} \exp[Ks_1 t_N + K(s_1 s_2 + s_1 s_{n_x+1} + t_N t_{N-1} + t_N t_{N-n_x})/2]. \quad (8)$$

Trial vectors of the form (6) were used, involving paths of maximum length 3. The estimates of the weights ω were obtained from the eigenvectors of small lattices for which the calculation was numerically exact. Results are displayed in Table 4. As was the case for the two-dimensional systems, we find a substantial decrease in variance for an increase of CPU time per flip on the order of 30%. The error bar of the 10×10 system seems too small, and should be regarded as preliminary.

TABLE 4. Results for the 3-dimensional Ising model at K=0.2 (roughly 10% above the critical temperature.) The fourth and sixth columns contain the absolute errors in the free energy followed by the respective one-flip errors for cases without and with variational guidance. The heads of the other columns are defined in the text. The number of random walkers was 2500 for the biggest system and 2000 for the others.

n_x	n_y	$-f_L$		e_2		e_2
10	10	2.137036	2×10^{-4}	10	4×10^{-6}	0.2
3	4	2.14682	1×10^{-4}	10	9×10^{-6}	1.
4	3	2.14661	1×10^{-4}	10	2×10^{-5}	2.

Acknowledgments

It is a pleasure to thank John Cardy for drawing our attention to the work of Fateev and Zamolodchikov on the Z(n) model, and to thank Henk Blöte for continuing discussions. This work, part of which was done at the Aspen Center for Physics, was funded at the University of Rhode Island by the National Science Foundation (NSF) under contract No. DMR-87-04730, and was conducted using the Cornell National Supercomputing Facility, a resource of the Center for Theory and Simulation in Science and Engineering at Cornell University, which is funded in part by the NSF, New York State, and the IBM Corporation.

References and Footnotes

1. V. A. Fateev and A. B. Zamolodchikov, Phys. Lett. **92A**, 35 (1982).
2. I. Affleck, Phys. Rev. Lett. **56**, 746 (1986); H.W.J. Blöte, J.L. Cardy, and M.P. Nightingale, ibid. p.742.

3. For a review see, J.L. Cardy, in "Phase Transitions and Critical Phenomena" Vol. **11** (eds. C. Domb and J.L. Lebowitz) (Academic Press, London,1987).
4. V.A. Fateev and A.B. Zamolodchikov, Zh. Eksp. Fiz. **89**, 380 (1985) [Sov. Phys. JETP **62**, 215 (1985)].
5. F.C. Alcaraz, J. Phys. A, **19**, L1085 (1986).
6. H.W.J. Blöte and M.P. Nightingale , Physica **112A**, 405 (1982).
7. M.P. Nightingale and H.W.J. Blöte, Phys. Rev. Lett. **60**, 1662 (1988).
8. D. M. Ceperley and M. H. Kalos, Monte Carlo Methods in Statistical Physics, edited by K. Binder (Springer, Berlin, 1979).
9. R.J. Baxter, J. Stat. Phys. **19**, 461 (1978).
10. M.P. Nightingale, Proceedings of the Third International Conference on Supercomputing, Boston, MA USA, May 1988.
11. C.J. Umrigar, K.G. Wilson, and J.W. Wilkins, Phys. Rev. Lett. **60**, 1719 (1988). Also see C. Umrigar, proceedings of this conference.

Finite Size Effects at First-Order Phase Transitions Revisited

P. Peczak and D.P. Landau

Center for Simulational Physics, University of Georgia, Athens, GA 30602, USA

Two years ago a phenomenological theory of finite size scaling at a temperature–driven first order phase transition was proposed by CHALLA et al. [1]. Approximating the energy distribution function by the superposition of two weighted Gaussian functions, the authors found that all finite–size effects, e.g. "rounding" and "shifting" of the transition, depend on the volume of the system L^d, in accordance with previous general scaling arguments. This prediction was successfully tested [1], by carrying out extensive Monte Carlo calculations, on the ten–state (Q=10) Potts model in two dimensions, defined by the Hamiltonian

$$\mathcal{H} = -J \sum_{(i,j)} \delta_{\sigma_i \sigma_j} \tag{1}$$

(where δ is the Kronecker delta, J is the interaction strength, the sum runs over all nearest–neighbor pairs of spins, and σ_j can have Q different values).

Recently, however, KATZNELSON et al. [2], studied the behavior of a weakly first order system, the Q=5 Potts model in two dimensions, and found yet another kind of finite size phenomenon: the existence for "small" lattices of false equilibrium states, which become metastable when the lattice size is increased beyond a critical value. This result was surprising because the previous simulations on the Q=10 Potts model [1], failed to detect any effect of this type.

Motivated by these unusual results, we decided to reexamine the behavior of the Q=5 Potts model on a L×L square lattice ($L \leq 240$) with periodic boundary conditions, by extensive Monte Carlo simulations on the Cyber 205 vector processor at the University of Georgia. The use of a memory–to–register swapping technique and a "checkerboard" algorithm [1] allowed us to obtain a speed 1.6 μsec. per update which was essential to obtain meaningful results. We concentrated our attention on the internal energy and order parameter distributions as functions of the lattice size L, following closely the method of analysis adopted by KATZNELSON et al [2]. We obtained extensive data for L between 32 and 240 at T_c. The results for all sizes studied exhibited very similar behavior: for short runs, the histograms for n, the fraction of configurations having an energy between E and E+ΔE, did, indeed, show more than two peaks, however the positions and sizes of the peaks varied from one part of the simulation to another. This was true even for runs as long as 10^6MCS for L=240 which still indicate multiple peaks in the distribution shown in Fig. 1. This histogram became smoother as the run was extended and by the time 5×10^6 MCS were averaged together the distribution of the energy (see Fig. 1) showed only a single peak with a shoulder suggesting the presence of another peak. (Since this simulation was not carried out at the effective finite lattice critical temperature $T_c(L)$, the two peaks are not expected to be of equal height.) Simulations at higher temperatures near $T_c(L)$clearly showed two peaks with half widths of the order of the peak separation. The shallow minimum between the peaks indicated that interface effects between domains of different phases play an important role; comparison with data obtained for smaller lattices showed that the minimum does, indeed, deepen with increasing lattice size. In contrast, for

Springer Proceedings in Physics Vol. 33: **Computer Simulation Studies in Condensed Matter Physics**
Editors: D. P. Landau · K. K. Mon · H-B. Schüttler

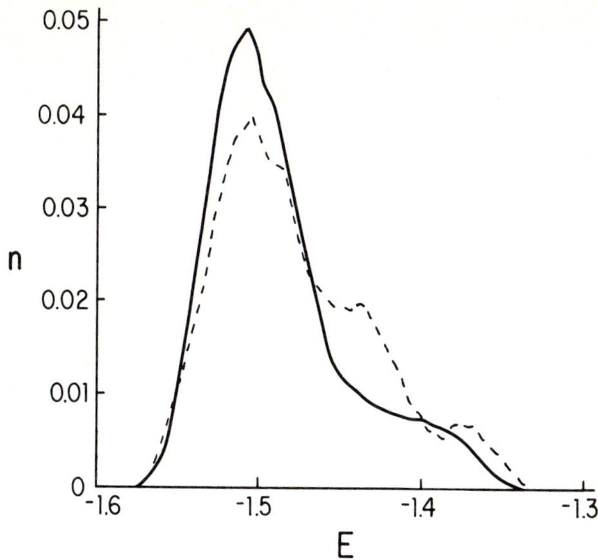

Fig. 1. The distribution n of the configurations which have internal energy per spin E in units of J for L=240, $T=T_c$ after 1.0×10^6 (broken line) and 5.0×10^6 (continuous line) Monte Carlo steps/site. Between two consecutive measurements, the lattice was updated 20 times. The internal energy bin width is 0.004

the Q=10 Potts model even the distribution for an L=12 lattice showed two well separated peaks with a very deep minimum between them.

The behavior of this weakly first order transition evokes the image of a continuous phase change. The time dependence of order parameter for L=90 lattice at T_c clearly showed how long the initially disordered system takes to transfer to the ordered phase, with several unsuccessful trials along the way. Also, pictures of the lattice in the transition region showed the features characteristic of a second order phase transition: irregular, ragged clusters of all sizes. The typical number of clusters n(s) vs. their size s showed a power law dependence characteristic for fractal shapes, not droplets: e.g. averaging over 3700 configurations on an L=90 lattice at T_c (i.e. in the ordered phase for a finite system), with order parameter smaller than 0.1 we obtained

$$n(s) \approx s^{-\tau} \quad \text{with } \tau = 1.94 \pm 0.07 \ . \tag{2}$$

This pseudo–second order behavior of the Q=5 Potts model resembles the phenomenon known in the phenomenology of liquid crystals as the pretransition effect [3]. Indeed, our analysis of the correlation function $\Gamma(r)$ for L=90 lattice calculated at several temperatures in the critical region($3 \times 10^{-3} \leq t \leq 10^{-1}$ where $t = |T-T_c|/T_c$), with a fit of the data to the Ornstein–Zernike form [4], revealed a second order transition–type scaling relation for the correlation length with $\eta = 0.24 \pm 0.02$ and

$$\xi \approx t^{-\nu} \quad \text{with } \nu = 0.59 \pm 0.03 \ . \tag{3}$$

Calculated from an independent log–log plot of susceptibility vs. reduced temperature the value of an exponent $\gamma = 1.02 \pm 0.03$ was used to check that the exponents γ, η and ν

satisfy the Fisher equality [4]; i.e., $\gamma = (2 - \eta)\,\nu$. One expects that for a certain value of t, when the system finally "realizes" its asymptotic first order type of behavior, the scaling relation will break down, but with the lattice size we used we couldn't see this effect. Apparently the size of the largest droplet is of order at least 100 lattice spacings.

KATZNELSON et al. [2] also reported in their paper that the energy distribution peaks $E_{\pm}$ are shifted in the volume–size dependent fashion but that the limiting values (for $L \to \infty$) do not correspond to those predicted by the theory. Unfortunately, the phenomenological theory of finite size scaling by CHALLA et al. assumes that there is no L–dependence of the energy maxima. However, we have reasons to believe that the lack of good statistics, as well as limited lattice size may also account for this apparent discrepancy with theory.

Although the results of our simulations generally confirm the validity of CHALLA et al. theory in the case of a weakly first order transition, it is necessary to investigate the behavior of larger lattices ($L \gtrsim 350$), for which a contribution of interface effects is less significant (and Gaussian energy distribution peaks are distinctly separated), to resolve the remaining problems.

This research has been supported in part by the National Science Foundation.

References

1. M. S. S. Challa, D. P. Landau, K. Binder: Phys. Rev. B34, 1841 (1986).
2. E. Katznelson, P. G. Lauwers: Phys. Lett. B186, 385 (1987).
3. See ref. in H. Kelker, R. Hatz: "Handbook of Liquid Crystals", Verlag Chemie, Weinheim; Deerfield Beach, Florida; Basel (1980), Ch. 8.
4. M. E. Fisher, R. J. Burford: Phys. Rev. 156, 310 (1967).

Correlation Time Measurements for the d=2 Ising Model

A.M. Ferrenberg and R.H. Swendsen

Department of Physics, Carnegie Mellon University,
Pittsburgh, PA 15213, USA

Three methods of extracting the correlation time from the decay of time-displaced-correlation-functions are discussed. One of the methods is used to determine correlation times for Monte Carlo simulations of the d=2 Ising model on square lattices with L=6,8,10 and 12. From these results, we estimate the dynamical critical exponent to be z = 2.121±0.008 in agreement with recent Monte Carlo results.

1. INTRODUCTION

Monte Carlo methods have been used extensively to study the equilibrium properties of many models. In particular, critical properties have been studied in great detail. However, the dynamic properties of even simple models are not well known compared to the static properties. The idea of universality, or universality classes, plays an important role in our understanding of static critical phenomena. An interesting question is whether dynamic universality classes exist. A recent paper by TANG and LANDAU [1] has addressed the question of dynamic universality in two-dimensional Potts models. They concluded that within their errors, a single dynamical critical exponent describes two, three and four state Potts models, although the errors are too large to permit too strong a statement about dynamic universality.

The dynamical critical exponent z which describes the divergence of the correlation time τ is defined by $\tau \sim \xi^z$ or $\tau \sim t^{z\nu}$ where $t = (T-T_c)/T_c$. In practice, it is convenient to use finite size scaling to determine z. Then $\tau \sim L^z$ where L is the linear dimension of the system. As with other scaling relations, we expect this behavior only for sufficiently large systems. To use finite size scaling to determine a critical exponent, one must be able to measure the scaled quantity accurately on sufficiently large systems. For directly measurable quantities (magnetization, energy, specific heat etc.) the uncertainty in the measured value of a quantity A decreases as the number of measurements increases

$$\frac{\delta A}{A} \approx \sqrt{\frac{\tau}{N}}$$

where N is the total number of measurements. The correlation time, however, is not a directly measurable quantity. It must be extracted from other quantities such as the relaxation of the energy or magnetization from an ordered state, or a time-displaced-correlation-function (TDCF f(t)). In this article, three methods for determining correlation times are discussed. The correlation times for the d=2 Ising model for four lattice sizes (L=6,8,10,12) are calculated and from this we calculate an estimate for z. Results from such small lattices will, of course, have an unknown systematic error. However, by performing very long runs on small lattices using a MicroVAX, we have reduced the statistical errors by a factor of seven without resorting to a supercomputer for the calculation.

Springer Proceedings in Physics Vol. 33: **Computer Simulation Studies in Condensed Matter Physics**
Editors: D. P. Landau • K. K. Mon • H-B. Schüttler

The dynamic properties of a Monte Carlo simulation are governed by the Master Equation

$$\frac{\partial P(\sigma,t)}{\partial t} = \sum_{\sigma}(P(\sigma',t)W(\sigma'\to\sigma) - P(\sigma,t)W(\sigma\to\sigma')) = LP(\sigma,t) \quad , \tag{1}$$

where σ and σ' represent configurations of spins, $P(\sigma,t)$ is the probability that the system is in configuration σ at time t, and $W(\sigma\to\sigma')$ is the probability per unit time of the transition from σ to σ'.

The general solution to (1) can be written as

$$P_{\sigma o}(\sigma,t) = P_{eq}(\sigma) + \sum_i a_i(\sigma_o)P_i(\sigma)e^{-t/\tau_i} \quad , \tag{2}$$

where $P_{eq}(\sigma)$ is the equilibrium probability distribution (detailed balance guarantees that $P_{eq}(\sigma)$ is an eigenvector of L with eigenvalue zero) and $P_i(\sigma)$ and $-1/\tau_i$ are the ith eigenvector (eigenvalue) of L. The coefficients $a_i(\sigma_o)$ are determined by the condition that $P(\sigma,0) = \delta(\sigma,\sigma_o)$. The correlation time τ is defined as the largest of the τ_i. Because of the exponential decay of the probability distribution to the equilibrium distribution, time-dependent quantities also exhibit exponential decay to equilibrium values

$$f(t) = \sum_i f_i e^{-t/\tau_i}.$$

In principle, τ can be determined easily. Since the form for the decay of f(t) is known, it is straightforward to extract τ with standard non-linear curve fitting routines. One can either fit the long-time part of f(t) to a single exponential or fit the entire function to a sum of several exponentials. In practice, however, the accuracy with which one can calculate the TDCF places severe limits on the determination of τ. f(t) is not itself a measured quantity, but rather is calculated from the measured quantity $\langle E(0)E(t)\rangle$ in the following way:

$$f(t) = \frac{\langle E(0)E(t)\rangle - \langle E\rangle^2_{eq}}{\langle E^2\rangle_{eq} - \langle E\rangle^2_{eq}}.$$

In models where $\langle E\rangle_{eq}$ is known exactly it is common to use this exact value in calculating f(t). Surprisingly, we found that more accurate results are obtained when the calculated averages are used rather than the exact results. In Fig. 1, plots of f(t) vs t demonstrating this effect are shown.

At long times,

$$\frac{\delta f(t)}{f(t)} \sim \sqrt{\frac{\tau}{N}} \exp(t/\tau).$$

This exponential increase in the error in f(t) makes it impractical to use only the long-time part of f(t) to extract τ. At short times, the error in f(t) is small, but more than one exponential is involved in the decay. If too many parameters are used in a non-linear curve-fitting algorithm, the results obtained are unreliable. The three methods of extracting τ from a TDCF investigated here are

- Method 1: The short-time part of the TDCF is cut off and then f(t) is fit to a single exponential.

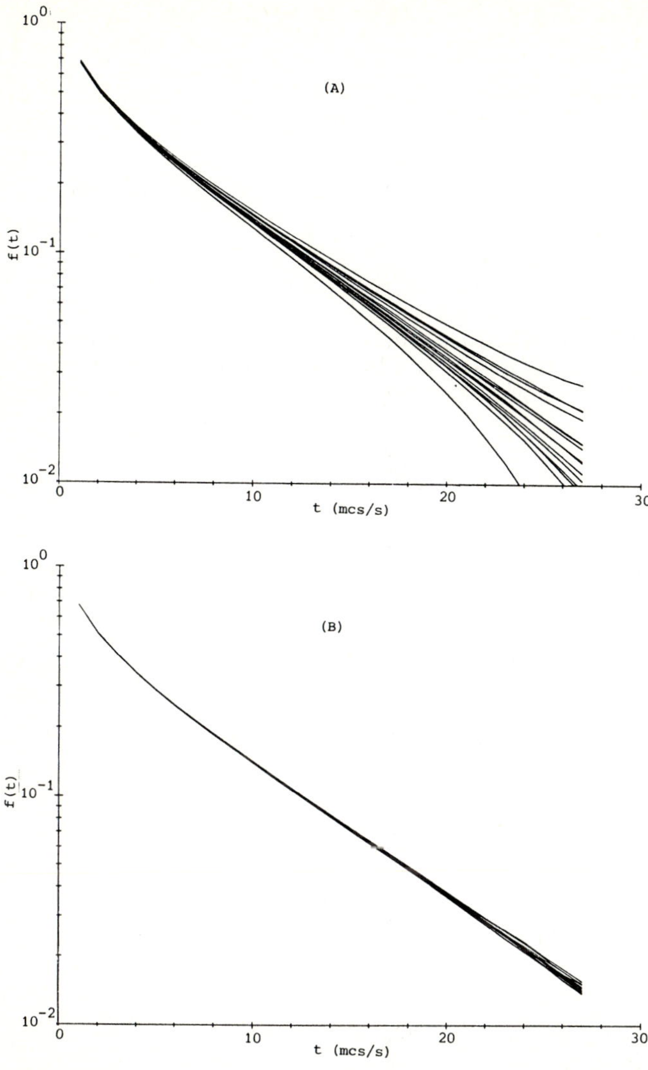

Fig. 1. Semi-log plot of f(t) vs t for fifteen MC runs of 2.0 x 10^7 each with (A) the exact energy and (B) the measured energy subtracted

- Method 2: The long-time part of the TDCF is cut off and then f(t) is fit to a sum of several exponentials.
- Method 3: Some long-time and short-time parts of the TDCF are cut off and then f(t) is fit to a sum of several exponentials.

2. RESULTS

Fifteen Monte Carlo simulations of the d=2 Ising model were performed on an 8^2 lattice at the infinite lattice critical temperature using a red-black spin update algorithm. The length of each simulation was 2 x 10^7 lattice sweeps. The Energy-Energy TDCF was calculated for each run, and the three methods indicated above were used to calculate τ. The range of times and number of exponentials used for the three methods were

- Method 1: $10 < t < 41$ one exponential
- Method 2: $0 < t < 20$ three exponentials
- Method 3: $6 < t < 20$ two exponentials

The results for τ and its amplitude in f(t) for the three methods are

Method 1: $\tau = 7.540 \pm 0.036$ amplitude = 0.529 ± 0.017
Method 2: $\tau = 7.590 \pm 0.018$ amplitude = 0.5211 ± 0.0015
Method 3: $\tau = 7.589 \pm 0.015$ amplitude = 0.5213 ± 0.0018

The third method of analysis was then applied to different sized systems to calculate z. Between 8 and 13 runs of 2×10^7 lattice sweeps were performed for lattices with L=6,10 and 12. The correlation times for these systems are

L	τ
6	4.159 ± 0.015
8	7.589 ± 0.015
10	12.328 ± 0.048
12	17.994 ± 0.093

A plot of these correlation times is shown in Fig. 2. The errors shown are five times the actual errors. The value obtained from these four points is $z = 2.121 \pm 0.008$ which is in good agreement with Tang and Landau's result $z = 2.13 \pm 0.06$, obtained for system sizes from L=12 to L=64. Work is currently in progress to extend our calculation to larger systems to examine the systematic errors in our results.

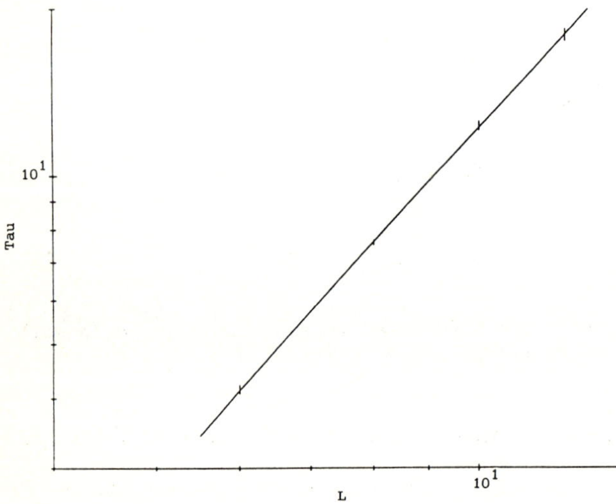

Fig. 2. Log-Log plot of τ vs L. The error bars shown are five times the actual errors. The solid line is the best fit used to determine z.

3. CONCLUSIONS

We have calculated the correlation time for the d=2 Ising model on an 8^2 lattice using three different extraction methods. From our results, we conclude that the correlation time for this system can be determined most accurately by fitting the TDCF over early times where the error in f(t) is small. Methods 2 and 3 agree very well in their values for both τ and the amplitude of the decay. We found that f(t) can be determined more accurately when measured averages, rather than exact averages, are used in its calculation. We also found that for the d=2 Ising model, it is possible to accurately determine the dynamical critical exponent using very small lattices.

This work has been supported by NSF Grant No. DMR-8613218.

4. REFERENCES

1. S. Tang and D.P. Landau, Phys. Rev. B36, 567 (1987) and references therein.

Monte Carlo Study of the Critical Dynamics at the Surface of an Ising Model

P.A. Slotte, S. Wansleben, and D.P. Landau

Center for Simulational Physics, University of Georgia, Athens, GA30602, USA

The model studied is the ferromagnetic 3D–Ising model on a simple cubic lattice with a free surface, i.e., the hamiltonian

$$-\beta\mathcal{H} = K \sum_{<ij>_b} s_i s_j + K_s \sum_{<ij>_s} s_i s_j, \qquad s_i = \pm 1, \tag{1}$$

where the second sum runs over spin pairs where both spins are in the surface, the first sum runs over all other nearest–neighbor spin pairs, and $\beta = 1/kT$, with k the Boltzmann constant and T temperature. The static properties of this model are well known [1–4]. For temperatures below the bulk critical temperature, $K_c \approx 0.221654$ [5], both the surface and the bulk are ferromagnetically ordered. For strong surface coupling, i.e., ratios of the surface to bulk coupling, $R = K_s/K$, above a tricritical value, $R_t \approx 1.52$ [2], the surface orders at a higher temperature than the bulk. We have studied the critical dynamics of the surface layer in this model at the ordinary transition ($R = 1.0$) and at the tricritical point ($R = 1.52$, also called the special transition).

The kinetics used is Metropolis Monte Carlo and the model simulated has D layers of L×L cross section with periodic boundary conditions parallel to the free surfaces. The spin–layers are updated sequentially, and each layer is updated using a two sublattice checkerboard decomposition. This updating scheme was implemented on a CYBER 205 with a vectorized multispin coding program which, on a 64×64×36–lattice, generates 75 MC–steps per spin each second. The dynamics of the model is described in terms of autocorrelation functions:

$$C_m(t) = \frac{<m(t_0)m(t_0+t)> - <m>^2}{<m^2> - <m>^2}. \tag{2}$$

Here m is the magnetization (energy .....) of the surface layer, the time, t, is measured in MC–steps per spin (MCS), and the expectation values, <...>, are found by taking the mean values over all initial times, t_0.

Near a critical point one expects critical slowing down [6]. This implies that the autocorrelation functions are characterized by a time scale, τ, which according to the dynamic scaling hypothesis diverges as

$$\tau(K) \sim [\xi(K)]^z \sim (K_c - K)^{-z\nu} \tag{3}$$

where ξ is the correlation length which diverges with static critical exponent ν and z is the dynamical critical exponent. The dynamical exponent is expected to be universal for several types of kinetics so that our MC results may also be applicable to other models. An unambiguous definition of the characteristic time scale is found by observing that for sufficiently long times the correlation functions fall off exponentially:

$$C(t) \sim \exp(-t/\tau). \tag{4}$$

Springer Proceedings in Physics Vol. 33: **Computer Simulation Studies in Condensed Matter Physics**
Editors: D. P. Landau • K. K. Mon • H-B. Schüttler

For a finite system (3) gives the finite size scaling form of the time scale at the critical temperature:

$$\tau(K_c) \sim L^z, \tag{5}$$

where L is the linear size of the system. This relation has been successfully used in earlier work on the critical dynamics of bulk transitions [7,8]. However, in the current problem there are two finite size length scales, the system size parallel to the surface, L, and the thickness of the slab, D. With a small D (5) cannot be applied and we have chosen to measure the characteristic time as a function of temperature and apply (3) directly.

Very long runs, typically at least 10^6 MCS, are needed to obtain good statistics for the long time behavior of C(t). Typical correlation functions are shown in Fig. 1. Near the ordinary transition (R = 1.0), C(t) differs significantly from exponential behavior for short times, and it is very small (≈ 0.05) when it reaches the asymptotic behavior (4). Since C(t) in general can be expressed as a sum over exponential terms, $a_i \exp(-t/\tau_i)$, where τ_i are the inverse eigenvalues of the Liouville operator, we have used a two exponential fit to improve the estimates for the leading behavior (4). This fits the data well for C(t) ≲ 0.7. Near the tricritical point (R = 1.52) C(t) has a qualitatively different behavior, and can be described by a single exponential even for short times. The characteristic time itself is much larger than for R = 1 though, and again runs of the order of 10^6 MCS are necessary to get good statistics. The bulk autocorrelation function, which is shown for comparison, falls off slower than the surface correlations for either value of R.

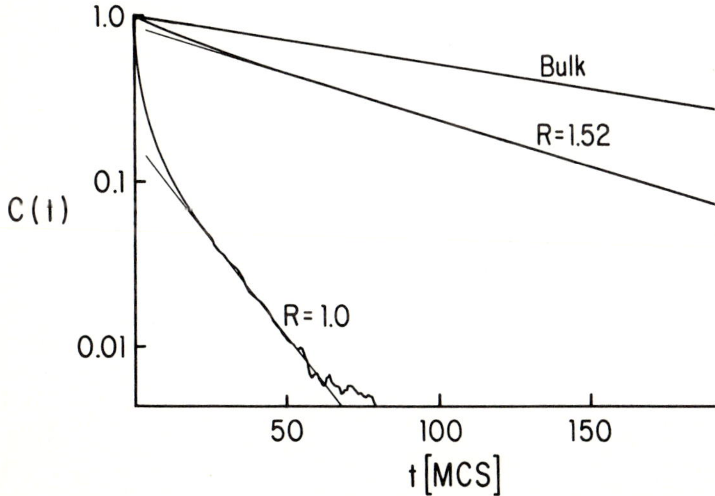

Fig. 1 Autocorrelation functions for the surface layer magnetization near the ordinary transition (R = 1.0) and near the tricritical point (R = 1.52) for $(K_c-K)/K_c = 0.0165$ and 64×64×36–lattice. The bulk magnetization at a slightly different coupling, $(K_c-K)/K_c = 0.0156$, and for a 48×48×48–lattice, is also shown [7]. The straight lines show the asymptotic exponential decay.

We have measured the characteristic time for several values of the reduced coupling, $(K_c-K)/K_c$, in the range 10^{-3}–10^{-1}. According to (3) the data should lie on a straight line with slope $z\nu$ in a log–log plot. For small reduced couplings, where the bulk correlation length is comparable with D, finite size effects are important. The bulk correlation length is ≈ 10 lattice spacings at reduced couplings of ≈ 10^{-2}, and we have studied films with D = 18, 36, 72 to estimate the finite size effect. For the largest values of the reduced coupling, the asymptotic form (3) is not valid, but for intermediate values the data fall on straight lines. We have extensive data near the ordinary transition (R = 1) and the results are consistent with $z\nu \approx 1.27$, where z is the bulk dynamical exponent [7,9], and ν is the bulk correlation length exponent [10]. That there is no true surface dynamic behavior different from the bulk in this case was predicted by ϵ–expansions [11]. A MC–study of the relaxation to equilibrium in the present model also gave this result [12]. Work to study the tricritical relaxation more quantitatively is in progress.

This research was supported in part by the National Science Foundation.

1. K. Binder: In Phase Transitions and Critical Phenomena, ed. C. Domb and J. Lebowitz, Vol. 8 (Academic Press, London 1983).
2. K. Binder and D. P. Landau: Phys. Rev. Lett. 52, 318 (1984).
3. H. Diehl and S. Dietrich: Phys. Rev. B24, 2878 (1981).
4. M. Kikuchi and Y. Okabe: Prog. Theor. Phys. 73, 32 (1985), Prog. Theor. Phys. 74, 458 (1985), Y. Okabe, M. Kikuchi and K. Ohno: Prog. Theor. Phys. 75, 496 (1986).
5. G. S. Pawley, R. H. Swendsen, D. J. Wallace and K. G. Wilson: Phys. Rev. B29, 4030 (1984).
6. P. C. Hohenberg and B. I. Halperin: Mod. Phys. Rev. 49, 435 (1977).
7. S. Wansleben and D. P. Landau: J. Appl. Phys. 61, 3968 (1987).
8. S. Tang and D. P. Landau: Phys. Rev. 36, 567 (1987).
9. R. Bausch, V. Dohm, H. K. Janssen and R. K. P. Zia: Phys. Rev. Lett. 47, 1837 (1981).
10. J. C. LeGuillou and J. Zinn–Justin: Phys. Rev. B21, 3976 (1980).
11. S. Dietrich and H. W. Diehl: Z. Phys. B51, 343 (1983).
12. M. Kikuchi and Y. Okabe: Phys. Rev. Lett. 55, 1220 (1985).

A New Model of Interactive Percolation

S.R. Anderson and F. Family

The Department of Physics, Emory University, Atlanta, GA 30322, USA

The most widely used model for the study of disordered systems is the percolation model [1], in which the only parameter is the concentration p. Percolation can be illustrated by a simple irreversible cluster-growth process, in which sites on an initially empty lattice are selected at random and then occupied. Nearest-neighbor sites are considered to be connected, forming clusters of varying sizes and geometries. The concentration increases with time, and above a critical concentration p_c a cluster extends across the entire system. Its formation signals a geometric phase transition, and the region around p_c can be analyzed for scaling behavior and critical exponents.

In real systems, however, there are interactions, and the assumption that the occupation is independent of the state of the neighboring sites is not valid. A simple generalization can therefore be made [2] in which two probabilities, p_0 and p_1, are used to decide if a site should be occupied, according to whether the site has no neighbors or at least one neighbor, respectively. The properties of the system, such as the threshold p_c, will then depend only on the concentration p and the ratio $r = p_1/p_0$ (modulo the time scale). Figure 1 shows the distribution of clusters at the percolation threshold for several values of r.

When $r = 1$, random percolation is recovered. When $r > 1$, the site occupation will be "ferromagnetic" in nature, since sites that have occupied neighbors will be preferred over those that don't. Cluster nucleation will therefore be small, resulting in widely separated clusters which grow in an Eden-like manner [3]. Eventually, these compact Eden clusters, or "blobs" [4], will link together and form larger, fractal clusters [5]. When $r < 1$, the site occupation will be "antiferromagnetic", since the presence of occupied neighbors will be inhibitory. Initially, therefore, there will be many single-particle blobs, which will only link together with difficulty. They will tend to lie on one of two sublattices, forming next-nearest-neighbor-connected "domains".

At first glance, the percolative structures shown in Fig. 1 appear to be very different from each other. However, suppose that, instead of the original particles, the blobs are taken to be the basic

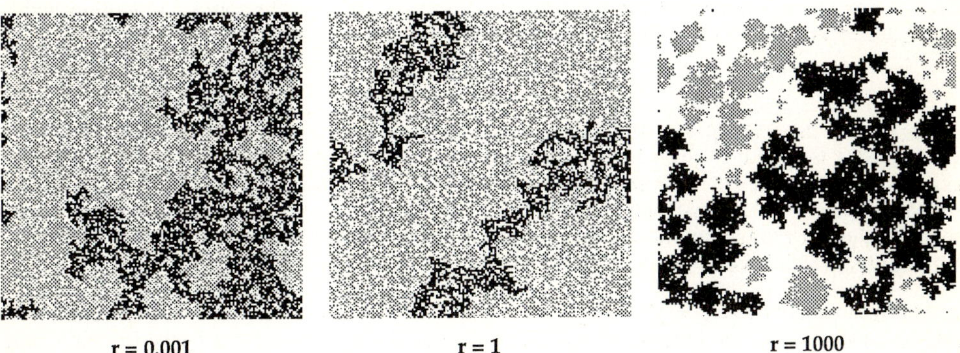

Fig. 1 Sample cluster distributions at the percolation threshold $p_c(r)$, on a lattice of size $L = 160$. The spanning cluster is black; all others are gray

Springer Proceedings in Physics Vol. 33: **Computer Simulation Studies in Condensed Matter Physics**
Editors: D. P. Landau • K. K. Mon • H-B. Schüttler

elements which join together to form the clusters. This makes sense, since they are the result of the interactions, and their connections with each other are due to their random placement in space and time on the lattice. Now, the critical behavior in the percolation model is due to the divergence of the connectivity length ξ [1], so that the blob size, a local quantity, should not then be important. It is therefore anticipated that the critical behavior will be independent of r, and the universality class will remain that of random percolation.

To verify these expectations, simulations of this cluster-growth model were carried out using standard Monte Carlo techniques [6] on a square lattice with periodic boundary conditions. The number of runs (i.e. configurations) that were averaged over varied from 32,000 for $L = 16$ to 2000 for $L = 512$. The results presented here are based on a percolation threshold p_c defined by the existence of a cluster which spans the lattice in both directions.

To determine p_c as a function of r, the finite-size scaling relation [1,7]

$$p_c(\infty) - p_c(L) \sim L^{-1/\nu} \tag{1}$$

is used. Nonlinear fits to this equation are shown in Fig. 2(a) for each value of r. It is clear from this figure that the percolation threshold has a strong dependence on r, with a range $0.5 \leq p_c(r) \leq 0.65$. It may be noted that the data for $r = 100$ has a fair amount of curvature present for the smaller lattice sizes. This is due to the large blob size at this value of r, and is a type of finite-size effect (the fit shown for $r = 100$ does not make use of the smaller-lattice data).

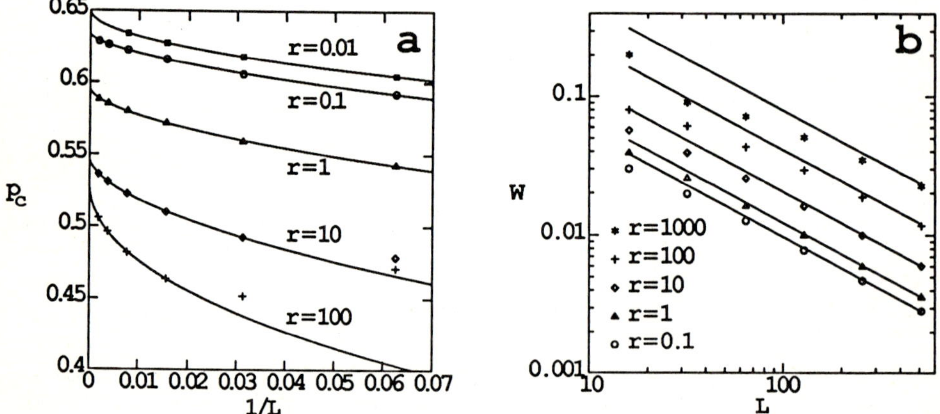

Fig. 2 (a) Finite-size-scaling analysis for the percolation threshold $p_c(r,L)$. The extrapolations to $1/L = 0$ provide an estimate of the true threshold $p_c(r,\infty)$. (b) The threshold distribution width $W(L)$, for several values of r. The slopes of the data give the exponent $1/\nu$

Because of the finite size of the lattices used, the percolation threshold has a distribution of values about its average, given by a width $W(L) = (\langle p_c(L)^2\rangle - \langle p_c(L)\rangle^2)^{1/2}$. This quantity is also expected [8] to scale as $L^{-1/\nu}$, so that the data for W should provide a better determination of the critical exponent ν than that obtained from (1). Figure 2(b) shows W as a function of L for several values of r. Asymptotically, these curves all appear to approach the straight lines on this log-log plot, which have slopes equal to the random percolation value $1/\nu = 3/4$ [1].

The mean cluster size $S = \sum s^2 n_s / \sum s\, n_s$ is shown in Fig. 3(a) as a function of p for several different values of r. In addition to the decrease in peak height as r, and hence the blob size, decreases, it can also be seen that the peak position increases, in accordance with the results of Fig. 2(a). Finite-size scaling [1] predicts that, at the threshold p_c, S will diverge with the lattice size L as $S(p_c) \sim L^{\gamma/\nu}$. Figure 3(b) shows $S(L)$ at the percolation threshold; again, all of the

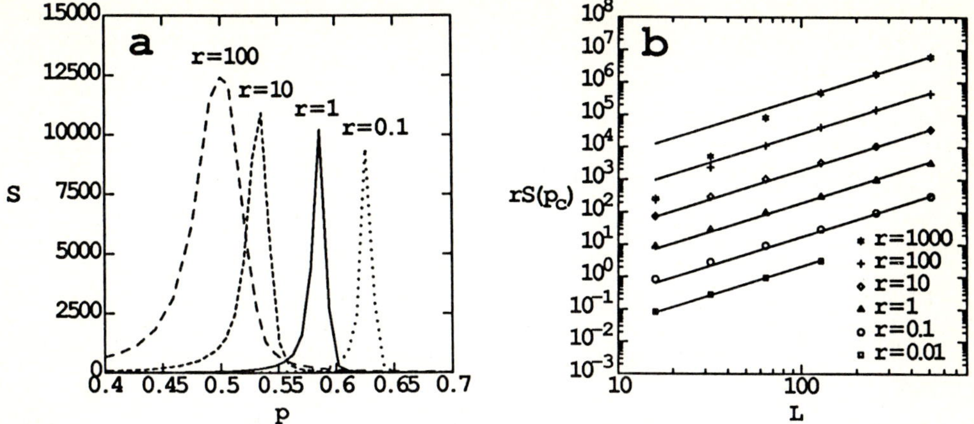

Fig. 3 The mean cluster size S for several values of r: (a) as a function of p, for $L = 512$; (b) as a function of L, for $p = p_c(r)$, and scaled by r. The slopes of the data give the exponent ratio γ/ν

curves asymptotically approach the straight lines, which have a slope given by the random percolation value $\gamma/\nu = 43/24 \approx 1.79$ [1].

A similar analysis can be performed for the spanning cluster density P. In a finite system, it is expected that $P(p_c) \sim L^{-\beta/\nu}$, and this relation is verified for all values of r, with an exponent consistent with the random percolation value. Note that, because of the scaling relation $D = d - \beta/\nu$, where D is the fractal dimension [1,5], this indicates that the latter is also independent of r.

As a final indication of the unimportance of the blob size, the cluster size distribution n_s is shown at the percolation threshold in Fig. 4. It is apparent from this figure that, as $s \to \infty$, n_s becomes *independent* of r. In other words, the number of very large clusters does not depend on the size of the blobs which make them up. This shows the relative unimportance of the blob size at larger length scales, and in particular at the critical point where ξ is diverging.

In conclusion, the percolating clusters in this model can be thought of as being built up out of compact clusters, or blobs, whose size increase with r. For length scales much larger than the blob size, the value of r is unimportant. Hence, the critical behavior is not expected to depend on r. For the range of r studied, the measured exponents are consistent with those of ordinary

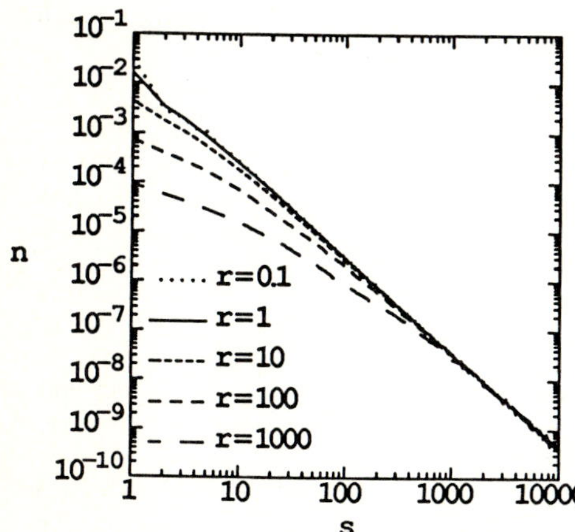

Fig. 4 The cluster size distribution n_s, for several values of r, for a lattice of size $L = 512$

random percolation. This is strong evidence, therefore, that, despite the presence of interactions in the model, the universality class is unaffected.

Acknowledgements. The authors wish to thank Daniel E. Platt, Daniel C. Hong, and M. Daoud for many helpful conversations. This work was supported by the Office of Naval Research and the Petroleum Research Fund, administered by the American Chemical Society.

References

1. D. Stauffer: *Introduction to Percolation Theory* (Taylor and Francis, London, 1985)
2. J.W. Evans, J.A. Bartz, D.E. Sanders: Phys. Rev. A *34*, 1434 (1986)
3. M. Eden: In *Proceedings of the Fourth Berkeley Symposium on Mathematical Statistics and Probability*, ed. by F. Neyman (University of California Press, Berkeley, 1961)
4. P.G. de Gennes: *Scaling Concepts in Polymer Physics* (Cornell University Press, Ithaca, NY, 1979)
5. B.B. Mandelbrot: *The Fractal Geometry of Nature* (W.H. Freeman, New York, 1983)
6. K. Binder (ed.): *Monte Carlo Methods in Statistical Physics* (Springer, Berlin, 1986)
7. M.E. Fisher: In *Proceedings of the Intern. Summer School Enrico Fermi, Varenna, Italy*, ed. by J.S. Green (Academic Press, New York, 1971)
8. P.J. Reynolds, H.E. Stanley, W. Klein: Phys. Rev. B *21*, 1223 (1980)

MD Simulation of 2D Rb Liquid and Solid Phases in Graphite

J.D. Fan[1], *O.A. Karim*[2], *G. Reiter*[1], *and S.C. Moss*[1]

[1]Physics Department, University of Houston, Houston, TX77004, USA
[2]Chemistry Department, University of California at Berkeley, Berkeley, CA94720, USA

We present here initial results of our molecular dynamics (MD) study of an intercalated two dimensional (2D) Rb liquid modulated by its periodic graphite host/1-4/. The circular average of the in-plane liquid structure factor S(q) has been experimentally determined by X-ray scattering for stage 2 Rb/4/, K/5/ and Cs/6/ in highly oriented pyrolytic graphite (HOPG) and there is considerable X-ray photographic data on the anisotropic liquid scattering from intercalated graphite single crystals/6-9/. There is also extensive information on the in-plane alkali dynamics from inelastic/10/ and quasielastic/11/ neutron scattering that may be directly compared to MD calculations of $S(q,\omega)$.

In the present simulation, we fit the measured, circularly averaged, X-ray structure factor for stage 2 liquid Rb in HOPG in which each Rb layer, separated by 2 graphite planes, is uncorrelated with its neighboring layers. We also calculate the full anisotropic S(q) replicating all of the features predicted by theory/12/ and observed in the X-ray single crystal photos of Parry/7/ and Rousseaux et al/9/. Finally, we cool the 2D liquid on the computer to give a diffraction pattern in excellent agreement with results on the incommensurate ordered state. A crucial input in our simulations is the graphite-Rb modulation potential which is also derivable from X-ray scattering/12,13/ through a careful measurement of the (modulated) Rb contribution to the HK.L graphite Bragg peaks. The graphite-K potential has been evaluated in a similar fashion/14/ and both Rb and K results have recently been summarized in a short review together with a preliminary MD comparison/15/.

For the MD we chose a rhombic box of 36 x 36 graphite unit cells, with edges parallel to the cell axes $|a_1| = |a_2| = 2.46$Å. Given two carbon atoms per cell, this box contains 216 Rb atoms at a planar density of $C_{12}Rb$. [With this composition a good fit could be made to the position of the first peak in the liquid structure factor which is often used as a measure of nearest neighbor distance, or number density, for close-packed systems.] The repulsive part of the Rb-Rb potential , $V_p(r)$, was taken from the treatment of Visscher and Falicov/16/ as adapted for the graphite intercalation compounds by Plischke/17/ in his calculation of the 2D liquid state. The only unknown in this potential is the dielectric constant in the layer which we treat as an adjustable parameter.

The other contribution to the potential energy is the graphite modulation potential, $V_m(r)$, induced in the Rb liquid. This takes the form/12/,

$$V_m(\vec{r}) = \sum_{HK} V_{HK}\, e^{i\vec{q}_{HK}\cdot\vec{r}} ,$$

where $\vec{q}_{HK}$ is a reciprocal lattice vector of the graphite. This periodic potential is impressed on the Rb liquid, acting to pull the ions into hexagon centers/12,18/, and it competes for registry against $V_p(r)$. Values for βV_{HK} ($\beta = 1/kT$) are experimentally available for Rb from the work of Thompson/19/ and

Springer Proceedings in Physics Vol. 33: **Computer Simulation Studies in Condensed Matter Physics**
Editors: D. P. Landau · K. K. Mon · H-B. Schüttler

from a re-analysis of the data of Ohshima et al./4/, both of which are supplied in Moss et al./13/.

The MD simulation was carried out at 300K in the 216-atom array with periodic boundary conditions using both the AS9000 computer at the University of Houston and the CRAY at the National Center for Supercomputing Applications, University of Illinois at Urbana-Champaign. The time step in the simulation was a constant $\Delta t = 0.04$ ps. The total number of configurations or snapshots, NS, over which the statistical average of the static structure factor was calculated is NS = 280 with a time separation between any two adjacent shots of 100 Δt.

We wish to calculate both $S(\vec{q})$, for comparison with the scattering from intercalated single crystals/6-9/, and S(q), which is its circular average appropriate to the Rb data from HOPG/4,15,19/. Because the S(q) from HOPG has been quite accurately determined/15,19/, this was fit first. The MD result may be circularly-averaged to give a radial correlation function g(r) from which S(q) is obtained via a 2D Bessel function transform /5,6/. With the imposed $V_m(r)$, however, g(r) oscillates about 1.0 at large r and must be cut at the end of our periodic box. To remove this oscillatory part of g(r), and the attendant ringing in S(q), some care had to be taken so that the modulated liquid S(q) remained with only its Bragg-like piece eliminated.

Our final fit, shown in Fig. 1, was obtained by matching S(q) at the first sharp peak ($q = 1.18\text{Å}^{-1}$) to give an in-plane density of $C_{12}Rb$ and an effective dielectric constant of of 2.35, (compared to a free space value of 1.0). We have labelled selected features in Fig. 1, for reference later: (a) is the first sharp diffraction peak whose position scales directly with the in-plane density and whose width yields a correlation range, $2\pi/\Delta Q$, of ~23Å or nearly four Rb diameters; (b) is a modulation shoulder on the second liquid diffuse peak at (c); (d) is the $\{10.0\}$ Bragg position of graphite, $\{HK\} = \{10\}$; (e) and (f) compose a split third diffuse peak from the liquid.

In calculating the anisotropic structure factor, $S(\vec{q})$, a second method was employed which did not involve the removal of the periodic component in g(r) but which otherwise gave essentially the same S(q) for HOPG. In this method we treat the periodic 216-atom box used in the simulation as one supercell of an infinite lattice. Each snapshot then gives us the coordinates $\vec{r}_i$ of the 216-atom basis set of the supercell. In this case, however, $\vec{q}$ is evaluated only at $\vec{q}_{hk} = h\vec{B}_1 + k\vec{B}_2$ where $\vec{A}_1 = 36\vec{a}_1$, $\vec{A}_2 = 36\vec{a}_2$ and $\vec{B}_{1,2}$ are reciprocal to $\vec{A}_{1,2}$. This procedure subdivides the normal graphite reciprocal lattice into (h,k) subcells at each point of which $S(\vec{q}_{hk})$ is evaluated; $S(\vec{q})$ is then given by an interpolation procedure.

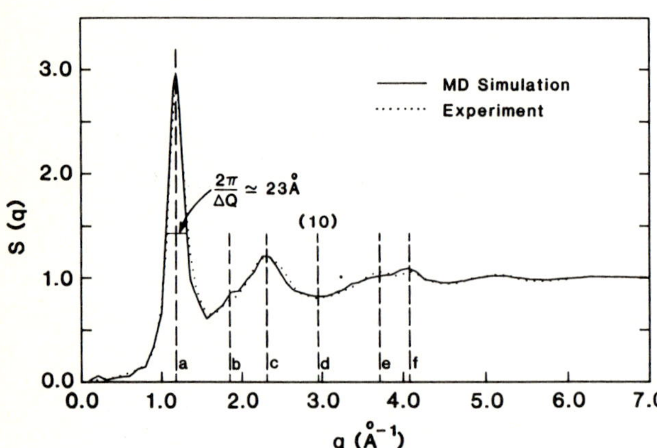

Fig. 1 Comparison of the MD simulation with the circularly averaged structure factor, S(q), for Stage 2 Rb in HOPG. A correlation range of ~ 23 Å is indicated as are the prominent features in S(q) at q values of a - f. The experimental graphite (10.0) peak at 2.94 Å^{-1} (d) has been removed along with the (11.0) peak at ~ 5.0 Å^{-1}.

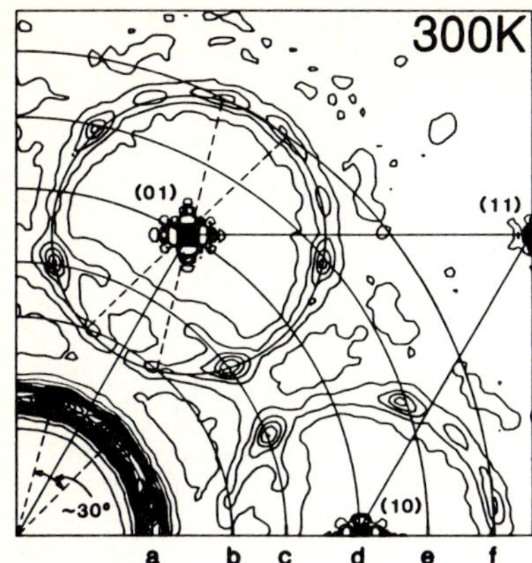

Fig. 2 The anisotropic liquid $S(\vec{q})$ for Rb as it would appear in an HK.0 section of a graphite single crystal. Note: the sharp Rb contributions at the {10} and {11} graphite Bragg positions; the anisotropic modulated liquid scattering; and the development of halos about all {10} positions. [Note also the absence of a halo about (11).] The origins of a - f in Fig. 1 are indicated. The ripple at the Bragg point is noise.

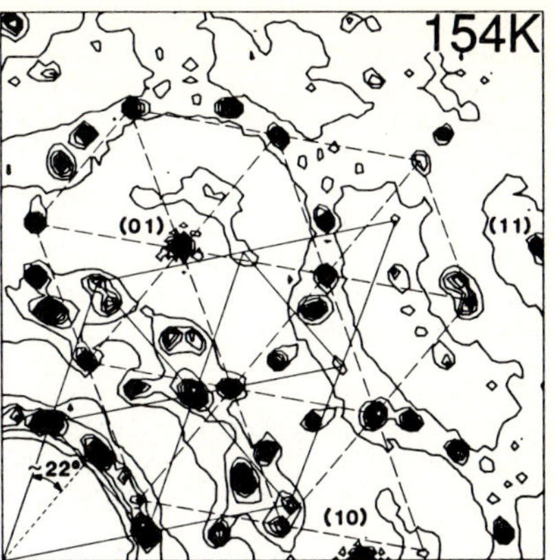

Fig. 3 $S(\vec{q})$ for the 2D Rb of Fig. 2 cooled on the computer to 154 K, below the measured ordering temperature of 165 K. Note the development of a hexagonal alkali reciprocal lattice [———] which is then reproduced as a modulation pattern [----] about the (01), and (10), positions. There are two rotationally split alkali lattices separated from [10] by ~ 11° (compared to ~ 15° in the liquid).

Fig. 2 shows $S(\vec{q})$ for liquid Rb in single crystal graphite using the supercell method, obtained by first averaging the calculation over three irreducible 30° segments and then displaying this averaged segment, via reflection and rotation, over the 360° pattern. There is a clear anisotropic splitting of the first sharp liquid diffraction peak into lobes ~ ±15° off the [10] axes, for which the ratio $I_{max}/I_{min} \simeq 1.7$. In the ordered incommensurate solid phase (see Fig. 3), the alkali Bragg spots that develop out of these split diffuse peaks appear rotated off the graphite [10.0] directions by ~ 10° - 15°, depending on in-plane density/8,9/. Clearly we are seeing this modulation effect in the 2D Rb liquid much as it has been observed experimentally by Rousseaux et al./9/ and predicted by Reiter and Moss/12/. [While the prediction of an anisotropic $S(\vec{q})$ is perhaps intuitively obvious, its actual analytical description is rather complex, involving multiparticle correlations and cumbersome integrals/12/.]

Fig. 2 reveals halos, for which $S(\vec{q}) \Rightarrow S(\vec{q} + \vec{q}_{HK})$, about the {10} positions, both as observed/6-9/ and as predicted by theory/12/. Also displayed are the origins of the circularly averaged features, a-f in Fig. 1. For instance, the diffuse splitting of intensity on the a-circle ($q = 1.18$ Å^{-1}) is weakly reproduced in the halos about (10) or (01) as lobes, noted on the (01) halo (and on the b-circle and f-circle as the main contribution to those features in Fig. 1). The halos are not, however, symmetrical about each {10} Bragg point because there are interferences with the normal 2D (unmodulated) liquid scattering which is pronounced around the entire c-circle but enhanced on the halo. These halo fine features are quite interesting because the wealth of detail in $S(\vec{q})$ from the ordered phases/7-9,18,20/ emerges directly out of what appear to be rather faint modulation features in the liquid pattern. This is clearly shown in Fig. 3 in which the 300 K liquid of Fig. 2 has been cooled on the computer to 154 K, or 11 degrees below the measured ordering temperature/9,11/. Here the modulation features are well developed and the calculated diffraction pattern, whose indexing scheme is given in Refs. 8, 9, 18 and 20, reproduces in detail the measured pattern/9/. It should also be noted that there is no halo about the (11) reciprocal lattice point in Fig. 2. Theory/12/ indicates that the intensity in a liquid halo will be proportional to $|V_{HK}|^2$ and from Refs. 13 and 15 we note that $\beta V_{10} = -0.45$ while $\beta V_{11} = -0.06$. We therefore expect only a very weak halo about $\{11\}$ and none about the other $\{HK\}$ positions. In summary, then, our simulation appears to give the major features of the modulated liquid and solid phases. The dynamical aspects are also being studied/21/.

This research was supported by the NSF on DMR-8603662 and by the National Center for Supercomputing Applications, University of Illinois at Urbana-Champaign to whom we express our thanks. We also thank Dr. Adrian Wright for suggesting the supercell method, J. L. Robertson for assistance in data reduction and X. B. Kan for helpful discussions of the calculation. The work of Omar A. Karim was done in the Chemistry Department, University of Houston, and we thank Prof. J. A. McCammon for his support.

References

1. For a recent review see: "Intercalation in Layered Materials," M. S. Dresselhaus, ed., NATO ASI Series, Series B - Physics 148 (1986).
2. G. S. Parry and D. E. Nixon: Nature 216, 909 (1967).
3. H. Zabel, S. C. Moss, N. Caswell and S. A. Solin: Phys. Rev. Lett. 42, 1552 (1979).
4. K. Ohshima, S. C. Moss and R. Clarke: Synth Met. 12, 125 (1985).
5. H. Zabel, Y. M. Jan and S. C. Moss: Physica (Utrecht) 99B, 453 (1980).
6. Roy Clarke, N. Caswell, S. A. Solin and P. M. Horn: Phys. Rev. Lett 43, 2018 (1979).
7. G. S. Parry: Mater. Sci. Eng. 31, 99 (1977).
8. M. Mori, S. C. Moss, Y. M. Jan and H. Zabel: Phys. Rev. 25, 1287 (1982); F. Rousseaux, R. Moret, D. Guerard, P. Lagrange and M. Lelaurain: J. Phys. (Paris) Lett. 45, L111 (1984).
9. F. Rousseaux, R. Moret, D. Guerard, P. Lagrange, and M. Lelaurain: Synth. Met. 12, 45 (1985).
10. W. A. Kamitakahara and H. Zabel: Phys. Rev. B32, 7817 (1985).
11. H. Zabel, S. E. Hardcastle, D. A. Neumann, M. Suzuki and A. Magerl: Phys. Rev. Lett. 57, 2041 (1986).
12. G. Reiter and S. C. Moss: Phys. Rev. B33 , 7209 (1986).
13. S. C. Moss, G. Reiter, J. L. Robertson, C. Thompson, J. D. Fan and K. Ohshima: Phys. Rev. Lett. 57, 3191 (1986).
14. X. B. Kan, J. L. Robertson, S. C. Moss, K. Ohshima and C. J. Sparks: unpublished.
15. S. C. Moss, X. B. Kan, J. D. Fan, J. L. Robertson, G. Reiter and Omar A. Karim: In "Competing Interactions and Microstructures: Statics and Dynamics," Proceedings of a Los Alamos Workshop, May 5-8, 1987, R. LeSar, ed., Springer-Verlag (New York), in press.
16. P. B. Visscher and L. M. Falicov: Phys. Rev. B3,2541 (1971).
17. M. Plischke: Can. J. Phys. 59, 802 (1981).
18. M. J. Winokur and R. Clarke: Phys. Rev. Lett. 54, 811 (1985).
19. C. Thompson: Thesis for the Ph.D., University of Houston, May 1987, unpublished.
20. M. Suzuki: Phys. Rev. B33, 1386 (1986).
21. J. D. Fan: Thesis for the Ph.D., University of Houston (in progress).

Index of Contributors

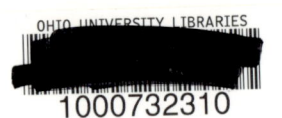